Advances in Geographic Information Science

More information about this series at http://www.springer.com/series/7712

Francis Harvey · Yee Leung
Editors

Advances in Spatial Data Handling and Analysis

Select Papers from the 16th IGU Spatial Data Handling Symposium

Springer

Editors
Francis Harvey
Department of Geography, Environment and Society
University of Minnesota
Minneapolis, MN
USA

Yee Leung
Department of Geography and Resource Management
The Chinese University of Hong Kong
Hong Kong SAR
China

ISSN 1867-2434 ISSN 1867-2442 (electronic)
Advances in Geographic Information Science
ISBN 978-3-319-19949-8 ISBN 978-3-319-19950-4 (eBook)
DOI 10.1007/978-3-319-19950-4

Library of Congress Control Number: 2015941501

Springer Cham Heidelberg New York Dordrecht London

Printed on acid-free paper

Springer International Publishing AG Switzerland is part of Springer Science+Business Media (www.springer.com)

Preface

The 16th Spatial Data Handling (SDH) conference was a special event held in Toronto from October 6 to 8, 2014, with the 2014 ISPRS Technical Commission II Midterm Symposium. We coordinated SDH with colleagues from ISPRS once again, which led to a particularly dynamic range of papers and discussions at the joint conference. Our thanks go out to Songnian Li and Suzana Dragicevic from ISPRS for their help and unflappable collegiality. This robust collaboration has always been central to assuring the vitality of the GIScience field. The submissions in the SDH proceedings you have in this volume have gone through a double-blind review and intensive editing. They represent some of the most engaging work going on connected to GIScience. As always was the orientation of SDH, the chosen contributions connect theory and applications. This continues to be important as it helps researchers connect their work to important recipients and see how the field is changing. Whether you are new to the field and have been active for many years, the following chapters have a great deal of insights and innovative developments to offer.

The International Symposium on Spatial Data Handling (SDH) continues as the biennial international research forum for Geographic Information Science (GIScience), again co-organized by the Commission on Geographic Information Science and the Commission on Modeling Geographical Systems of the International Geographical Union (IGU). The conference brings together scholars and professionals from a wide range of disciplines, including (but not limited to) geography, computer science, information science, engineering, statistics, and geostatistics, with a broad range of applications in others sciences that contribute to the development of the theory of GIScience. The 2014 SDH conference jointly organized with the ISPRS Technical Commission, offered additional possibilities for exchange during the conference. The SDH proceedings are, however, independent and reflect a distinct focus on traditional SDH contributions that connect science with applications.

Indeed, since the first conference in 1984, the SDH conference has developed through a long tradition and evolved in parallel with the discipline of GIScience. As a well-known international meeting place for GIScience, it remains a leading scientific meeting in the field. After the conference's third appearance in Australasia 2 years ago (held in Hong Kong, after Sydney 1988 and Beijing 2000), the 2012 SDH in Bonn, Germany, the 2014 SDH conference took place in Toronto, Canada. While retaining tradition, again, following the lead of the 2012 SDH organizers, we again opened two different submission tracks making the distinction between long and short paper formats that reflect different publication cultures.

Even with these changes, the SDI review process continues to retain high standards. Over 52 paper proposals were submitted to the initial abstract review. 28 papers were submitted to a second double-blind peer review for publication in the SDH proceedings. Only 17 of the papers were revised and went through further editorial processing in preparation for the proceedings you have now.

Contents

Introduction

The change since SDH 2012 in Bonn, Germany, in accepting long and short papers, reflects larger changes in the field that has grown by leaps and bounds. If one counts the millions of Google Earth and other online mapping software users to the traditional number of the GIS users, GIScience is at the forefront of developments that change the way millions of people around the world every day find their way, understand their vicinities, share plans, and discover the world. As Dan Sui points out in his contribution to SDH 2014, the ways in which GI is used are changing. His valuable insights offer a number of reflective comments on these developments. The remaining chapters offer innovative research that point to a number of important changes. Reflecting developments noted in the SDH 2012 proceedings. We can point to mixing of old and new topics in some general characteristics of the SDH 2014 proceedings' papers. First, four fundamental GIS functions [(1) data capture; (2) modeling, storage, and retrieval; (3) manipulation and analysis; and (4) presentation and sharing] are important in all contributions. Second, fundamental Spatial Data Handling topics (e.g., modeling topological relations and modeling terrain) are complemented with new and emerging challenges (spatio-temporal and mobile GIS, new geodata sources) that mirror changes in the field.

In the first section of the SDH 2014 proceedings, Web and Crowd Sourcing GI, Monika Sester and Sagi Dalyot describe an approach to enrich conventional routing instructions with data coming from conventional digital topographic maps and volunteered geographic information (VGI). They describe a taxonomy of relevant features and relations with methods for their automatic extraction and evaluation. The concept of a taxonomy is also evident in the following contribution from Claus Rinner and Victoria Fast. They propose criteria to distinguish types of user-generated data and contents and relate these to types of online applications. Rob Feick and Collin Robertson, in their chapter, offer a framework for mining information from georeferenced user-generated content (UGC) and volunteered geographic information (VGI) to gain new insight into how people describe and delimit urban places. Their chapter aims to extend recent efforts to explore semantic similarity in information coming from georeferenced user-generated content (UGC) and volunteered geographic information (VGI) with the aim of gaining new insights into how people describe and delimit urban places such as neighborhoods, vernacular landmarks, and locales. It computes measures of tag importance using both a naïve aspatial approach and a method based on spatial relations. Obviously, the boom in the creation of VGI data is leading to the awareness of its potential and the challenges in finding effective strategies to ensure its reliable use. Haydn Lawrence and co-authors consider how to evaluate feasible VGI study areas and the appropriateness of different aggregation unit sizes through three different components of data quality: coverage, density, and user-heterogeneity. In real time, or when assessing large amounts of crowd-sourced data, another challenge is distinguishing events. Khatereh Polous and co-authors explore this and develop a generic and extensible programming framework to find meaningful patterns out of heterogeneous and unstructured online data streams in their chapter. The chapter from Xiogang Zhou and co-authors take on these issues and examine a central issue in GIScience uncertainty in the context of the identification of activity centers and vagueness in geographic context problem.

The second section, Network Analysis, points to significant changes in the ways in which GI is used now as sensors that provide location data become increasingly ubiquitous. The chapter from Jing Tian and co-authors ties into issues from the proceeding section and introduces a robust approach for mining the co-location patterns between network spatial phenomena demonstrated in a study in Shenzhen City, China. Picking up on innovative spatial statistical approach, Xiang Zhang and Tingua Ai describe an approach used to automatically assess OSM road tags for errors arising from widely applied to detect inconsistent tags and provide intelligent suggestions for data correction. Christopher Brown and Xiaobai Yao describe an example of the processing of large network sets that demonstrates innovative approaches to data analysis related to their creation of a comprehensive road change database of the USA working with a compilation of US Census Bureau TIGER/Line GIS datasets from 2008 to 2012. Addressing the complexity of dealing with large databases of road networks, Jing Tian and others describe an approach to classifying road segments based on the measures of four parameters: length, degree, closeness, and betweenness. Wei Tu and co-authors develop a spatio-temporal decision support framework for logistics distribution in the metropolitan area. It integrates a traffic information database, a spatio-temporal heuristic algorithm, a cloud GIS-based logistics server, and intelligent mobile apps for logistic planning.

Turning to the third section of the proceedings, Spatial Modeling and Reasoning, it is evident that contributions continue to come from enhancements to modeling approaches based on innovative and insightful spatial statistical analysis.

Gev Ben-Haim, Sagi Dalyot, and Yerach Doytsher propose a method for spatial regional inner-accuracies analysis that is based on terrain characteristics and error-propagation computation via a simultaneous relative DTMs comparison mechanism. It focuses on the application of a multiple wide-coverage DTM comparison method using weighted LSA to perform local analysis and calculation of absolute vertical accuracies of DTMs. The next paper in this section, from Matthew Quick and Jane Law, provides insights into the social disorganization theory, related social processes, and the spatial dynamics of crime that reflect spatial demographic changes in Toronto at the neighborhood scale. The study uses exploratory spatial data analysis to locate crime hot spots against concentration of immigrants, ethnic heterogeneity, and residential mobility. The final paper in the section explores the values of spatial data handling approaches in public health, which regularly involve large amounts of data. Bouden and Moulin propose an approach that can generate an informed virtual geographic environment (IVGE) composed of a set of cells in which the evolution and interaction of the involved populations can be simulated plausibly. Since the number of these cells is huge, they propose an interesting threshold-based merging algorithm that creates spatial subdivisions with the maximal size and suitability for a given biological phenomenon.

The final section of the proceedings, Statistical and Spatial Analysis, contains contributions that also speak plainly to the continued centrality of spatial statistics and analysis in enhancing our abilities to make sense of the world around us. Zhuang He and co-authors describe an innovative development of spatial statistical techniques for defining rules to assess building development. The rules consider 3D building layouts that comply with the rules and have optimized indicators employing a marked point process (MPP) of 3D boxes, whose probability distribution can be defined through Gibbs energy with regard to a reference process. Yilun Lin and Francis Harvey extend previous spatial statistical analysis in an empirical analysis of the spatial variation of individual uniqueness in the US census data given the simple demographics four states and three scales. The study contributes to geographic information privacy relevant to reverse geocoding and related spatial aggregation contributes. Patrick Kirby and Scott Mitchell return to Monte Carlo methods with an innovative analysis of thee effects of positional and thematic uncertainties in a set of landscape maps on model-averaged regression coefficients. Their results indicate that the uncertainty estimates from model averaging outweigh the effects of positional and thematic uncertainties in the landscape maps. The results also point to the need for further simulation approaches that account for spatial autocorrelation. In the final paper of this section, Ikuho Yamada and Atsuyuki Okabe also return to a mainstay in spatial statistical analysis of spatial autocorrelation, Moran's *I*, and describe results from intensive simulation experiments that suggest that discrepancies from the normality are more serious for target variables with larger skewness and kurtosis. Their work proposes to carry out multiple sets of simulations and turn to the mean of simulated percentage point values as the critical values in Moran's I to reduce the instability inherent in simulation-based approaches.

In concluding, we hope that these papers will encourage and contribute to the continuation of research in all areas of Geographic Information Science. We also wish to give expression to our thanks to many people who made SDH 2014 possible: all those who submitted their research and participated in the meeting, the program committee members for reviewing and sharing their experience, the steering committee for their support, and the local organizing committee.

January 2015

Francis Harvey
Yee Leung

Alternative GIS (alt.gis) and the Six Senses of the New Mind: Is alt.gis Transforming GIS into a Liberation Technology?

Daniel Z. Sui

Abstract The past twenty years have witnessed the emergence of alternative GIS (alt.gis) practices that are quite different from conventional GIS activities. Intellectually tied to the critical cartography pioneered by J.B. Harley in the late 1980s and early 1990s, alt.gis has evolved from early discussions on GIS & Society, critical GIS, PPGIS, feminist GIS, map stories/geonarratives, deep maps to qualitative GIS, neogeography, crowdsourcing, volunteered geographic information (VGI), geoweb, post-colonial computing, geogames, geodesign, and spatial humanities. This paper develops a preliminary framework to loosely knit together the diverse intellectual threads for Alt.gis. This paper argues that Daniel Pink's framework for the six senses of the whole new mind (design, story, symphony, empathy, play, and meaning) captures the six major Alt.gis practices remarkably well. Mainstream GIS development has been predominantly concentrating on automated cartography/map-making, spatial modeling, geo-computation, and database development with the goal for efficiency, which tend to be closely associated with the left-side/slow thinking capabilities of the human brain. In contrast, Alt.gis focuses more on geo-narratives, qualitative/mixed methods, story-telling, and synthesis with the goal for achieving equity and social justice, which tend to be more closely associated with the right-side/fast thinking capabilities of the human brain. Evidently, Alt.gis has enabled GIS users to transcend what Heidegger called the enframing nature of technology and has led them explore new territories with greater sensitivities. In this sense, Alt.gis has become an enabling technology that continues to empower GIS users in their quest for a better, more equitable and sustainable world.

Keywords alt.gis · Critical GIS · Geodesign · Map story · Synthesis · Geogames

D.Z. Sui (✉)
The Ohio State University, Columbus, Ohio, USA
e-mail: sui.10@osu.edu

F. Harvey and Y. Leung (eds.), *Advances in Spatial Data Handling and Analysis*, Advances in Geographic Information Science,
DOI 10.1007/978-3-319-19950-4_1

1 GIS and alt.gis

I first heard about the term alt.gis from Nadine Schuurman in late 2012. I was told that Schuurman coined the term by inspirations from the music genre—alt.country. Also known as alternative country or insurgent country, alt.country is a catch-all phrase that includes loosely defined sub-genre of country music that differs significantly in style from mainstream or pop country music. Those alt.country music bands and artists often have incorporated eclectic styles including roots rock, bluegrass, rockabilly, honky-tonk, alternative rock, folk rock, and sometimes punk (wikipedia.org). Although closely related to country music, alt.country has become its own genre and enriched the country music scene.

By analogy, alt.gis may also serve as a general umbrella term for all those GIS research and application activities that are different from the main stream/conventional GIS research and applications. Upon further reflection, I think the neologism alt.gis is an intriguing idea that can be used to describe the multiple exciting developments in GIS during the past 20 years.[1] The goal of this paper is to conduct a preliminary survey of the diverse intellectual threads for alt.gis practices. By tying the resulting alt.gis concepts to Pink's (2006) framework for the whole new mind, a preliminary framework is developed to understand the diverse alt.gis activities. The take-home message of this paper can be summarized as this: mainstream GIS and GIScience development has been predominantly concentrating on cartographic, quantitative, computing, and database approaches with the goal for efficiency, which tend to be closely associated with the left-side/slow thinking capabilities of the human brain; alt.gis, in contrast, focuses more on narrative, qualitative, story-telling, and synthesis approaches with goals orientated towards equity and justice, which tend to be more closely associated with the right-side/fast thinking capabilities of the human brain.

2 Six Emerging alt.gis Practices

2.1 alt.gis I—Changing Geography by Design: From Description to Prescription and Geodesign

Among the multiple alt.gis practices, geodesign may be the latest to arrive on the scene. The growth momentum of geodesign and its impacts, however, have been quite breath-taking in recent years (www.geodesignsummit.com). Geodesign formalizes and expands the role of geospatial technologies in the entire life cycle of design flows. It is a framework as well as a technology to leverage geographic

[1]See Wilson (2012) for a more comprehensive list of bibliography on critical GIS, which considerably overlaps with the literature of alt.gis as reviewed in this paper.

information, resulting in designs that more closely follow natural systems. Geodesign is designed with nature in mind (Dangermond 2009) with the goal of changing geography (Steinitz 2012), which is conceptually consistent with the change geographers (Castree et al. 2010) have been aiming for. We now have not only the concepts, but the tools and technologies to change the world as we desire through geodesign.

Goodchild (2010) argues that GIS is about what *is*, geodesign is about what *could be*. Indeed, as George Bernard Shaw once famously said, "some men see things as they are and say 'why?'—I dream things that never were and say 'why not?' (http://www.bartleby.com/73/465.html)." For quite some time GIS, like some men in the Shaw quote, concentrated predominantly on descriptive and analytical functions—and consequently GIS has been quite a successful technology in "describing" where things are located on or near earth surfaces and (to some extent) why. Admittedly, GIS technologies have advanced significantly as descriptive and analytical tools (de Smith et al. 2011), and yet, despite its deep roots in environmental design, beyond the naive GIS conceptual work (Egenhofer and Mark 1995), GIS until recently has been ill-equipped to enable users to imagine things that do not exist, and ask *why not*—the primary task for design.

2.2 alt.gis II—Giving Voice to Geographic Lore: From Maps to Map Stories and Geo-Narratives

One creative way to practice qualitative GIS in recent years is through map stories or location-based story-telling more generally. This is one of the interesting ways of using GIS to give voice to geographic lore (ESRI 2012). As a result, today's maps are not simply used as illustrations. Instead, they are increasingly used as a medium to tell stories, even in full narrative atlases (Wood 2010; Elwood and Mitchell 2012; Mennis et al. 2013). Such map-based geospatial storytelling has empowered contemporary map makers to describe reality in a way unattainable by traditional approaches (Caquard 2011; http://mapstory.org; http://storymaps.esri.com/home). Story maps often integrate text, multimedia, and interactive functions to inform, educate, entertain, and inspire people. Story maps are capable of empowering a global community to organize knowledge about the world spatially and temporally. With platforms such as mapstory.org or storymaps.esri.com, people of all kinds can turn into story tellers who are capable of creating, storing, and collaborating on map stories and improve on understanding of global dynamics. Incorporating four Cs (countries, cities, companies, and communities) as integral components of geostrategy, GIS-based maps story-telling helps people in all walks of life better understand the world. With more and more platforms for location-based story-tell coming into existence, such as echolocation, storieseverywhere.org, geographic lore will capture people's attention and imagination.

2.3 alt.gis III—Mashing up for Synthesis

With the explosive development of volunteered geographic information (VGI), geoweb, crowdsourcing, social media, and user-generated content, mashing up has rapidly evolved as a new Alt.gis practice. This alt.gis emphasis on synthesis through mashup is quite different from the earlier focus on analysis through statistics or mapping (Sui 2009). Many forms of synthesis in the context of VGI applications can be described as mashups. Mashup is perhaps one of the most important of the frequently mentioned buzz words nowadays. Indeed, many of the rapidly expanding web 2.0 applications can be broadly described as mashups. Borrowed from the music industry, a mashup originally refers to a song or composition created by blending two or more songs—a task that can be achieved by digitally overlaying the vocal track of one song seamlessly over the music track of another (Wikipedia.org). In the context of web-based applications, a mashup may have multiple meanings. At the functional/service level, a mashup may refer to a web page or application that combines data or functionality from two or more external sources to create a new service. In terms of the actual content, a mashup can be a digital media file that contains a combination of text, maps, audio, video, and animation, which recombines and modifies existing digital works to create a derivative work. The term mashup implies easy, fast integration, frequently using open APIs (application programming interfaces) and data sources to produce something new. A growing number of industry leaders such as Google, Microsoft, Yahoo, and MapQuest have developed new products and services that enable their users to develop their own APIs for a variety of mashups. Undoubtedly, mashup capabilities have enabled millions of web users to develop a variety of creative applications. It is definitely one of the major technical breakthroughs since WWW came into existence, but mashup is much more than a technical advance. I believe that the true significance of mashup lies in its potential promotion of new habits of mind towards synthesis. As cognitive psychologists have shown us (Wallace and Gruber 1989), synthesis, rather than analysis, is the defining characteristics of human creativity. GIS and Geography are uniquely positioned to lead this new wave of synthesis.

2.4 alt.gis V—Empowering the Powerless: From Efficiency to Empathy

In an effort to move critical GIS from dense theoretical discussions into meaningful practices, we have witnessed growing alt.gis practices focusing on social goals (Harris et al. 1995), ranging from human right abuses (http://shr.aaas.org/geotech), sexual harassment (harassmap.org), to gender inequality (Stephens 2013), and an ever enlarging digital divide (Graham 2011; Haklay 2013). In sharp contrast to earlier GIS applications with an emphasis on efficiency (“doing the thing right”), the growing alt.gis applications stress doing the right thing, thus contributing to our growing sense of empathy for the dire situations people face in various part of the world. For example, as demonstrated by AAAS's analysis of the on-going conflicts in Syria, geospatial technologies are playing increasingly key roles in revealing to the world the devastating destruction wrought by war in the city of Aleppo in Syria (Davis 2013). Furthermore, geospatial technologies have also significantly broadened the ability of NGOs to rapidly gather, analyze, and disseminate authoritative information, especially during times of crisis.

Despite the progress made along this direction, we should be keenly aware that this world remains a deeply divided one among multiple boundaries—both physical and digital (Sui et al. 2012). In the context of geographic information (and to some extent other types of data as well), the biggest irony remains that Murphy's law is still at work—information is usually the least available where it is most needed. We have witnessed this paradox unfolding painfully in front of our eyes in the Darfur crisis in northern Sudan (2006), the aftermath of the Haiti earthquake (2010), and the BP explosion in the Gulf of Mexico (2011). Undoubtedly, how to deal with big data in a shrinking and divided world will be a major challenge for Alt.gis researchers in the years ahead. The strengths, weaknesses, opportunities, and threats (SWOT) of VGI for improving the spatial data infrastructure (SDI) are quite different in the two global contexts of North and South (Genovese and Roche 2010). Furthermore, as Gilbert and Masucci (2011) show so clearly in their recent work on uneven information and communication geographies, we must move away from the traditional, linear conceptualization of a digital divide, concerned primarily with physical access to computers and the Internet. It is abundantly clear that Alt.gis practices have vastly increased our sensitivity to the human conditions in various parts of the world. By cultivating a more empathetic state of mind, Alt.gis potentially can espouse more positive changes toward a more just, equitable, and fair world.

2.5 *alt.gis IV—Let's Play: From Geoinformatics to Geogames*

It is a well-known fact among psychologists that humans rarely succeed at anything unless they are having fun doing it. Homo Sapiens, on top of multiple other traits and characteristics, are really Homo Luden (the player). Indeed, the human quest for various causes can be made a lot easier when they are never bored. Being a bit playful and lighthearted is key to preventing boredom and dullness. In fact, the play ethic has been proposed as the new way of living in the digital age (Kane 2004).

Geogames are gaining more recognition for their importance in both Alt.gis research and education (Alhqvist et al. 2012). The interactions between GIS and the on-line virtual world Second Life (SL) is particularly noteworthy (Sui 2008). The interactions between GIS and SL so far have been a two-way street. First, geospatial technologies have been increasingly used to map out the geography of SL. Google map/earth mashups have been created to show the spatial distributions and linkages of continents and islands in SL. This is a continuation of Martin Dodge and Rob Kitchin's work on mapping cyberspace (www.mappingcyberspace.com). Since SL is a virtual world, mapping SL using GIS will further promote the use of spatialization as a method to understand the non-spatial world.

The rapid convergence of Google Earth and SL has promoted Wade Roush to envision the emergence of a Second Earth (SE), in which geospatial technologies and 3-D simulation-based social networking technologies are seamlessly integrated. In SE, not only everything and everybody in SL (more generally on the Web as well as on Earth) are fully georeferenced or geotagged (thus fully mappable using GIS), but also SL's avatars, controls, and modeling tools are becoming an integral part of the tool box for users to process geospatial data. The development of SE is happening at a much faster rate than I originally anticipated. Both DataGlobe and GeoGlobe are currently being developed as a prototype of SE. Furthermore, both Google Earth Street View and the latest version of Microsoft Virtual Earth have implemented some features of Roush's SE. This new development will certainly make geospatial technologies accessible by a much larger audience. Nowadays, even the characters in the Simpsons use Google Earth to plan their daily activities.

2.6 alt.gis VI—Seeking Deeper Meaning: From Space to Place

Among the six aptitudes, the most elusive is perhaps the last one—meaning. How can we enrich people's lives and give them deeper meaning through our research and educational activities in GIS and Alt.gis? It is indeed very ironic for human existence that the means to live are often inversely proportional to the meanings to live for. Furthermore, as Austrian psychologist and Holocaust survivor Victor Frankl (n.d.) so eloquently summarized, "… the meaning of life differs from man to man, from day to day and from hour to hour. What matters, therefore, is not the meaning of life in general but rather the specific meaning of a person's life at a given moment." More than ever, now that we have so much better means to collect data about individual lives and behaviors with detailed spatial and temporal tags, new Alt.gis practices have empowered us with a better chance to flesh out the meaning of life.

Broadly casted as humanistic GIS (Sui 2004, 2005), the affective stream of Alt.gis practices draw conceptual inspirations from the arts and humanities instead of traditional computing and geodesy fields (Kwan 2007). Instead of honing GIS users' analytical skills, humanistic/affective GIS practices sharpen users' artistic and emotional sensitivity. Inspired by techniques commonly used in cinematography, Aitken and Craine (2009) argued that affective GIS visualization should move beyond the representational space of the final maps, aiming to touch people at the visceral level, not simply the intellectual level. According to Aitken and Craine (2009), affective geovisualizations are "soulful" and "they tug at our hearts to the extent that we may be mobilized to action (page 140)." By reframing GIS applications to the recent broad literature on affect (feelings and emotions), Young and Gilmore (2013), once again, demonstrated the power of affective participatory GIS and emotion-charged visualizations in land disputes involving the Maijuna people in the Peruvian Amazon.

The shift from space to place in Alt.gis practices also signifies a major change that facilities and contributes to people's keen sense of place, thus making their lives more meaningful. In contrast to space, place can be considered as 'bottom-up' and contingent, representing the outlooks and actions of more typical folks (Tuan 1977). Geography practiced from the place tradition favors a slant/side perspective (e.g. Google Streetview or regular photos), qualitative, and multi-sensory. According to Agnew (2011), the concept of place usually include three pillars: **location** (as defined latitude and longitude), **locale** (as defined by both physical/environmental and socioeconomic/cultural context), and a **sense of place** (as defined by human subjective perception/attachment to a particular location/locale). Apparently, place is semantically a much richer concept than space.

Until recently, GIS has been dominated by locational perspectives using Cartesian coordinates according to Euclidean geometry. The massive amounts of volunteered geographic information (VGI) in general, and geo-tagged or location-based social media data in particular, seem to revive our approach to the world from the perspective of place, almost reaching the point of hyperlocalism dominated by "the tyranny of place" (Haklay 2010).

3 alt.gis and the Six Senses of the New Mind: Further Discussions

To succeed in this high concept and high touch[2] era, Pink (2006) highlights six essential aptitudes (termed "the six senses") that will define professional success and personal satisfaction including: design, story, symphony, empathy, play, and meaning. Although written for a different audience, Pink's main arguments resonate well with Howard Gardner's (2006) *Five Minds for the Future* published in the same year. In fact, Pink's six senses of the whole new mind are artfully interwoven into Gardner's five minds for the future. Both Pink's six senses and Garnder's five minds have been expanded and updated in Nobel Laureate Daniel Kahneman's (2013) best seller *Thinking, Fast and Slow*. Instead of using a spatial lateralization (right vs. left), Kahneman shifted the discussion of the human mind into a temporal

[2]According to Pink (2006), "high concept" refers the functions of the left side of the brain and "high touch" is related to the functions of the right side of the brain, which constitutes the whole new mind. We need both to succeed.

domain (fast vs. slow), with the right side of the brain corresponding to the fast thinking (emotional/intuitive) and the left side of the brain corresponding to the slow thinking (rational/logical). As we shall see, these maps of the mind, although highly simplified and even distorting at times, are quite helpful for us to better understand the diverse streams of alt.GIS activities. Mainstream GIS development has been predominantly concentrating on automated cartography/map-making, spatial modeling, geo-computation, and database development with the goal for efficiency, which tend to be closely associated with the left-side/slow thinking capabilities of the human brain. In contrast, alt.GIS focuses more on geo-narratives, qualitative/mixed methods, story-telling, and synthesis with the goal for achieving equity and social justice, which tend to be more closely associated with the right-side/fast thinking capabilities of the human brain. In this, the six emerging alt. gis practices reflect Pink's six senses of the new mind, with profound implications for both research and education in the years to come.

Evidently, alt.gis has enabled GIS users to transcend the enframing nature of technology and led them explore new territories with greater sensitivities—including defending human rights, improving governance, empowering the poor, protecting women and children, promoting economic development, and pursuing a variety of other social goods. In this sense, alt.gis has become an enabling technology (Diamond and Plattner 2012) along multiple fronts as it truly liberates GIS from its previous technical, theoretical, and application restraints. Consequently, it also empowers and liberates people in their quest for a better, more equitable and sustainable world. Digital media in general and location-based social media in particular are contributing to the more personalized collective action process that maintains high levels of engagement, agenda focus, and network strength (Bennett and Segerberg 2011). By implication, Alt.gis has pointed out interesting directions to reform our GIS curriculum. Moving away from the previous models of teaching **about** versus teaching **with** GIS, we perhaps need to realign GIS education efforts with Alt.gis practices focusing on the six senses of the new mind—a task more urgent in the context of the big data deluge and ubiquity of geo-tagged data (Crampton et al. 2013; Dodge and Kitchin 2013).

4 Summary and Conclusions

Obviously, diverse alt.gis practices have emerged during the past twenty years, which are quite different from the traditional mainstream GIS work. Although I must also admit that the boundaries between alt.gis and mainstream GIS are some sometimes not so clear cut, as demonstrated by earlier works by Schuurman (2006), Kwan (2008), and Kwan and Ding (2008). The emerging literature on alt.gis nonetheless has developed its own distinct flavor. This paper has found that the diverse alt.gis practices are remarkably consistent with the six senses of the new mind. This paper develops a preliminary framework to loosely knit the diverse intellectual threads of alt.gis. It suggests that Daniel Pink's framework for the six senses of the whole new mind (design, story, symphony, empathy, play, and meaning) captures the six major alt.gis practices remarkably well. Mainstream GIS development has predominantly concentrated on cartographic, quantitative, computing, and spatial database-oriented with the goal for efficiency, which tend to be closely associated with the left-side/slow thinking capabilities of the human brain. In contrast, alt.gis focuses more on narrative, qualitative, story-telling, and synthesis-oriented with the goal of equity, which tends to be more closely associated with the right-side/fast thinking capabilities of the human brain. In this sense, alt.gis has become an enabling technology that continues to empower GIS users in their quest for a better, more equitable and sustainable world. Moving forward, an engaged pluralism, as advocated by Barnes and Sheppard (2010), may be the most productive approach.

Acknowledgments Critical comments by Francis Harvey and Nadine Schuurman on an earlier draft of this paper are gratefully acknowledgement. The author is solely responsible for any remaining problems.

References

Agnew J (2011) Space and place. In: Agnew J, Livingston DN (eds) The sage handbook of geographical knowledge. Sage, Los Angeles, CA, pp 316–330

Ahlqvist O, Ramanathan J, Loffing T, Kocher A (2012) Geospatial human-environment simulation through integration of massive multiplayer online games and geographic information systems. Trans GIS 16(3):331–350

Aitken S, Craine J (2009) Into the image and beyond: affective visual geographies and GIScience. In: Cope M, Elwood S (eds) Qualitative GIS: a mixed methods approach. Sage, Thousand Oaks, CA, pp 139–153

Barnes TJ, Sheppard E (2010) 'Nothing includes everything': towards engaged pluralism in anglophone economic geography. Prog Hum Geogr 34(2):193–214

Bennett WL, Segerberg A (2011) Digital media and the personalization of collective action: social technology and the organization of protests against the global economic crisis. Info Commun Soc 14(6):770–799

Caquard S (2011) Cartography I: mapping narrative cartography. Prog Hum Geogr. doi:10.1177/0309132511423796

Castree N, Chatterton P, Heynen N, Larner W, Wright M (eds) (2010) The point is to change it. Antipode book series. Wiley-Blackwell, Oxford and Malden

Crampton JW, Graham M, Poorthuis A, Shelton T, Stephens M, Wilson MW, Zook M (2013) Beyond the geotag: situating 'big data' and leveraging the potential of the geoweb. Cartograph Geogr Inf Sci 40(02):130–139

Dangermond J (2009) GIS: designing our future. Available on-line at: http://www.esri.com/news/arcnews/summer09articles/gis-designing-our-future.html (last accessed 5 June 2013)

Davis C (2013) Aleppo satellite images reveal destruction wrought by Syria War. Available on-line at: http://www.huffingtonpost.com/2013/08/06/aleppo-satellite-images-reveal-destruction_n_3713640.html

de Smith M, Longley P, Goodchild MF (2011) Geospatial analysis: a comprehensive guide, Kindle edn. Troubador Publishing, London

Diamond L, Plattner M (eds) (2012) Liberation technology: social media and the struggle for democracy. JHU Press, Baltimore

Dodge M, Kitchin R (2013) Crowdsourced cartography: mapping experience and knowledge. Environment and Planning A 45(1):19–36
Egenhofer MJ, Mark DM (1995) Naive geography. In: AU Frank, W Kuhn (eds) Spatial information theory: a theoretical basis for GIS, Lecture Notes in Computer Sciences No. 988. Springer, Berlin, pp 1–15
Elwood SA, Mitchell K (2012) Mapping children's politics: spatial stories, dialogic relations and political formation. Geografisker Ann Series B Hum Geogr 94(1):1–15
ESRI (2012) Telling stories with maps: a white paper. Available on-line at: http://storymaps.esri.com/downloads/Telling%20Stories%20with%20Maps.pdf
Gardner H (2006) Five minds for the future. Harvard Business School Press, Boston
Genovese E, Roche S (2010) Potential of VGI as a resource for SDIs in the North/South context. Geomatica 64(4):439–450
Gilbert M, Masucci M (2011) Information and communication technology geographies: strategies for bridging the digital divide. Praxis (e) Press—University of British Columbia, Vancouver
Goodchild MF (2010) Towards geodesign: repurposing cartography and GIS? Cartogr Perspect 66 (Fall):55–69
Graham M (2011) Time machines and virtual portals: the spatialities of the digital divide. Progress in Development Studies 11(3):211–227
Haklay M (2010) The tyranny of place and OpenStreetMap. http://povesham.wordpress.com/2010/07/10/the-tyranny-of-place-and-openstreetmap/
Haklay M (2013) Neogeography and the delusion of democritisation. Environ Plan A 45(1):55–69
Harris TM, Weiner D, Warner TA, Levin R (1995) Pursuing social goals through participatory geographic information systems. In: Pickles J (ed) Ground truth: the social implications of geographic information systems. Guilford Press, New York
Kahneman D (2013) Thinking, fast and slow. Farrar, Straus, and Giroux, New York
Kane P (2004) The play ethic: a manifesto for a different way of living. MacMillan, New York
Kwan M (2007) Affecting geospatial technologies: toward a feminist politics of emotion. Prof Geogr 59(1):27–34
Kwan M (2008) From oral histories to visual narratives: re-presenting the post-September 11 experiences of the Muslim women in the USA. Soc Cult Geogr 9(6):653–669
Kwan M, Ding G (2008) Geo-narrative: extending geographic information systems for narrative analysis in qualitative and mixed-method research. Prof Geogr 60(4):443–465
Mennis J, Mason MJ, Cao Y (2013) Qualitative GIS and the visualization of narrative activity space data. Int J Geogr Inf Sci 27(2):267–291
Pink DH (2006) A whole new mind: why right-brainers rule the world. Riverband
Schuurman N (2006) Formalization matters: critical GIScience and ontology research. Ann Assoc Am Geogr 96(4):726–739
Steinitz C (2012) A framework for geodesign: changing geography by design. ESRI Press, Redlands
Stephens M (2013) Gender and the GeoWeb: divisions in the production of user-generated cartographic information. GeoJournal (forthcoming)
Sui DZ (2004) GIS, cartography, and the third culture: geographical imaginations in the computer age. Prof Geogr 56(1):62–72
Sui DZ (2005) Beethoven, picasso and GIS: is spatial really special? GeoWorld 9:22–24
Sui DZ (2008) Geography and GIS in second life. GeoWorld 9:19–21
Sui DZ (2009) Mashup and the spirit of GIS and geography. GeoWorld 12:15–17
Sui DZ, Goodchild MF, Elwood S (2012) Volunteered geographic information, the exaflood, and the growing digital divide. In: Sui DZ, Elwood S, Goodchild MF (eds) Crowdsourcing geographic knowledge: volunteered geographic information in theory and practice. Springer, Berlin, pp 1–20
Tuan Y (1977) Space and place: the perspective of experience. University of Minnesota, Minneapolis

Wallace DB, Gruber HE (1989) Creative people at work. Oxford University Press, New York
Wilson MW (2012) Critical GIS reading list. Available on-line at: www.criticalGIS.com (last accessed 23 Aug 2013)
Wood D (2010) Everything sings: maps for a narrative atlas. SIGLIO, Los Angeles
Young J, Gilmore M (2013) The spatial politics of affect and emotion in participatory GIS. Ann Assoc Am Geogr 103(4):808–823

Part I
Web and Crowd-Sourcing GI

Enriching Navigation Instructions to Support the Formation of Mental Maps

Monika Sester and Sagi Dalyot

Abstract Navigation systems are widely used today. These systems usually provide turn-by-turn origin-to-destination navigation instructions via visual and audio guidance. Whereas the systems most of the times work successfully, experiments have shown that users have problems in developing survey knowledge of the environment they passed through. This is due to the fact that users simply follow the instructions, without actively mentally mapping their current location in respect to the environment. The research presented in this paper aims at enriching conventional routing instructions with additional information (survey knowledge), which puts the route into the general spatial context and thus facilitates the building up of the users' mental map of the environment. In this paper, first investigations are presented in terms exploring two different data sources for enrichment, namely topographic and VGI data. A taxonomy of relevant features and relations is given, and methods for their automatic extraction and evaluation are described.

Keywords Wayfinding · Mental map · Car navigation · Survey knowledge · Landmarks

1 Introduction and Overview

Navigation devices provide a technical support for efficient wayfinding and navigation, mainly by guiding a user from location *A* to location *B*. This is carried out by turn-by-turn instructions including relative distances and road landmarks, often accompanied by a map overlay, which gives the user an idea about the spatial

M. Sester (✉)
Institute of Cartography and Geoinformatics, Leibniz Universität Hannover,
Hannover, Germany
e-mail: monika.sester@ikg.uni-hannover.de

S. Dalyot
Transportation and Geo-Information Engineering, Technion, Haifa, Israel
e-mail: dalyot@tx.technion.ac.il

F. Harvey and Y. Leung (eds.), *Advances in Spatial Data Handling and Analysis*, Advances in Geographic Information Science,
DOI 10.1007/978-3-319-19950-4_2

context around the route. The users mostly rely on listening and following the instructions. Thus, an explicit acquisition of the spatial knowledge of the area traversed by the route is not needed—and thus usually not done.

Whereas the systems in most of the time successfully solve their main task of guiding users to their destination, experiments have shown that users have problems in developing overview knowledge of the environment they passed through. This is due to the fact that users simply follow the instructions given by the navigation system, without actively mentally mapping their current location in respect to the environment (e.g., acquiring and using spatial survey knowledge). This is different when users analyze a map and structure the route before their trip: they scan the map, identify important locations and landmarks in the surroundings, and more specifically in relation to their planned route, consequently obtaining a general impression of the layout of the environment and the route they have to follow. This leads to the formation of a mental map of the environment. The lack of environmental overview knowledge, as in car navigation, is a deficiency, and as such it may also lead to problems when it comes to react on deviations, e.g. traffic jams or road constructions, and in worst cases it might lead to unsafe driving and accidents.

The research hypothesis described in this paper is that communicating augmented information about objects (and their spatial context) in their environment will help the user to memorize routes better and also explore an unknown area better, thus, aiming to mimic the map study process. This information is in addition to landmarks, which are crucial for wayfinding Fig. 1. schematically visualizes the additional information when going from mere turn-by-turn-instructions, over to a landmark-based description to a description using survey knowledge.

In this way, the navigation process is complemented by an educational experience, which informs users and makes them aware of supplementary spatial information. This process involves the retrieval, extraction and integration of environmental information, and its adequate selection and communication to the user. In this paper, first investigations are presented in terms of spatial data interpretation and retrieval exploring two different data sources for enrichment, namely topographic data and crowdsourced geo-tagged information. Whereas the first provides conventional environmental information, the latter has the potential to

Fig. 1 Turn-by-turn-instructions (*left*); use of landmarks (*middle*); enrichment by survey knowledge (*right*)

supplement with additional environmental information and being better suited to the interests and needs of the users, and thus provides better user adaptability.

The paper is organized as follows: the next section outlines the state of the art relevant for this research, followed by the presentation of a taxonony of important features and relations necessary for conveying survey information. Subsequently, methods for the automatic extraction of these features from geographic databases are sketched and how to link them to a route description via relations. Based on this, examples for route enrichments are shown, which are compared with routes derived from navigation systems. A discussion and an outlook on future work conclude the paper.

2 State of the Art

The research addresses different research fields in the domain of spatial cognition and spatial data processing. The state of the art in these domains will be briefly discussed in the following.

2.1 Conceptualization of Space

The spatial representation of the environment in the human mind is formed from a variety of geographic features. According to Lynch (1960) these can be divided into paths, districts, edges, landmarks and nodes. They include the locations of these features in the environment, the distances among them and the knowledge necessary to orientate oneself in the environment. The primary function of this spatial representation is to facilitate location and movement within the larger physical environment and to prevent oneself from getting lost (Siegel and White 1975). According to Thorndyke (1981) there are three different types of spatial knowledge:

- *Landmark knowledge*: memory of prominent geographic features or objects.
- *Procedural knowledge*: knowledge of route representation, action sequences that connect separate locations, and procedures for navigating between points.
- *Survey knowledge*: knowledge of the configuration of features and the global organisation of those objects and the relationship between different routes.

The author also states that these three types of knowledge are build sequentially; that a person's knowledge typically progresses from landmark to procedural to survey knowledge with increasing familiarity with the environment (Thorndyke and Hayes-Roth 1982). People traditionally acquire environment knowledge through direct exploration of an environment, and/or through indirect methods, such as the study of formal maps, informal sketch maps, verbal instructions, and so on (Burnett and Lee 2005).

2.2 Landmarks and Survey Knowledge—Cognitive Aspects of Wayfinding

The important role of landmarks and survey knowledge for the effectiveness of route directions and descriptions was studied and confirmed by several researchers (Deakin 1996; Denis et al. 1999; Michon and Denis 2001; Tom and Denis 2003). The conclusion made by these researchers showed that survey knowledge having a more absolute reference frame, and not merely in relation and relative to the specified route, produces more broad and comprehensive mental and cognitive understanding of the space and orientation in it. There have been many studies on the role of landmarks on wayfinding, especially on the automation of the process of defining and selecting appropriate landmarks (Raubal and Winter 2002; Elias 2003). Several studies showed that navigation systems contribute much less to the development of cognitive spatial models, as compared to the use of printed maps (Burnett and Lee 2005; Dickmann and Kestermann 2011; Leshed et al. 2008) This has the negative effect that users are lost if they cannot use their navigation systems. Also, it is difficult for them to evaluate alternative routes, e.g. in case of roadblock (Hipp et al. 2010).

Daniel and Denis (1998) analyzed propositions used in route descriptions and came up with 5 classes, two of which only relate to local landmarks in the vicinity (class 3) and their properties (class 4). Schwering et al. (2013) extended this concept by additional classes indicating orientation information, namely orientation using local landmarks, orientation using global landmarks, turning/non-turning movements using local landmarks. In their study with human subjects they revealed that global landmarks were only rarely used in route descriptions. However, they claim that those landmarks would support global orientation.

2.3 VGI—Volunteered Geographic Information

VGI describes a collective and collaborative accumulation of spatial data through a web platform (Goodchild 2007). Although typically VGI data are not as complete in coverage, thematic granularity, homogeneity and quality as authoritative data, they can provide a rich source of complementary information compared to authoritative data (Jackson et al. 2010; Sester et al. 2014). VGI data sources provide large amounts of personal knowledge, and thus hold a great potential for improving the quality of existing spatial databases used for navigation and Location Based Services (LBS). The exploitation of VGI for different purposes is currently actively researched, e.g. in the context of catastrophe management (Dittrich and Lucas 2014; Fuchs et al. 2013). Furthermore, VGI can be used as data source for the retrieval of more informal information, such as vernacular places or local knowledge, which often is not represented in maps, although heavily used by people in their daily lives. The automatic extraction of such information has been investigated by several researchers (e.g. Jones et al. 2008; Paelke et al. 2012). VGI have the potential to provide timely and immediate information, because the update process of such data is usually event based (and not time based as in authoritative data).

3 Taxonomy of Survey Information for Formation of a Mental Map

The type of information required for enriching conventional routing has two components: one that relates to general environmental information; the other that refers to individual information, and thus takes individual preferences into account. The assumption is that the former information gives orientation concerning visible topographic objects ("crossing the railway line"), while the latter supports memorizing, as it includes emotions and thus is of relevance for the individual ("the junction where the famous/your favorite restaurant is").

Besides conveying specific objects in the local environment of users on their routes, also basic fundamental concepts should be provided, which embed these features in a broader spatial context, such as cardinal directions, relations to other (known) objects. The abstraction and integration of both delivers with survey knowledge, such that the mental representation of this knowledge can be considered as "cognitive map" (Sholl 1987).

When selecting features, saliency plays a crucial role: this has been investigated for landmarks e.g. by Raubal and Winter (2002). In this way, characteristics of landmarks were identified, which support the immediate perception but also visual distinction of different objects. In the following, the investigation of Elias (2002) is used as a basis for defining relevant survey information, which is extended by additional relational features.

3.1 Information from Topographic Maps

The following list of objects from a digital topographic map is examined for features and attributes, which can enhance a route description by providing orientation information. These are objects directly crossing or bordering the route, objects in the vicinity, or objects which are important due to their size, their regional or global relevance and also visibility. Importance can be measured by a certain attribute (size, form) or derived from the proper name (e.g. church).

In the topographic data set of Germany (ATKIS), the following object classes have been identified:

- Transportation:
  - road traffic (e.g. roads)
  - rail traffic (e.g. tram rails, railroads)
  - traffic infrastructure (e.g. tunnels, bridges)
- Water:
  - sea, lakes, ponds
  - streams, rivers, creeks
- Settlement:
  - built-up areas
  - industrial areas
  - open areas with specific function (sports facilities, zoos, ..)
  - special buildings and facilities (towers, monuments, ..)
- Vegetation:
  - forests
  - gardens, parks

From cadastre, in particular the following information is relevant:

- Buildings:
  - Large buildings
  - buildings with special, extraordinary, remarkable shape
  - buildings with specific use (name)
- Important infrastructures:
  - railway stations
  - airports.

3.2 Information from VGI

Since VGI data is maintained by the public and various sources exist, no single and consistent catalog is maintained, as in ATKIS data. VGI platforms considered here are not characterized as GIS ones (such as OSM), such that they are not geographic platforms in nature. Therefore, VGI datasets considered at this stage are the ones explicitly storing position tags (geo-tagged); these are later geo-registered to the extent of the topographic maps, emphasizing on the retrieval of complementary semantic attractions that might not exist in authoritative datasets.

Two volunteered data sources are considered and analyzed here: *Wikipedia* (wikipedia.org) and *foursquare* (foursquare.com). Wikipedia is used here for the retrieval of cultural, architectural and historical objects, as well as other geographic features (such as water bodies). Foursquare, maintained and ranked by the public, delivers a list of attractions that are less touristic-oriented and more recreational; e.g., coffee houses, art galleries, restaurants. The update-rate of foursquare is very frequent (updates are aggregated as '*check-ins*'), offering the possibility of turning all features existing in the spatial extent into a 'venue' (feature), thus the density of spatial features is very high.

3.3 Overall Basic Concepts in Cities

In the conceptualization of the environment, people are using also overall concepts and vernacular descriptions, e.g.:

- city center
- old town, historical center
- urban districts
- ring roads

Such implicit information can be extracted from VGI data (e.g. Jones et al. 2008; Lüscher and Weibel 2013) or from topographic data based on spatial data interpretation (Heinzle and Anders 2007).

3.4 Spatial Relations

It is very important to link the different concepts with each other, with the route, and the current position of the user. In the following, some important relation concepts are listed:

- directions and cardinal directions (e.g. North, South, Northwest)
- in front of
- right of, left of
- towards (north, south, …, object)
- close to/near/next to
- passing through (e.g. city center, harbor area, …)
- passing at the border of
- passing at the northern border of
- between x and y (e.g., between river and city center)
- crossing (e.g. route crosses railway line).

4 Extraction of Survey Information from Spatial Data

The information defined in Sect. 3 can be retrieved from existing spatial information in different ways. On the one hand, some of the information is explicitly modeled in the spatial data sets and thus it can directly be selected and used. In our case, the topographic data is retrieved based on this approach. On the other hand, some information has to be aggregated and/or interpreted from the given information, e.g. the size of an industrial area has to be computed after all the individual industry objects have been merged. Also, additional attributes can be calculated or determined, which allow attaching a measure of salience or relevance to the objects.

In the following, it is assumed that topographic data can be selected according to given criteria, whereas the selection of relevant information from large collections of VGI data is a challenge.

4.1 Extraction of Survey Information from VGI

For the retrieval of Wikipedia features (entries), the Wikipedia API[1] is used based on SQL format. Two options are available for the retrieval of geographic entries: either by a bounding box or by a radius search. The use of *Location*, *Geographic coordinates*, and *Category* tags of the Wikipedia entry are used to retrieve the semantic attractions desired. In order to be able to assign a ranking weight (significance measure), i.e., the potential a feature has in contributing to a better cognitive appreciation and understanding of the space the user is in, the following entry attributes (tags) are considered:

- *Date of page creation*—the "older" the entry is, the more significant and important it might be
- *Total number of edits* and *Date of latest edit*—more edits exist meaning that the entry is more significant and being updated
- *Total number of distinct authors*—more distinct editors exist meaning that the entry is more significant, important, and reliable
- *What links here*—more links exist mean that the entry is relevant to various categories and other entries, i.e., its hierarchy might be high
- *Page view statistics*—high entry views implies that an entry is of interest

All attractions existing in foursquare have a *rating* (on a scale of 0–10). Foursquare entries are retrieved with the help of the foursquare API,[2] according to the entry position (coordinates). Additional parameters can be used to tune the query, whereas here the use of *section* was implemented, retrieving features categorized as *outdoors* and *sights*. Other attributes used for significance measure are:

[1]https://www.mediawiki.org/wiki/Extension:GeoData.

[2]https://developer.foursquare.com/docs/venues/explore.

- *checkinsCount*—total number of users "visits", i.e., check-in, to the feature; the higher the value implies the features is of an importance
- *usersCount*—total number of unique users "visited" (checked-in) the feature; the higher the value implies the feature is of an importance
- *tipCount*—total number of tips given to the feature
- *numPhotos*—total number of photos uploaded by users to the entry page.

4.2 Determination of Spatial Relations Between Features

The relevant features are not merely listed, but they have to be linked to the route, which the user is taking. This is achieved using spatial relations. They have to be extracted using spatial operations in a GIS or in a spatial database. Some of them are directly available as implementations (e.g. topological relations using 9I-model); some have to be developed and implemented. In the following, a preliminary list of operations is given for the realization of the above mentioned relations.

Directions can be determined as heading angle towards an object with respect to the north direction; thus the operator takes two points as input, as well as a direction angle and a tolerance angle (depicted in Fig. 2).

Cardinal directions can be determined by a variant of the direction operator using appropriate values for the angles (e.g., direction north: $\boldsymbol{\beta} = 0\,°$, $\boldsymbol{\alpha} = 45\,°$).

Van Kreveld and Reinbacher (Van Kreveld and Reinbacher 2004) describe an algorithm for partitioning a polygon into compass directions, to determine the cardinal directions within a polygon object. In this way, information such as "in western part of the city" can be created.

Towards < cardinal direction or object > can be calculated using the current heading intersected with a target object or target direction. It can be realized with the direction operator.

In front of can also be realized with a variant of the direction operator (depicted in Fig. 3):

Right of, left of can be determined based on a given orientation of an object and an analysis of direction angles.

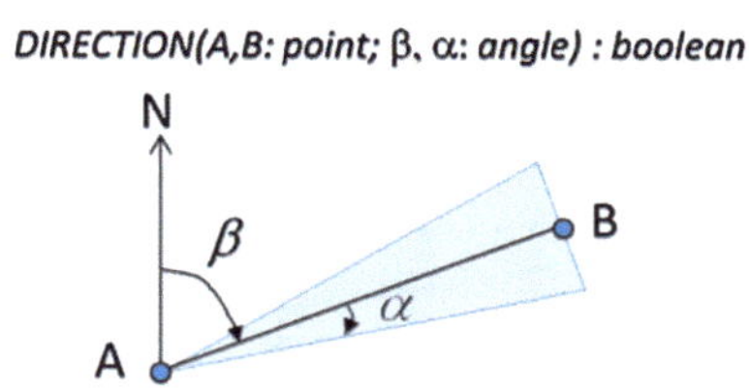

Fig. 2 Direction operator: input two points (A, B), direction angle β and tolerance angle α

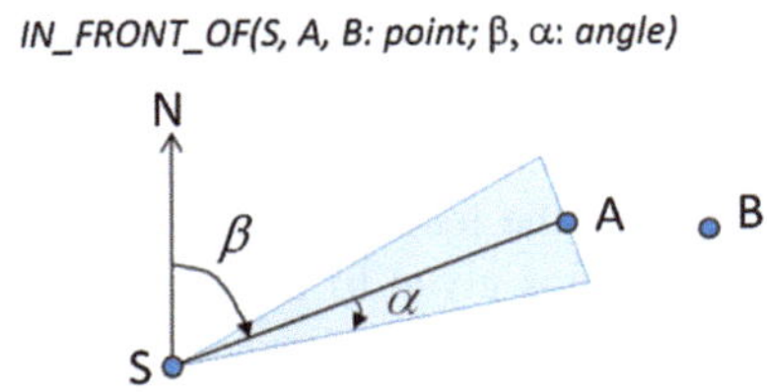

Fig. 3 In front of operator: related to start point S and destination point A that lies in front of point B

Near, close to, next to is an operation that has to take the distance of two objects as well as their importance into account, e.g. for an airport, a larger distance can be used to be considered as near as opposed to a small coffee-shop. This can be formalized using an exponential function of the distance (Arampatzis et al. 2006), depicted in Eq. 1 and Fig. 4.

$$near(A, B) = \exp(-L * D(A, B)) \quad (1)$$

where A and B are the centroids of the two (point) objects, D(A,B) is their Euclidian distance. Thus, proximity scores decay exponentially from 1 to 0 with increasing distance. L controls the rate of decay and has to be set according to the importance of the object: a large object like an airport has a small L-value, leading to a slow decay.

Passing through is an operation on a line and a polygon feature and can be realized using the overlay or intersection function from the 9I-model. It can be refined by comparing the relative sizes of the areas that are partitioned by the linear object into "passing through the middle", or "passing in the northern part of". The relation "passing at the border of" can be realized using the touch relation.

Crossing is an operation which relates to two linear features and is realized with an intersection operation (e.g. crossing the railway line).

Between: x and y (e.g. a road between river and city center) can be determined by analyzing the relative ordering in the neighborhood of three objects.

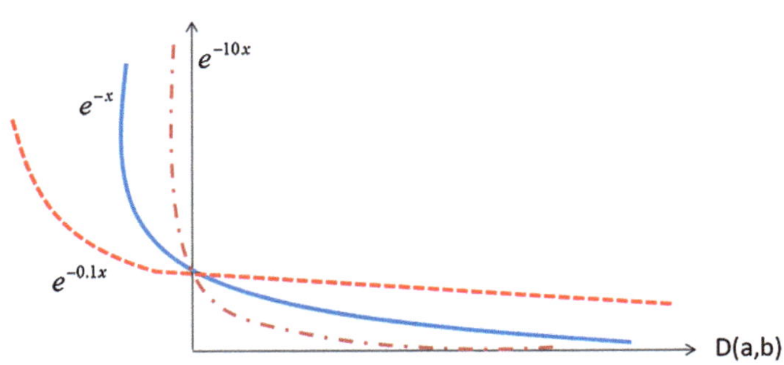

Fig. 4 Near-Operation: function of distance and decay factor; low value (e.g. 0.1) leads to slow decay, corresponding to a large neighborhood

5 Combination of Routes and Survey Information

In order to enrich a given route with survey information, an appropriate selection of relevant objects has to be made; relevant objects do have a relation with the route, namely:

1. they are adjacent to or cross the route (e.g. railroad)
2. they have a certain importance
3. they are of local perceptual significance, as they are directly visible from the route (e.g. city hall on right hand side of the route)
4. they have regional or global relevance and thus are put into a spatial context with the current location (position and heading) on the route

The regional or global importance of an object can be determined from the function or name of the object (e.g. Opera house, TV tower). Winter et al. (2008) propose an alternative by creating a hierarchical ordering of (landmark) objects based on a saliency criterion (Elias 2003). In VGI data, as explained above, the notion of relevance can be determined using the ranking or rating measures.

The concept for combining the survey information with a given route is as follows:

1. Identification of survey knowledge elements in the local environment around the route (buffer).
2. Determination of spatial relations between the route and the extracted elements.
3. Identification of global features in a larger environment of the route.
4. Determination of spatial relations between those features and the route.

As a result, the route is enriched with additional survey information. Besides only naming or mentioning the objects, also further information about the objects can be communicated, e.g. when the historical church was built and by whom.

6 Examples

6.1 Data Retrieval from VGI

Entries and places from in the area of Hannover city, Germany, were retrieved based on the concepts detailed in Sect. 4.1. In Wikipedia, two main data sources were considered: the German Wikipedia-DE, and the English one-EN. Since route examples analyzed are in Hannover, more DE entries were retrieved than EN ones: 721 and 124, respectively, detailed in Table 1. Though the DE data is much more complete and has much more details, it was found that all EN entries retrieved had DE ones as well, and perhaps more important—the majority of significant features (i.e., high count of entry parameters used for ranking) existed in the retrieved data. Such as historic, transportation (main train stations and airport), cultural and architectural objects, university. It was hard to use the DE *category* tag, since no consistent classes

Table 1 Number of entries retrieved for DE and EN Wikipedia

Feature type	EN (total = 124)	DE (total = 721)	Description
City	26	21	Different districts existing in the area
EDU	2	–	Universities
Landmarks	64	596	All types (churches, museums, statues, …)
Mountain	4	2	–
Railway station	12	12	Mainly high speed train stations (not trams)
River	3	1	–
Water bodies	4	4	lakes
Null	9	94	–

were used (for example: Marktkirche in the city center is categorized under "church" in the EN entry, whereas in the DE entry it is categorized under 10(!) different sub-categories). Also, the use of native-language ("local") Wikipedia holds a great challenge, since additional language barrier exists in building the required taxonomies. Moreover, the DE data consisted of many entries that could be considered as irrelevant, for example: local shops and entries having a geo-tag but present only general knowledge (football games took place in Hannover in 2003). Consequently, only the EN entries were considered at this stage.

In the analysis of foursquare data, the focus here was given to two main categories considered of having more relevance to wayfinding: *outdoors* and *sights*. In contrast to Wikipedia, the type of venues (named categoryName), was fairly easy to retrieve, and was clear and easy to identify into taxonomies (for example: Plaza, City Hall, Monument/Landmark, Garden, Church). A first impression on the retrieved places seemed very promising, since it was found that many geographic features, which we did not expect to find, did exist in foursquare, namely city districts (*Neighborhood*), together with other features that did not exist in the EN Wikipedia. However, *rating* parameter was found unusable since it did not exist in most entries. Alternatively, the other parameters suggested in Sect. 4.1 were used, which had a very high correlation, depicted in Table 2 (*summary* in Table 2 represents the number of "likes" users-comments for a venue).

Table 2 Feature statistics of different relevance parameters in foursquare

Name	Category name	Checkin count	Users count	Num photos	Summary
HDI Arena	Soccer Stadium	3899	1435	254	109
Kröpcke	Plaza	3517	1199	113	16
Maschsee	Lake	3208	1265	254	48
Leibniz Universität Hannover	University	2652	503	58	11
Zoo Hannover	Zoo	2340	1186	175	115
Ernst-August-Platz	Plaza	2058	676	25	9
Neues Rathaus	City Hall	1843	934	265	18

Fig. 5 VGI features retrieved for the centre of Hannover superimposed on the road network and topographic features (*polygons*). *Circle* depicts the features' ranking (some features are not visible having low ranking), *blue* Wikipedia, *red* foursquare

Figure 5 gives a general impression of all features extracted by all available sources. It is interesting to observe that, while there is an overlap of features in the data sets, as expected VGI data contributes with local information specifically about places, and—in principle, but not analyzed here—with additional information such as shopping centers, restaurants, leisure places. The large pink area in the East symbolizes a large forest (Eilenriede), whereas the rectangular polygon in the West is the famous Baroque Garden; next to it, in green, is the George's Garden.

6.2 Route Examples

In the following, example routes have been selected and enriched with the survey information (Fig. 6 and Fig. 8). The descriptions are compared with a conventional routing instruction from Google (Fig. 7 and Fig. 9). It has to be noted, that conventional routing instructions do serve another purpose (only wayfinding and guiding), and thus are mainly based on directions and street names. Still, the comparison with the survey information shows the richness of the information conveyed.

6.2.1 Example 1

In Google Maps, Route 1 is described in Figs. 7:

Using the proposed concept, the following information will be conveyed along with the routing instructions:

1. your start is in the district Herrenhausen, which lies in the north west of the city; on your right you should see the new Herrenhausen castle; drive along Herrnhäuser Straße in eastward direction; you enter the district Nordstadt
2. continue on Nienburger Straße in direction south-east, the route heads all the way towards the Neues Rathaus (city hall)
3. you will drive along the north-eastern border of the Georgengarten with the Wilhelm-Busch Museum on your right
4. To your left, you see the Welfenschloss, the main building of Leibniz Universität Hannover
5. continue on Königsworther Platz and then turn right onto Königsworther Straße and drive southwards

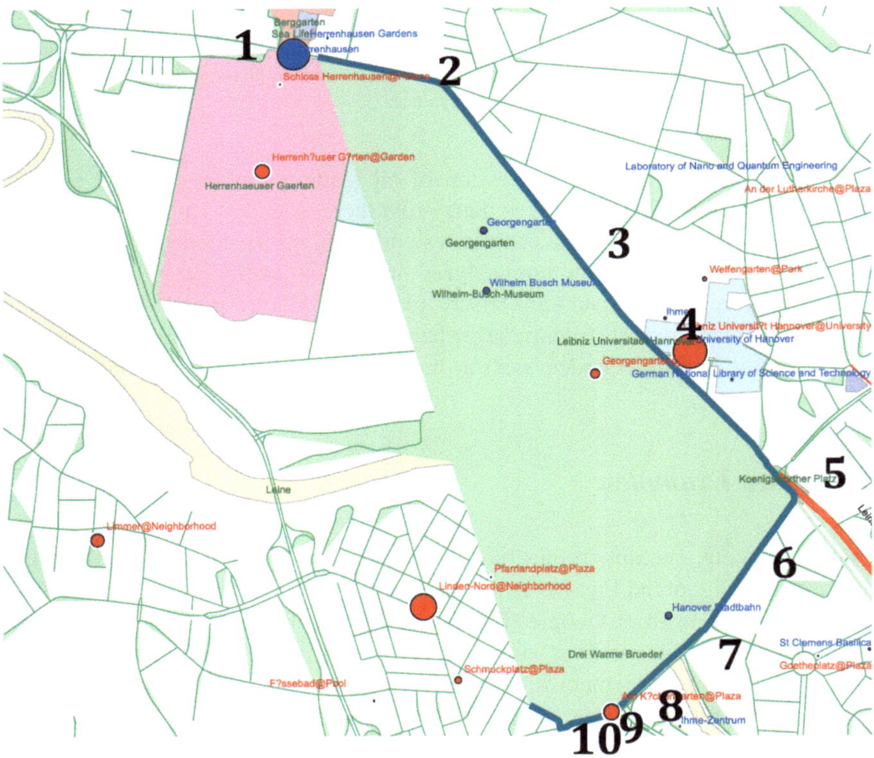

Fig. 6 Route 1: origin (*top-left*) to destination (*middle-bottom*) with route labels of enriched routing instructions

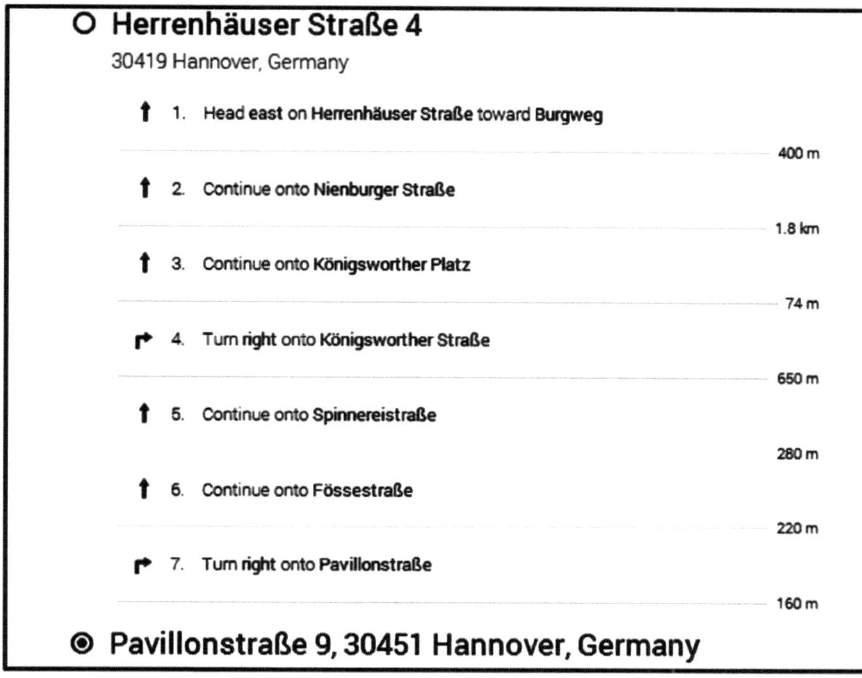

Fig. 7 Instructions for route 1 given in google maps (*source* google maps)

6. you are now in the district Calenberger Neustadt, which is located west of the city center
7. you will now cross the river Leine and reach the city district Linden-Nord; continue straight on Spinnereistraße
8. On your right hand side you pass the huge heating plant with its three towers; it is colloquially called “drei warme Brüder”—the three hot brothers
9. You now access Am Küchengarten plaza
10. turn right into the Pavillionstraße towards Linden-Nord centre.

In this way, both information about local points of interest is given, but more importantly, general survey information is conveyed, which locates the objects into a reference frame, e.g. with compass bearings, and with reference to other objects. According to Thorndyke and Hayes-Roth (1982), survey knowledge includes in addition also distances between the objects. This is not yet included in the proposed concept, and is subject to future work.

6.2.2 Example 2

In Google Maps, Route 1 is described in Figs. 9:

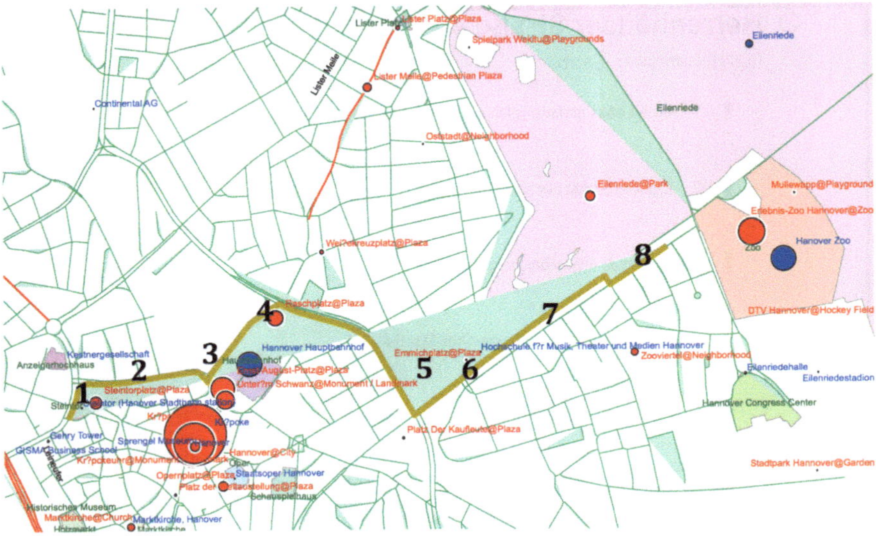

Fig. 8 Route 2: origin (*middle-left*) to destination (*top-right*) with route labels of the enriched routing instructions

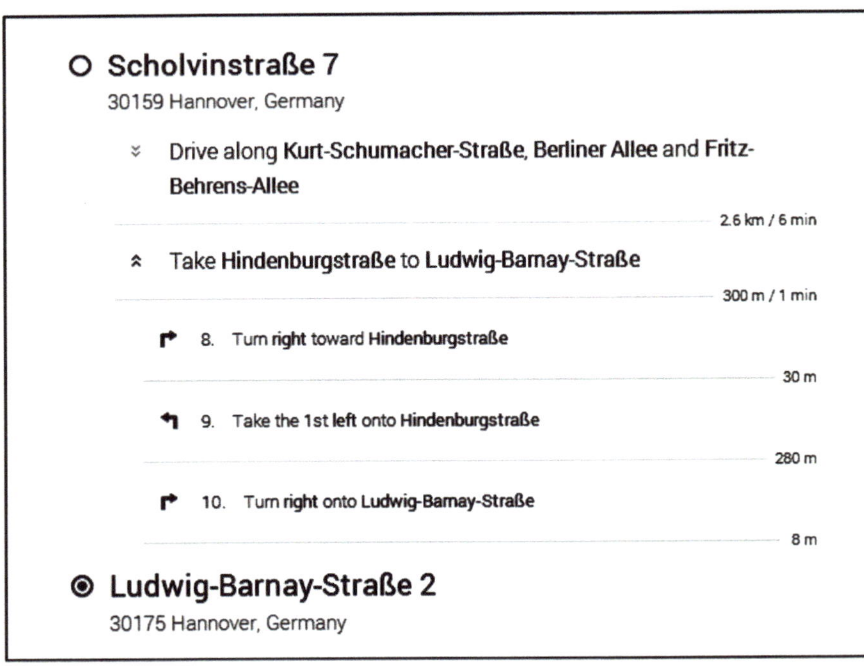
Scholvinstraße 7
30159 Hannover, Germany

Drive along **Kurt-Schumacher-Straße, Berliner Allee** and **Fritz-Behrens-Allee**

2.6 km / 6 min

Take **Hindenburgstraße** to **Ludwig-Barnay-Straße**

300 m / 1 min

8. Turn **right** toward **Hindenburgstraße**

30 m

9. Take the 1st **left** onto **Hindenburgstraße**

280 m

10. Turn **right** onto **Ludwig-Barnay-Straße**

8 m

Ludwig-Barnay-Straße 2
30175 Hannover, Germany

Fig. 9 Instructions for Route 2 given in Google Maps (*source* Google Maps)

Using the proposed concept, the following information will be conveyed along with the routing instructions:

1. Your start is in the quarter Steintor, which lies in the west of the city centre; on your right you should see the Steintor subway station and Georgstrasse which leads to Kroepke Plaza—the city centre; drive north along Am Steintor
2. continue right on Kurt-Schumacher in direction north-east, the route heads all the way towards the central train station (Hauptbahnhof)
3. Approaching central train station, before head left, you should see ahead of the Ernst-August-Plaza with the "Unterm Schwanz" monument—a statue of King Ernst-August I on horseback.
4. Drive north while the central-train-station is on you right on Raschplaz, and turn right when approaching on Raschplatz plaza; this is one of Hannover's main entertainment centres with a large cinema complex.
5. Drive eastwards along Berliner Allee until the next main intersection
6. Head left northwards towards Zooviertel neighborhood, which holds the Hannover Congress Center
7. Continue straight while you pass on your left the Eilenriede park—one of Germany's largest urban parks
8. While approaching your destination you will see the Hannover Zoo with its attractions.

7 Conclusion and Outlook

In the paper, a concept for the enrichment of routing directions with additional survey information is proposed. The features have been selected based on a literature survey. A selection of features is presented, which can be found either in digital topographic maps or in VGI data, and methods for their automatic selection are described. Exemplary route descriptions have been manually created using those features and linking them to the route with spatial relation expressions.

There are several issues that will be investigated in the future. An important topic is the conducting user experiments in order to verify the hypotheses set up in this paper concerning their influence on the human cognition. To this end, tests will have to be conducted, where a group of test people who used enriched route descriptions have to solve some tasks, as opposed to ones, who used conventional navigation instructions: e.g., draw a sketch map of the environment, describe the route in the reverse order, or answer questions concerning the overall structure of the environment. Based on these investigations, further refinements on the features and relations involved will be conducted. It is also expected that the type of features used will differ with increased familiarity of the users with the environment. An interesting topic will be to identify which type of feature is useful for which degree of familiarity.

Concerning VGI additional investigation is required into the retrieval of the most significant information, in terms of the feature itself, but also in terms of user-specifications (e.g., preferences or wayfinding purpose), and perhaps in terms of LBSN and recommendations from other users. Another aspect is the filtering-out of duplicated features retrieved both from the VGI and topographic datasets. Still, the use of VGI has important benefits, in that firstly it allows to present supplementary as well as user-adapted information. Secondly, from social media data, the notion of relevance can be directly extracted, e.g. the popularity of a place and its purpose.

All in all this process has great potential in addressing cognitive problems associated with the building up of a mental map, but also in terms of user-defined requirements and the exploration of new places.

Acknowledgments This research benefited from the input of Dipl.-Inform. Daniel Fitzner and Anna Walter who provided valuable ideas and assistance to the analysis and experiments.

References

Arampatzis A, van Kreveld M, Reinbacher I, Jones CB, Vaid S, Clough PD, Joho H, Sanderson M (2006) Web-based delineation of imprecise regions. Comput Environ Urban Syst (CEUS) 30 (4):436–459

Burnett GE, Lee K (2005) The effect of vehicle navigation systems on the formation of cognitive maps. In: Underwood G (ed) Traffic and transport psychology: theory and application. Elsevier, Amsterdam, pp 407–418

Daniel M-P, Denis M (1998) Spatial descriptions as navigational aids: a cognitive analysis of route directions. Kognitionswissenschaft 45–52

Deakin A (1996) Landmarks as navigational aids on street maps. Cartogr Geogr Info Syst 23 (1):21–36

Denis M, Pazzaglia F, Cornoldi C, Bertolo L (1999) Spatial discourse and navigation: an analysis of route directions in the city of Venice. Appl Cogn Psychol 13:145–174

Dickmann F, Kestermann E (2011) Der Einfluss von Stadtplänen und Fahrzeug-Navigationssystemen auf die Entwicklung mentaler Raummodelle. Kartographische Nachrichten 4:183–190

Dittrich A, Lucas C (2014) Is this twitter event a disaster? In: Proceeding of AGILE Conference Geographic Information Science. Castellón Spain

Elias B (2002) Automatic derivation of location maps. GeoSpatial theory process. Appl, ISPRS Archives 34/4

Elias B (2003) Extracting landmarks with data mining methods. In: Kuhn W, Worboys MF, Timpf S (eds) Spatial information theory. foundations of geographic information science. Lecture Notes in Computer Science. Springer, Heidelberg, pp 375–389

Fuchs G, Andrienko N, Andrienko G, Bothe S, Stange H (2013) Tracing the German centennial flood in the stream of tweets: first lessons learned. ACM Press, pp 31–38

Goodchild M (2007) Citizens as sensors: the world of volunteered geography. GeoJournal 69:211–221

Heinzle F, Anders K-H (2007) Characterising space via pattern recognition techniques: identifying patterns in road networks In: Mackaness W, Ruas A, Sarjakoski T (eds) Generalisation of geographic information: cartographic modelling and applications, Published on behalf of the International Cartographic Association by Elsevier, pp 233–253

Hipp M, Schaub F, Kargl F, Weber M (2010) Interaction weaknesses of personal navigation devices. In: Proceeding of the 2nd international conference on automotive user interfaces and interactive vehicular applications, pp 129–136

Jackson MJ, Rahemtulla HA, Morley J (2010) The synergistic use of authenticated and crowd-sourced data for emergency response. In: Proceedings of the second international workshop on validation of geo-information products for crisis management (VALgEO). Ispra, Italy, pp 91–99

Jones CB, Purves RS, Clough PD, Joho H (2008) Modelling vague places with knowledge from the Web. Int J Geogr Inf Sci 22(10):1045–1065

Leshed G, Velden T, Rieger O, Kot B, Sengers P (2008) In-car gps navigation: engagement with and disengagement from the environment. In: Proceedings of the SIGCHI conference on human factors in computing systems, pp 1675–1684

Lüscher P, Weibel R (2013) Exploiting empirical knowledge for automatic delineation of city centres from large-scale topographic databases. Comput Environ Urban Syst 37:18–34

Lynch K (1960) The image of the city. MIT Press, Cambridge

Michon P, Denis M (2001) When and why are visual landmarks used in giving directions? In: Montello D (ed) Proceedings of the international conference COSIT 2001, Spatial information theory. Springer, Verlag, pp 292–305

Paelke V, Dahinden T, Eggert D, Mondzech J (2012) Location based context awareness through tag-cloud visualizations, advances in geo-spatial information science. CRC Press, Florida

Raubal M, Winter S (2002) Enriching wayfinding instructions with local landmarks. In: Egenhofer MJ, Mark DM (eds) Geographic information science. Lecture Notes in Computer Science, vol 2478. Springer, Berlin, pp 243-259

Schwering A, Li R, Anacta VJA (2013) Orientation information in different forms of route instructions. In: Short paper proceedings of the 16th AGILE conference on geographic information science. Leuven

Sester, M, Arsanjani JJ, Klammer R, Burghardt D, Haunert JH (2014) Integrating and generalising volunteered geographic information. In: Abstracting geographic information in a data rich world. Springer International Publishing, pp 119–155

Sholl MJ (1987) Cognitive maps as orienting schemata. J Exp Psychol 13(4):615–628

Siegel A, White S (1975) The development of spatial representations of large-scale environments. In: Reese H (ed) Advances in child development and behaviour, 10. Academic Press, New York, pp 9–55

Thorndyke PW (1981) Spatial cognition and reasoning. In: Harvey J (ed) Cognition, social behaviour, and the environment. Lawrence Erlbaum Associates, USA

Thorndyke PW, Hayes-Roth B (1982) Differences in spatial knowledge acquired from maps and navigation. Cognit Psychol 14:560–589

Tom A, Denis M (2003) Referring to landmark or street information in route directions: what difference does it make? In: Kuhn W, Worboys M, Timpf S (eds) COSIT 2003, LNCS 2825. Springer, Berlin, pp 362–374

Van Kreveld M, Reinbacher I (2004) Good NEWS: partitioning a simple polygon by compass directions. Int J Comput Geom Appl 14:233–259

Winter S, Tomko M, Elias B, Sester M (2008) Landmark hierarchies in context. Environ Plan B Plan Des 35:381–398

A Classification of User Contributions on the Participatory Geoweb

Claus Rinner and Victoria Fast

Abstract "Volunteered geographic information" (VGI) is the term most widely used to describe a variety of user contributions on the participatory Geoweb. These contributions range from coordinate locations and geometries to categorical observations, attribute tags, numeric measurements, and content ratings, as well as complex narratives, photos, and videos. Although researchers are creating and studying Geoweb applications, different types of VGI, and the related phenomena of neogeography, citizen science, and crowd-sourcing, systematic characterizations of user-contributed local knowledge are scarce. In this paper, we propose criteria to distinguish types of user-generated data and contents, and relate these to types of Geoweb applications. The proposed classification provides a conceptual framework to examine the participatory Geoweb, facilitate the processing of user contributions, and identify possible gaps in the data/content types currently used. This approach could help improve the effectiveness of current Geoweb applications, and increase the uptake of the valuable geographic information they generate.

Keywords Crowd-sourced data · Geospatial web · Online participation · User-generated geographic content · Volunteered geographic information

1 Introduction

The Geospatial Web, or short "Geoweb", is a network of Web 2.0 applications that enable two-way communication of geospatial data among and between citizens and organizations, including government (Leszczynski and Wilson 2013). This participatory Geoweb consists of an ever-increasing number of mapping applications that collect user contributions (Johnson and Sieber 2012). These geographically

C. Rinner (✉) · V. Fast
Ryerson University, Toronto, Canada
e-mail: crinner@ryerson.ca

F. Harvey and Y. Leung (eds.), *Advances in Spatial Data Handling and Analysis*, Advances in Geographic Information Science,
DOI 10.1007/978-3-319-19950-4_3

referenced user contributions take on very different forms, including coordinate locations, geometries, categorical observations, attribute tags, numeric measurements, and content ratings, as well as complex narratives and multimedia items, such as photos and videos (Coleman 2010). To date, there is no agreed-upon terminology to distinguish these different types of user contributions, although the term "volunteered geographic information" (VGI) is widely used for a variety of contributions (Elwood et al. 2012). In addition, researchers have noted that the processes, from which VGI emerges, also differ from each other. This led to additional characterizations such as involuntary geographic information, ambient geospatial information, and facilitated VGI (Tulloch 2008; Seeger 2008; Stefanidis et al. 2011).

In an attempt to comprehensively classify VGI, Deparday (2010) distinguishes types of VGI along a continuum of scientific knowledge, local knowledge, and personal knowledge. He associates these three types with multiple criteria, including the technique used to capture location information; the supported geographic feature type; the structured versus unstructured nature of attribute data; subjectivity versus objectivity of contributed information; degree of "volunteeredness" of contributions; and quality of participation. The mechanism for capturing location information is viewed as a key determinant for the nature of user contributions. Sources of location information include GPS recording, cellular phone positioning, address geocoding, and manual drawing on a map (Deparday 2010). Some of these mechanisms (e.g., GPS recording) are automatic, while others (e.g., map drawing) require the user's intention to locate their contributions. The supported geographic feature type, including points, lines, and polygons, also determines the nature of user contributions. A fundamental distinction has to be made between contributions linked to geographic coordinates, contributions linked to user-defined shapes, and contributions linked to geographic features representing real-world objects (Rinner 2001).

In addition to the location component of VGI, Deparday (2010) reviews the "text component", which he also terms "attribute data". Deparday (2010) distinguishes structured from unstructured attribute data, where structured attributes are defined as those "that conform to a range of values on nominal, ordinal, interval or ratios scales" (p. 21). With reference to Tulloch (2008), Deparday (2010) makes another important distinction: that between subjective and objective information provided by users. Likewise, Rinner et al. (2011) separate observations (i.e., objective information) from opinions (i.e., subjective information). Finally, the degree of interaction in the VGI collection process can be characterized as a one-way or two-way flow of information according to Deparday (2010), although since the completion of his thesis, the two-way information flow has been identified as a critical component of participant engagement on the Geoweb (e.g., Walker and Rinner 2013).

In this paper, we attempt to systematically classify Geoweb contributions by their data type. We believe that Deparday's (2010) linear classification misses some combinations of classification criteria and does not consider volunteered quantitative data. We therefore review the user contributions to current Geoweb applications from a variety of domains, including citizen science applications in weather mapping and invasive species monitoring, collaborative basemap creation and maintenance, crisis mapping initiatives, geosocial media and business review Web sites, and map-based discussion forums in urban and regional planning. From these examples, we identify different types of contributions in terms of their spatial and attribute dimensions; their data measurement levels; and their content types.

The following Sect. 2 examines current research on the Geoweb, geospatial crowd-sourcing, VGI, and user-generated geographic content (UGGC). Section 3 embraces a systems perspective of VGI and discusses the core functional groups of data/content input, management, analysis, and presentation along with examples of existing VGI systems. Section 4 proposes a framework for types of user contributions on the Geoweb, while Sect. 5 discusses possible uses of the framework and concludes the paper with an outlook on future research.

2 Research Context: Geoweb, Crowd-Sourced Data, VGI, and UGGC

With the emergence of expandable Web mapping interfaces such as Google Maps, more people have become entrenched in using maps for personalized directions as well as mapping data of interest in "map mashups". These significant developments can be viewed from a number of perspectives, including the blended roles of producer and consumer of user contributions, the non-expert nature of contributors, their motivations, and the number of participants and quality of their contributions.

Coleman et al. (2009) and Sieber and Rahemtulla (2010) have discussed the dual role of "produsers" or "prosumers" in the context of interactive Web mapping. The term neogeography has been used to describe the new expanded range of citizen non-experts involved in mapping enterprises based on their local knowledge rather than formal training (Haklay et al. 2008). The underlying technical infrastructure was termed the "geospatial Web" (Scharl and Tochtermann 2007), or "Geoweb", although a clear definition of the Geoweb is still missing. An important distinction between the informational Geoweb and the participatory Geoweb is made by Johnson and Sieber (2012). The informational Geoweb enables a one-way flow of information from producers to consumers, whereas the participatory Geoweb enables two-way communication. The participatory Geoweb relates to earlier developments in Geographic Information Systems (GIS) research and practice, such as public participation GIS, participatory GIS, collaborative GIS, and Sieber's (2004) call for a second-generation, bottom-up GIS.

With respect to contributor motivation, the concept of volunteered geographic information (VGI) was coined by Goodchild (2007) as a type of user-generated content on the Web, which was also termed as user-generated *geographic* content (UGGC) (Goodchild 2008). Goodchild (2007) considers three types of sensors that are creating location-specific "information": static and mobile physical devices as well as human individuals themselves. Goodchild (2007) also discusses citizen science as a context, in which VGI requires some skill level of contributors. To refine the "humans as sensors" perspective, we argue that citizens rather act as conduits, aggregators, or interpreters of their local observations. In addition, we have to consider user contributions taking the form of opinions in addition to observations (Rinner et al. 2011).

Another perspective on user contributions on the participatory Geoweb is reflected in the concept of crowd-sourcing, and ultimately, big data. Crowd-sourcing implies the notion that the contributors' combined local knowledge (the "wisdom of the crowd") will guarantee the emergence of high-quality information (Zook et al. 2010). Conceptually quite different processes can result in crowd-sourced data, including "involuntary VGI" (Fischer 2012), "ambient geospatial information" (Stefanidis et al. 2011), and "facilitated VGI" (Seeger 2008). Geographers have noted recently that facilitated VGI settings often result in limited participation and therefore anything but big data. In contrast, the collection of ambient and thus involuntary geographic information can result in very large databases, such as millions of geographically referenced tweets or photographs. Consequently, big (geographic) data are increasingly discussed from the perspective of locational surveillance and spatial privacy.

Geo-tagged social media items are perhaps best characterized by Goodchild's (1997) "information with geographically determined interest". Goodchild (1997) noted that library items such as books or photographs often have a geographic footprint, which determines a user's interest, as we are typically most interested in material referring to nearby areas. The geospatial information in such independent items is only implicitly defined, such as through place references in a novel.

Overall, there is a range of ways, in which the location and attribute components of VGI can be connected. Figure 1 shows the traditional GIS data model on the left with tightly linked location and attribute information, as it occurs in both the vector and raster data model. The right-hand side of Fig. 1 shows loosely linked VGI consisting of geospatial and non-spatial elements that exist independently in their respective realms, but can be connected automatically (geo-coding) or by user interaction (geo-tagging). The centre of Fig. 1 characterizes an in-between situation, in which somewhat independent data/content items can readily be connected, as it occurs when an external data table is spatially joined to a feature dataset in GIS.

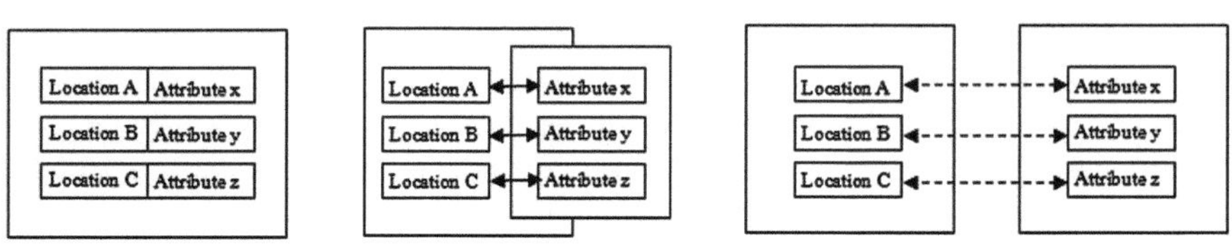

Fig. 1 Degree of connection between location and attribute components of VGI, associated with the link between location and attributes in traditional GIS data models

3 Functionality of VGI Systems

Fast and Rinner (2014) argued that the term "information" is improperly used in many publications on VGI, since information is a higher-order concept than the original user contributions suggest. In fact, many researchers are using the contradictory term "VGI data" (Coleman 2010; Cinnamon and Schuurman 2013; Sui et al. 2013). In analogy with Tomlinson's (2007) recommendations for successful GIS implementations, information should be regarded as the output of a VGI system rather than its input. Along the continuum of data-information-knowledge (e.g., Meeks 2007), VGI emerges from the processing of volunteered geographic data. This systems perspective still leaves the distinction of data from content up for debate. Therefore, this paper focuses on the handling of volunteered geographic data and volunteered geographic content as input to VGI systems.

Key to the definition of Geographic Information Systems (GIS) is the geospatial nature of the data being handled in these computer systems. By extension, key to VGI systems is the volunteered as well as the geospatial nature of their data/content. Examining the functionality of VGI systems serves to frame the sequence of tasks necessary for handling user contributions. Although VGI systems functions could be classified in different ways—for example, Turner discussed the production, consumption, analysis, visualization, and sharing of VGI (Wilson and Graham 2013)—we employ the functional groups of GIS as found in typical GIS definitions: input, management, analysis, and presentation, also known as the "IMAP" model (Bill and Fritsch 1999).

3.1 Input

VGI evolved around the concept of "citizens as sensors" (Goodchild 2007), using people as the primary input mechanism, and signaling a major shift from GIS input functions. Traditionally, the data stream began with either analogue or digital data, supplied by authoritative sources such as national mapping agencies. These data were captured using digitizing, scanning, data transfer, or key coding (Heywood et al. 2006). In the VGI systems framework, the data stream shifts from authoritative data sources to asserted data sources, where data capture mechanisms are often digital. In fact, while data had to be actively collected yesterday, some datasets are emerging as a byproduct of collaborative volunteer efforts today (Longley et al. 2011). For example, Google Maps for Mobile harnesses real-time traffic conditions from location data contributed by way of GPS-enabled smartphones (Barth 2009). In this case, the data input mechanisms are designed to be effortless on the part of the contributor; Google Maps for Mobile is "easy to install and use, […] making it easy for people to provide information about their own vehicle speed. There's no extra device to plug into your car and no extra software to buy" (Barth 2009). This example highlights the new mechanisms through which data can be collected from location-enabled devices.

While data input is becoming more automated, human intervention is still necessary and we still rely on data input tools (Goodchild and Li 2012). For example, anyone is invited to submit or edit the data of the OpenStreetMap initiative, but doing so still requires input devices (keyboard, mouse) and "ancient" input functions for geometric and attribute data, such as on-screen digitizing and categorization of geographic features. Nearly effortless input, as with the case of Google Maps for Mobile, could be the key to tapping into a broader source or user contributions.

3.2 Management

Similar to traditional GIS projects, VGI system require database management functions to facilitate the storage, organization, and retrieval of user contributions. A VGI system represents a multi-user environment, and requires a database management system that can accommodate its unique conditions. In addition to the spatial data and content, metadata are important to VGI systems. For example, contributor profiles stored in the metadata could assist project organizers in learning about the sample population contributing content, and lead to a more robust understanding of who is contributing, and why.

Of particular interest is the dynamic and sometimes real-time nature of the database construction. For timely updates on the map, especially in the case of emergency management where immediate response is necessary, the database management requires automatic storage and immediate retrieval. Databases can either be built for one project (e.g., emergency response) or be continuously maintained (e.g., OpenStreetMap). Another management consideration is verification procedures to control the input of UGGC, including contributor logins, trusted contributors, and approval mechanisms. Ushahidi, for example, requires system administrators to moderate and approve all contributions before they are published online. In this instance, users input geographic content but in turn, the system administrators manage, analyze, and present it as volunteered geographic information. Although data input is technically open to anyone in the Ushahidi platform, the management of the data gives administrators control over the information produced and shared (Baker and Neu 2013).

Exploring how VGI fits into existing spatial data infrastructures (SDIs) can assist in the management of VGI. Mooney and Corcoran (2011) asked, from a computer science perspective, whether VGI is ready to be a part of SDIs, stating that SDIs are typically institutionally sanctioned, top-down approaches to data dissemination. Initially, it seems difficult to translate this top-down systems planning approach to the realm of VGI, since VGI initiatives are commonly characterized as "grassroots" or community-based, emerging, and minimally constrained. Sui and Goodchild (2011, 1742) similarly asked: "What protocols and procedures can be developed to link asserted, crowd-sourced social-media data with authoritative data to fill gaps in spatial data infrastructure?" Budhathoki et al. (2008) explored the complementary nature of VGI and SDI, highlighting that future research is needed to reconceptualize the role of the user as the producer in SDI. Determining how VGI fits into existing SDIs could be the first step toward balancing both authoritative and assertive data frameworks (Elwood et al. 2012; Coleman 2010).

3.3 Analysis

Goodchild (2008) highlighted that the analytical functions of GIS is what made it such a powerful tool, revealing insights not otherwise evident. Information with geographically determined interest has "patterns that are well behaved and therefore amenable to modeling; and of sufficient variability to impact locational decisions" (Goodchild 1997, 387). However, data analysis in VGI systems is generally less developed than the other functional groups. Bowker (2005) advised that the mass amount of data being generated daily necessitates deciding what data are important and then paring them down to usable information, or risk all the data being useless. Due to the sheer amount of user contributions on the Geoweb, geospatial data mining is emerging as a popular method of classifying and consolidating predominantly qualitative contributions (Mennis and Guo 2009; Elwood et al. 2012). De facto however, VGI as the system output is constrained by the hardware and software used in a particular initiative, as well as by the number and engagement level of participants and the amount and quality of contributed data/content.

Some platforms are enabling the discovery of trends and relationships in emerging dataset. SwiftRiver, made available through Ushahidi, filters and verifies real-time data while providing some analysis capabilities, including semantic analysis to auto-classify contributions (Baker and Neu 2013). More advanced geospatial analysis functions, such as buffering or interpolation, are not widely implemented in VGI systems. On the one hand, the users of a VGI system are rarely trained in geographic problem solving, thus limiting their ability to properly apply such functions. On the other hand, many geospatial analysis functions require numeric data that are not the focus of typically qualitative VGI initiatives (Fogliaroni et al. 2010), or that would need to be generated from raw data (e.g., counts of contributions per area). An alternative to analysis within VGI system, particularly suited for those trained in spatial analysis, is exporting the dataset to be used in GIS or statistical software. Integrating the strengths of VGI system with existing GIS analysis capabilities has the potential to produce more complex VGI from user contributions.

3.4 Presentation

Effectively presenting the information derived from VGI systems is an important function, ultimately enhancing our spatial understanding of the world around us.

Similar to the information products resulting from the data-to-information transformation described by Tomlinson (2007), we need to consider information products as the output of VGI systems. In a Geoweb environment, these information products are extending beyond traditional static maps to dynamic and interactive methods for sharing and visualizing VGI (Wilson and Graham 2013). Elwood et al. (2012) define "VGI as geographic information acquired and made available to others." (p. 5) The Geoweb is not only enabling the collection of user contributions through increasingly simplified and interoperable mapping interfaces, but also leads to a wake of new options to make geographic information available to others (Sui and Goodchild 2011). The Geoweb provides an "intuitive view of spatial phenomena for a wider audience than conventional maps" (Li et al. 2011).

Although interactive maps extend visualization and presentation capabilities for communicating geographic information, we also need to consider outputs that are accessible, succinct, and easily integrated into existing information channels. Authoritative stakeholders (including government, academic, and private sectors) require information that can be incorporated into reports, publications, and policy. In this case, datasets should also be considered an information product derived from VGI systems, as they are more easily converted into traditional tabular, graphic, textual, and static map outputs. In particular, developing VGI systems derived datasets to integrate into various SDIs can be of most benefit, as the dataset could be the starting point for a host of research and development. For example, Johnson et al. (2012) focus on using UGGC as a data source to support tourism research. Creating a variety of information products, both interactive and static, to meet the information needs of a project and its stakeholders should be an area of attention moving forward with VGI systems.

4 Types of User Contributions on the Geoweb

With a view on the VGI systems outlined in the previous section, the following data types/formats of user contributions on the Geoweb can be identified:

(A) Locations (e.g., coordinates, geometries, geographic objects/features)
(B) Categorical observations (e.g., species; earthquake; R/S classification)
(C) Numeric measurements (e.g., temperature, flood water level, noise)
(D) Parameter settings in models (e.g., multi-criteria evaluation weights)
(E) Descriptions (e.g., annotations, narratives, stories)
(F) Media (e.g., audio, photos, video)
(G) Opinions (e.g., thumbs-up/thumbs-down, ratings, arguments, commentary)

Type A represents a purely geospatial type of user contributions while all other types represent the geographically referenced attribute component discussed in Sect. 2. Types B through G can be further distinguished in different ways. For example, types B, C, D, and part of type G are structured with fixed value ranges as defined by Deparday (2010), while types E, F, and arguments/commentary of type G are unstructured. Additionally, types B and C represent data at the nominal and numeric (including ordinal, interval, and ratio) levels of measurement. In contrast the remaining types are not easily understood as measurements but rather as the output of human assertive or creative activity. Finally, a pattern of increasing complexity of the information underlying types B through G can be observed, as well as within types C (from ordinal to ratio), E (from short annotations to extensive stories), F (from audio only to static images to animated images with audio), and G (from one-click ratings to extensive commentary).

On the basis of these observations, our proposed classification distinguishes two groups in each of the geospatial and attribute dimensions of user contributions on the Geoweb. In the attribute component (horizontal dimension of Fig. 2), we distinguish contributions akin to "data" in a narrower sense from those more appropriately described as "content". In the geospatial component (vertical dimension of Fig. 2), we distinguish contributions referring to coordinate locations from those referring to geographic features.

Within the "data" component, the different levels of measurement are found: quantitative (ordinal, interval, ratio) and qualitative (nominal). Within the "content" category, a similar distinction is not obvious. However, contents items can be distinguished by the complexity of their information, or the information needed to

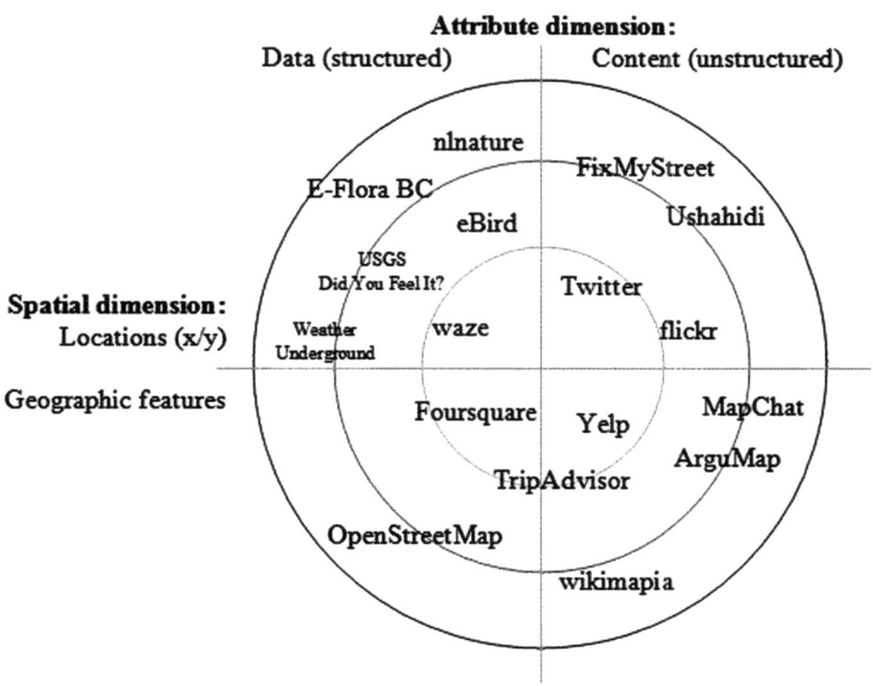

Fig. 2 Examples of Geoweb applications distinguished by the spatial and attribute dimension of user contributions

create them. Although the entire "content" component could be characterized as nominal, the scope of contents is different as seen in the abstract types discussed above. Content items such as audiovisual media are independent objects with only a loose geospatial link, as outlined in Fig. 1.

Some of these applications included in Fig. 2 collect data using a reporting feature via the Geoweb, while others rely on checklists, check-ins, and other input mechanisms, from which user contributions subsequently get added to a Geoweb environment. For example, OpenStreetMap, ArguMap, MapChat, and waze collect data directly on maps, while eBird, USGS Did You Feel It?, and E-Flora BC use forms, checklists, and check-ins to gather data.

The group of applications that collect structured data for x/y coordinate locations (top left in Fig. 2) includes NLNature, E-Flora BC, eBird, USGS Did You Feel it?, Weather Underground, and Waze. These are often characterized as citizen science applications, and typically require participants to actively contribute data. As an example, eBird builds on over a century of successful citizen science-based Christmas bird counts and collects data on where, when, and how long participants went birding, along with the bird species observed (seen or heard) via a checklist (Sullivan et al. 2009). The developers then map those observations, arranged with other more authoritative data. In the same class of applications, the electronic atlas of the flora of British Columbia, E-Flora BC, is compiled from a variety of (authoritative) databases with relevant flora information, and supplemented with mapped photo records from citizen scientists (Klinkenberg 2014); NLNature, or Newfoundland and Labrador Nature, encourages participants to observe wildlife and then post details and pictures of the plants, animals, and other interesting features (e.g., rocks, landmarks) they sighted to the online atlas (Lukyanenko et al. 2011); Weather Underground integrates measurements from over 34,000 personal weather stations to provide local weather forecasts; with its "Did You Feel It?" application, the United States Geological Survey (USGS) taps into information from people experiencing earthquakes. Organized by ZIP codes, the information collected includes a description of people's observations during the earthquake, the extent of the damage, and a questionnaire that aims to crowd-source the relative intensity of the event; and finally, Waze has been called a "social traffic app" that automatically collects travel times from users' smartphones and encourages users to manually submit information on other road conditions.

The class of Geoweb applications using unstructured content and x/y locations (top right in Fig. 2) includes Flickr, Twitter, FixMyStreet, and Ushahidi. Flickr is the world's largest photography community, allowing users to search, organize, and share photos, which includes an option for geotagging and organizing photos on a map. As an online social networking and microblogging service, Twitter allows particpants to upload short 140-character messages, or "tweets", which can be geolocated by the position of the user's mobile device. FixMyStreet allows residents to report, view, and discuss problems in their city (e.g., graffiti, potholes, garbage hotspots) by reporting the problem on a map. Finally, Ushahidi is a map-based platform to report incidents via mobile phones (short message service) or the Web. It supports textual observations, photos, and video uploads.

OpenStreetMap, Foursquare, and to some extent TripAdvisor represent a class of apps that collect structured data referring to geographic features (bottom left of Fig. 2). OpenStreetMap uses crowd-sourced GPS data and manual digitization to create a collaborative digital map of the world. Foursquare allows people to virtually check-into real-world locations. TripAdvisor is based on user reviews of travel amenities such as hotels. It includes both structured (e.g., price paid, star rating) and unstructured (e.g., free-text review) attribute data, and therefore also belongs to the following class of Geoweb applications.

A final class of Geoweb applications with unstructured content as their attribute dimension and geographic features as their spatial dimension (bottom right of Fig. 2) includes wikimapia, Yelp, MapChat, and ArguMap in addition to TripAdvisor. Wikimapia aims to describe the world by having users mark objects on a satellite image and provide textual and photographic documentation of the feature or area. Yelp collects reviews of local business from its users. A Google Map is one way of browsing the existing reviews, albeit not the primary point of access. Although the map uses markers at coordinate locations, these markers are geocoded from the addresses of business and therefore indirectly represent geographic features. In ArguMap (Rinner and Bird 2009) and MapChat (Hall et al. 2010), participant contributions in the form of annotations or commentary are linked directly with objects in the digital map environment.

5 Discussion and Conclusion

In reviewing the examples of Geoweb applications included in Fig. 2 and the corresponding types of user contributions, it seems that user contributions are determined to some extent by the purpose and the context (field of study) of the respective project. The schema in Fig. 2 suggests groups of applications that share similar contribution types as follows:

- **Crowd-mapping**: OpenStreetMap; wikimapia.
  These applications require users to refer to real-world geographic features when submitting attribute data/contents, which can be structured (OpenStreetMap) or unstructured (wikimapia).
- **Citizen sensing**: Weather Underground, waze, USGS Did You Feel It?
  In these applications, user contributions are in the form of structured data attached to x/y coordinate locations.
- **Citizen reporting**: eBird, E-Flora BC, nlnature; FixMyStreet, Ushahidi.
  Here, users provide structured (eBird, E-Flora BC, nlnature) or unstructured (FixMyStreet, Ushahidi) data/content linked to coordinate locations. The difference of the structured citizen reporting applications to the citizen sensing group above is the active nature of the users' contributions.

- **Map-based discourse**: TripAdvisor, Yelp, Argumap, MapChat.
  These applications operate on real-world geographic features, but unlike the crowd-mapping group above, their contents represents user opinions rather than facts (a distinction not currently captured in the diagram).
- **Geosocial media**: flickr, Foursquare, Twitter.
  The typical geosocial media applications (flickr, Twitter) use automatic geocoding of user contributions, the main interest in which is in the attribute dimension (photos, tweets). However, Foursquare distinctly focuses on the spatial dimension with references to real-world features.

Beyond purpose-specific determinants, additional characteristics of Geoweb applications seem to be associated with certain types of user contributions. With respect to the spatial dimension, Deparday (2010) already discussed the difference between manual geocoding (e.g., markers set by users in crowd-mapping or map-based discourse applications) and automatic geocoding (e.g., location recordings from cellphone or camera position in geosocial media). In addition, the different meanings of the location component are of interest: home location, current position when contributing, or location of the object of a contribution. While the user's home location is rarely captured explicitly due to privacy and safety concerns, citizen reporting applications would often locate contributions at the user's current location, while map-based discourse requires the location of the object of a contribution (e.g., restaurant being reviewed or urban planning project being commented on), irrespective of the user's location.

Brandusescu et al. (2015) discuss effects of temporality on the use of a crisis mapping platform, Ushahidi, in acute (short-term) and chronic (long-term) community development. The examples in Fig. 2 include large-scale, continuous projects (e.g. OpenStreetMap) as well as platforms (e.g., Argumap, MapChat, Ushahidi) for project-specific applications. Project-specific applications appear to allow for more complex, unstructured contents. This could be owing to a typically limited geographic scope, and a smaller number of users and contributions, which allows users to make sense of complex contributions that would not be possible to digest in a larger-scale application. In contrast, larger scale applications with more homogenous contents (e.g., photos on flickr) and those with structured data are limited to automatic summaries and reports, such as on the number of contributions per user. The influence of the type of user-generated data/content on its processing within the VGI system is an important observation towards developing more effective Geoweb applications.

The types of user contributions also have a relationship with the degree of user involvement in the Geoweb application. With respect to citizen science, Haklay (2013) proposes a new ladder of participation that culminates in "extreme citizen science", where users are involved in all steps from problem definition to data collection and analysis. A related distinction can be made between Geoweb applications that only collect original input versus those that allow for maintenance, corrections, and updates in the wiki sense (notably the above crowd-mapping group). Additionally, some applications allow for direct communication between users about contributions (map-based discourse group above, as well as Twitter). Both of these types of Geoweb applications tend to be associated with more complex types of user contributions, such as unstructured content referring to geographic features.

In order to guide the development of future Geoweb applications, the fit of the types of user contributions with geospatial data models should be examined. Tentatively, volunteered geographic *data* do fit with traditional spatial data handling in GIS, while volunteered geographic *content* requires new models and tools in order to become compatible with GIS. A participatory Geoweb data model would need to represent anything from points with attached narratives to complex geometries with attribute tags, and from isolated map annotations to structured geographically referenced argumentation. Such a data model could help with accuracy assessment of VGI and quantifying data quality on the Geoweb. It would also highlight any gaps in current Geoweb applications, where some type of user contribution may not be supported yet but present opportunities for citizen engagement.

Further examination of the types of user contributions on the participatory Geoweb should also refer to open data and e-government and their impact on citizen-government interactions. Spatial data infrastructures may increasingly rely on user-generated (or user-verified, -updated, -maintained) information. Ultimately, this research can assist with valuing VGI for societal decision-making.

Acknowledgments This research was partially supported by the GEOIDE Network of Centres of Excellence, Project PIV-41on "Participatory Geoweb for Engaging the Public on Global Environmental Change", and by the Social Sciences and Humanities Research Council of Canada, with a Partnership Grant on "How the Geospatial Web 2.0 is Reshaping Government-Citizen Interactions". We thank the co-investigators and students of both projects for inspiring discussions that have helped to frame this paper.

References

Baker R, Neu C (2013) Ushahidi: frameworks for effective platform management. Course description, TechChange—The Institute for Technology and Social Change. http://techchange.org/online-courses/ushahidi-frameworks-for-effective-platform-management/. Accessed 6 Sept 2014

Barth D (2009) The bright side of sitting in traffic: crowdsourcing road congestion data. Google Official Blog. http://googleblog.blogspot.ca/2009/08/bright-side-of-sitting-in-traffic.html. Accessed 6 Sept 2014

Bill R, Fritsch D (1999) Grundlagen der Geoinformationssysteme. Band 1: Hardware, Software and Daten. Wichmann, Heidelberg

Bowker GC (2005) Memory practices in the sciences. MIT Press, Cambridge

Brandusescu A, Sieber RE, Jochems S (2015) Confronting the hype: the use of crisis mapping for community development. Convergence 1–17. doi: 10.1177/1354856515584320

Budhathoki N, Bruce B, Nedovic-Budic Z (2008) Reconceptualizing the role of the user of spatial data infrastructure. GeoJ 72(3–4):149–160

Cinnamon J, Schuurman N (2013) Confronting the data-divide in a time of spatial turns and volunteered geographic information. GeoJ 78(4):657–674
Coleman D (2010) Volunteered geographic information in spatial data infrastructure: an early look at opportunities and constraints, Chap. 10. In: Rajabifard A, Crompvoets J, Kanantari M, Kok B (eds) Spatially enabling society: research, emerging trends and critical assessment. Leuven University Press, Leuven, pp 131–148
Coleman D, Georgiadou Y, Labonte J (2009) Volunteered geographic information: the nature and motivation of produsers. Int J Spatial Data Infrastr Res 4:332–358
Deparday V (2010) Enhancing volunteered geographical information (VGI) visualization with open source Web-based software. Master's thesis, University of Waterloo, Waterloo, Ontario, Canada. https://uwspace.uwaterloo.ca/handle/10012/5709. Accessed 6 Sept 2014
Elwood S, Goodchild M, Sui D (2012) Researching volunteered geographic information: spatial data, geographic research, and new social practice. Ann Assoc Am Geogr 102(3):571–590
Fast V, Rinner C (2014) A systems perspective on volunteered geographic information. ISPRS Int J Geo-Inf 3(4):1278–1292
Fischer F (2012) VGI as big data—a new but delicate geographic data-source. GEO-Informatics 15(3):46–47
Fogliaroni G, DeFelice J, Wallgrun O (2010) A qualitative perspective on volunteered geographic information. In: GIScience 2010 workshop on role of volunteered geographic information in advancing science, Zurich, 14 Sept 2010
Goodchild M (1997) Towards a geography of geographic information in a digital world. Comput Environ Urban Syst 21(6):377–391
Goodchild M (2007) Citizens as sensors: the world of volunteered geography. GeoJ 69:211–221
Goodchild M (2008) Assertion and authority: the science of user-generated geographic content. In: Proceedings of the Colloquium for Andrew U. Frank's 60th Birthday. GeoInfo 39. Department of Geoinformation and Cartography, Vienna University of Technology. http://www.geog.ucsb.edu/%7Egood/papers/454.pdf. Accessed 6 Sept 2014
Goodchild M, Li L (2012) Assuring the quality of volunteered geographic information. Spat Stat 1:110–120
Haklay M (2013) Citizen science and volunteered geographic information: overview and typology of participation. In: Sui D, Elwood S, Goodchild M (eds) Crowdsourcing geographic knowledge: volunteered geographic information (VGI) in theory and practice. Springer, Berlin, pp 105–122
Haklay M, Singleton A, Parker C (2008) Web mapping 2.0: the Neogeography of the Geoweb. Geogr Compass 2(6):2011–2039
Hall B, Chipeniuk R, Feick R, Leahy M. Deparday V (2010) Community-based production of geographic information using open source software and Web 2.0. Int J Geogr Inf Sci 24 (5):761–781
Heywood I, Cornelius S, Carver S (2006) An introduction to geographical information systems, 3rd edn. Pearson Education, Harlow
Johnson P, Sieber R (2012) Motivations driving government adoption of the Geoweb. GeoJ 77 (5):667–680
Johnson P, Sieber R, Magnien N, Ariwi J (2012) Mining the Web: user-generated content as a data source for tourism research. Curr Issues Tourism 15(3):293–299
Klinkenberg B (ed) (2014) E-Flora BC: Electronic Atlas of the Flora of British Columbia. Lab for Advanced Spatial Analysis, Department of Geography, University of British Columbia, Vancouver. http://ibis.geog.ubc.ca/biodiversity/eflora/. Accessed 6 Sept 2014
Leszczynski A, Wilson M (2013) Guest editorial: theorizing the Geoweb. GeoJ 78(6):915–919
Li S, Veenendaal B, Dragicevic S (2011) Advances, challenges and future directions in web-based GIS, mapping services and applications. In: Li S, Dragicevic S, Veenendaal B (eds) Advances in WebGIS, mapping services, and applications. Taylor and Francis, London, pp 3–12
Longley P, Goodchild M, Maguire D, Rhind D (2011) Geographic information systems and science, 3rd edn. Wiley, Hoboken

Lukyanenko R, Parsons J, Wiersma YF (2011) Citizen science 2.0: data management principles to harness the power of the crowd. In: Proceedings of 6th international conference on service-oriented perspectives in design science research, Milwaukee, WI, USA, 5–6 May 2011. Lecture Notes in Computer Science No. 6629, pp 465–473

Meeks WL (2007) The utility of geospatial data and information used in geographic information systems (GIS): an exploratory study into the factors that contribute to geospatial information utility. Doctoral Dissertation, The George Washington University, Ann Arbor, MI. http://pqdtopen.proquest.com/pubnum/3291997.html. Accessed 6 Sept 2014

Mennis J, Guo D (2009) Spatial data mining and geographic knowledge discovery—an introduction. Comput Environ Urban Syst 33:403–408

Mooney P, Corcoran P (2011) Can volunteered geographic information be a participant in eEnvironment and SDI? In: Hřebíček J, Schimak G, Denzer R (eds) Environmental software systems: frameworks of eEnvironment. IFIP Advances in Information and Communication Technology, vol 359. Springer, Berlin, pp 115–122

Rinner C (2001) Argumentation maps—GIS-based discussion support for online planning. Environ Plann B 28(6):847–863

Rinner C, Bird M (2009) Evaluating community engagement through argumentation maps—a public participation GIS case study. Environ Plann B 36(4):588–601

Rinner C, Kumari J, Mavedati S (2011) A geospatial Web application to map observations and opinions in environmental planning. In: Li S, Dragicevic S, Veenendaal B (eds) Advances in WebGIS, mapping services, and applications. Taylor and Francis, London, pp 277–291

Scharl A, Tochtermann K (eds) (2007) The geospatial web: How geobrowsers, social software and Web 2.0 are shaping the network society. Springer, London

Seeger C (2008) The role of facilitated volunteered geographic information in the landscape planning and site design process. GeoJ 72(3–4):199–213

Sieber RE (2004) Rewiring for a GIS/2. Cartographica 39(1):25–39

Sieber RE, Rahemtulla H (2010) Model of public participation on the Geoweb. Paper presented at the 6th international conference on geographic information science, Zurich, Switzerland, 14–17 Sept 2010

Stefanidis A, Crooks A, Radzikowski J (2011) Harvesting ambient geospatial information from social media feeds. GeoJ 78(2):319–338

Sui D, Goodchild M (2011) The convergence of GIS and social media: challenges for GIScience. Int J Geogr Inf Sci 25(11):1737–1748

Sui D, Elwood S, Goodchild M (eds) (2013) Crowdsourcing geographic knowledge: volunteered geographic information (VGI) in theory and practice. Springer, Berlin

Sullivan B, Wood C, Iliff M, Bonney R, Fink D, Kelling S (2009) eBird: a citizen-based bird observation network in the biological sciences. Biol Conserv 142:2282–2292

Tomlinson R (2007) Thinking about GIS: geographic information system planning for managers, 3rd edn. ESRI Press, Redlands

Tulloch D (2008) Is VGI participation? From vernal pools to video games. GeoJ 72(3–4):161–171

Walker B, Rinner C (2013) A qualitative framework for evaluating participation on the Geoweb. URISA J 25(2):15–24

Wilson M, Graham M (2013) Neogeography and volunteered geographic information: a conversation with Michael Goodchild and Andrew Turner. Environ Plann A 45(1):10–18

Zook M, Graham M, Shelton T, Gorman S (2010) Volunteered geographic information and crowdsourcing disaster relief: a case study of the Haitian earthquake. World Med Health Policy 2(2):6–32

Identifying Locally- and Globally-Distinctive Urban Place Descriptors from Heterogeneous User-Generated Content

R. Feick and C. Robertson

Abstract Place, which can be seen simply as space with meaning, has long been recognized as an important concept for understanding how individuals perceive, utilize and value their surroundings. There is increasing interest in mining information from geo-referenced user-generated content (UGC) and volunteered geographic information (VGI) to gain new insights into how people describe and delimit urban places such as neighbourhoods and vernacular landmarks and locales. In this paper, we aim to extend recent efforts to explore semantic similarity in these data by examining differences in place descriptors through georeferenced photo tags across multiple scales for selected cities in the USA. We compute measures of tag importance using both a naïve aspatial approach and a method based on spatial relations. We then compare the results of these methods for understanding tag semantics, and reveal to what degree certain characterizations as represented in tag-space are also spatially structured. Tag metrics are computed for multiple fixed resolutions that approximate typical urban place sizes (e.g. city, block, neighbourhood) and a simple extension of a well-known tag-frequency metric is proposed to capture differences in locally distinctive and globally distinctive tags. We present this analysis as an adaptation of traditional text analysis methods with ideas from spatial analysis in order to reveal hidden spatial structure within UGC.

Keywords GIS · Internet/Web · Urban · Data mining · Understanding · Scale · Method · Multiresolution

R. Feick (✉)
School of Planning, University of Waterloo, Waterloo, ON, Canada
e-mail: rdfeick@uwaterloo.ca

C. Robertson
Department of Geography and Environmental Studies,
Wilfrid Laurier University, Waterloo, ON, Canada
e-mail: crobertson@wlu.ca

F. Harvey and Y. Leung (eds.), *Advances in Spatial Data Handling and Analysis*, Advances in Geographic Information Science,
DOI 10.1007/978-3-319-19950-4_4

1 Introduction

People regularly use terms and concepts (e.g. neighbourhood, city center, tourist district, near, far) that are vague, context-specific and ambiguous to varying degrees when they communicate about the places they value, inhabit and interact within. Understanding how place is sensed and communicated is complex since place is a personal and socio-cultural construct that is influenced by individuals' cognitive processes, experiences and the dynamic context in which place-sensing is situated (Cresswell 2004). Traditional methods of gathering data related to how people sense and perceive place (e.g. photo-elicitation interviews, participant observation) provide rich qualitative data, however they are typically labour-intensive (Manzo 2005). The increased availability of geo-referenced user-generated content (UGC), such as photographs, videos and social media posts, offers a complementary avenue to explore how place is sensed and characterized across larger populations and geographic extents.

The volume and nature of this UGC or volunteered geographic information (VGI) varies substantially across sources and geographic regions. Generally, these data are comprised at least of: (a) a geographic reference or object (e.g. coordinates of a social media post, toponym reference, GPS track) and, (b) an associated set of largely unstructured text keywords, phrases or "tags". Through joint and separate analysis of these spatial and tag components, new insights relevant to our understanding of place use and perception have been gained. For example, Hollenstein and Purves (2010) and Li and Goodchild (2012) demonstrated how the spatial extents of vernacular regions, such as city cores and place references, can be interpreted from geotagged photos (GTPs). Others have illustrated how landmark preferences, urban movement patterns, and place semantics can be inferred from these data (Jankowski et al. 2010; Mackaness and Chaudhry 2013; Purves et al. 2011).

Place semantics, described by Rattenbury and Naaman (2009) as socially-defined locations associated with tag terms, offers a promising approach for identifying meanings and uncovering otherwise opaque tag and spatial structures in UGC. Inferring shared place meaning from multiple users' UGC is challenging in part because of the absence of common ontologies, the often incomplete and inconsistent nature of UGC, and the presence of idiosyncratic abbreviations, colloquialisms and conflations of natural language terms (Shelton et al. 2014; Li et al. 2013; Janowicz et al. 2011). Recently, more attention has been directed at examining spatial structure in place references and descriptors across multiple scales (Rattenbury and Naaman 2009; Mackaness and Chaudhry 2013; Feick and Robertson 2014). This vein of inquiry recognizes that individuals' perceptions of place are often comprised of locationally-specific, overlapping and/or hierarchical elements. In some cases, these perceptions may be anchored to explicit and formal entities (e.g. city-province-country), while many others are more personal or related to ephemeral events and experiences (e.g. "my neighbourhood", "safe areas", "music festival").

Our interest here lies in complementing recent methodological advances in the joint handling of space and semantics for the analysis of place meanings and descriptions encoded in UGC. In particular, we aim to build upon earlier work that examined tag dominance in GTPs across multiple place scales within a single city by investigating a method to uncover place references that are distinct within multiple urban centres (e.g. city hall) or are unique within a single urban entity (e.g. Empire State Building). We propose an adaption of the well-known tag frequency-inverse document frequency (tf-idf) that searches for relative uniqueness of terms (tags) across both global and local extents.

2 Methods

Recent work by Hollenstein and Purves (2010) examined how people characterize urban centres through certain place-oriented keywords in geotagged photographs across the USA and used these place references to derive the spatial extent of city boundaries. Similarly, Feick and Robertson (2014) explored the spatial dynamics of tag-space in terms of neighbourhood similarity and dissimilarity, and how this changes with level of spatial aggregation. However, we can envisage these works as two ends of a spectrum of spatial-semantic analysis, where in the former, the semantic focus is determined a priori (i.e., only looking at place-oriented tags) and in the latter only similarity and dissimilarity are examined and not the inherent meaning of the tags themselves. Here, we are interested in unearthing place semantics from UGC without prior manual identification of candidate place tags using methods developed for text search and analysis (tf-idf). We explore this idea through a case study of a large dataset of Flickr geotagged photograph metadata (GTPs).

3 Data and Study Area

Data were obtained from the Flickr API that covered the years 2001–2013 for a selection of 14 census urban areas identified by the US Census Bureau in 2012. Urban areas (UAs) are a census unit defined to represent the actual urban footprint comprised of built-up residential, commercial and industrial urban land uses. Officially defined UAs are stratified into urban clusters (2501–49,999 people) and urban areas (50,000+ people) (Fig. 1). Here we consider only the larger of these urban area classifications. Additional census geography was obtained including census tracts and census blocks in order to investigate spatial semantics across multiple spatial scales within UAs.

GTPs for 14 UAs of varying sizes and locations were obtained from the Flickr API using the python library flickrapi (http://stuvel.eu/flickrapi). A python script queried the API using a grid of reference points spaced 5 km apart for each UA. At each grid

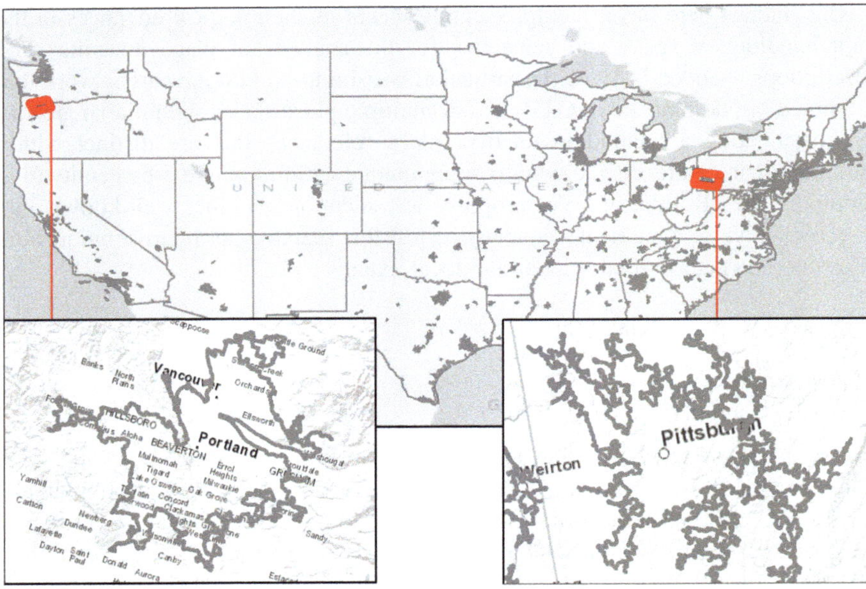

Fig. 1 Urban areas of the United States with two sample UAs highlighted

point, a radius search was used to obtain nearby photographs. This provided more consistent search results from the API than more direct single point-radius and bounding box searches. Once duplicate photographs were removed, 669,099 photos with unique Flickr photo IDs and valid latitude and longitude values were used for the analysis (Table 1).

Table 1 GTP, unique users and tag counts per urban area (UA)

Urban area	# of photos	# of unique tags	Unique users	Untagged photos
Boston	83,360	83,686	7138	7028
Chicago	110,859	106,717	8184	12,133
Dallas-Fort Worth-Arlington	36,147	34,028	3362	5595
Denver-Aurora	14,230	24,685	1705	1624
Detroit	41,493	44,455	3396	5290
Los Angeles-Long Beach-Anaheim	64,041	82,212	7477	6375
Minneapolis-St. Paul	33,166	42,483	2963	3546
New York-Newark	158,107	165,651	16,008	14,263
Philadelphia	48,440	47,171	4191	5905
Pittsburgh	16,308	24,535	1640	1560
San Jose	26,973	31,543	3185	2264
St. Louis	13,580	16,972	1507	1686
Tampa-St. Petersburg	22,395	27,627	2593	2995

4 Spatial Data Processing

To support spatial-semantic analysis over multiple spatial scales, GTPs were aggregated using spatial joins over three levels of census geography: UAs, Census Tracts and Census Blocks. With each aggregation, both the count of the number of GTPs and their full tag-sets were attached to each census geography unit. The number of tags associated with an individual GTP varied considerably with 70,264 of the 669,099 photos having no tags and a mean tag count of 98. To permit the relative uniqueness of tag words to be examined easily, the comma-separated tag array associated with each photo record was decomposed into *n* normalized rows in PostgreSQL database tables for each level of census geography. The one-to-many relationships established between tag records and their respective census geography tables enabled calculation of the metrics described in the following section and also supported examination of results across the hierarchy of UAs, census tracts and census blocks. It was found that census block geography was too fine to support meaningful analysis of tags except in specific localized areas. As a result, only results from the UA and CT level are reported here.

5 Analysis of Space, Scale and Semantics

The growth of the participant-web has led to vast amounts of text data being generated and a heightened need for methods such as sentiment analysis that can aid understanding of the meaning of users' content. Analysis of tag-sets attached to GTPs can therefore leverage text analysis tools to better understand embedded semantics and their spatial arrangement (Mackaness and Chaudhry 2013). Much of the text modelling literature is devoted to learning characteristic descriptions of documents within large collections (corpora) to support tasks such as classification, similarity assessment, relevance analysis and detection of anomalies or unusual features—largely in the context of information retrieval (Janowicz et al. 2011; Baeza-Yates and Ribeiro-Neto 1999; Blei et al. 2003).

A widely used method for identifying characteristic words from a document is term-frequency/inverse-document frequency, which is a normalized measure of the occurrence of a specific term within a document, relative to the number of documents containing that term (Salton and McGill 1983). The notion behind *tf-idf* is that distinctive terms will be mentioned frequently within a document, whilst being relatively infrequent across other documents in the corpus.

To extract place-semantics, we altered the *tf-idf* to compare distinctive tags local to a specific area, to those found to be distinctive across all locations. By considering each local unit of geography (i.e., census block, census tract, UA) as a document, we count the number of times each tag in the tag-set occurs in each area, relative to the inverse count of occurrences in other geographical units. For what we term the global-*tf-idf*, the set of eligible areas (i.e., the full tag-set) for the inverse tag-counts was set to all corresponding areas (i.e., blocks or census tracts) in the 14 UAs. In contrast, local *tf-idf* uses only the census units within a given UA. Global *tf-idf* therefore reflects the relative importance of a tag across all UAs, while local *tf-idf* reflects importance within a single urban context.

Tf-idf measures are conceptualized as reflecting local and global geographies in tag-space as presented in Fig. 2. Low values of global *tf-idf* are seen to indicate either within-unit (e.g. within census tracts) infrequency and/or across-unit frequency, while high values indicate within-unit frequency and/or across-unit infrequency. Constraining ourselves to urban settings, we might expect low global *tf-idf* to be generic descriptors characteristic of urban settings (e.g., 'street', 'park') and high values to be landmark-type place-tags, with high within-unit and low across-unit frequencies. Local *tf-idf* differs only in the reference set to which within-unit tag frequencies are compared which, in this case, are spatially constrained to geographic units in a common UA. As such, low values of local *tf-idf* may indicate either low within-unit frequency or high across-unit frequency in the city. In contrast, high values indicate high within-unit tag counts and low across-unit tag counts. Ultimately, we aim to compare values in both local and global tag-space in order to reveal characteristic place-semantics.

Following recent research in urban place semantics (Mackaness and Chaudhry 2013), we examine the tag-space relationships across spatial scale. Using census block and census tract geographies as spatial units, we compare the trajectories of randomly selected tags in both global *tf-idf* and local *tf-idf*. Within this paper, the

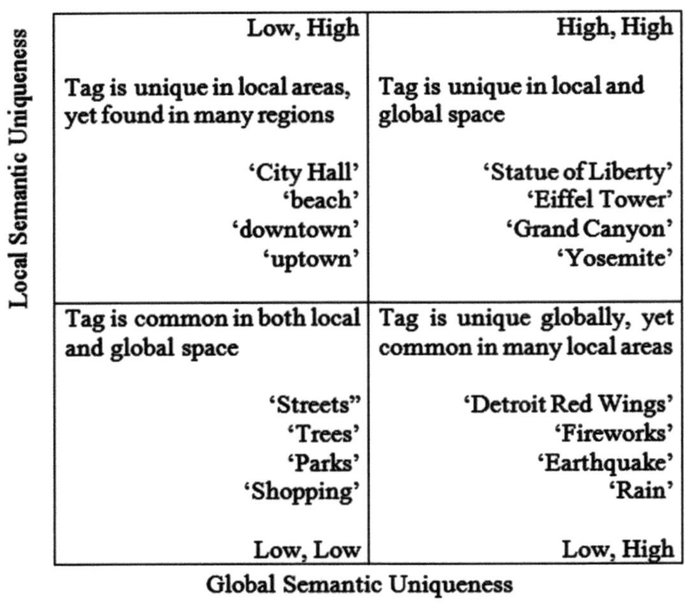

Fig. 2 Partitioning tag-space into local and global dimensions

ratio of local *tf-idf* to global *tf-idf* is used to operationalize Fig. 2 and to explore similarity in place descriptors across space. Finally, we examine spatial patterns of local and global semantics using local measures of spatial autocorrelation to visualize spatial clustering in tag-space. With little theoretical basis upon which to determine the spatial weights matrix required for local measures of spatial analysis, we apply an iterative neighbourhood optimization modification of the local Gi* statistic (Getis and Ord 1992) described by Aldstadt and Getis (2006).

6 Results

The distribution of the local/global *tf-idf* ratio (LGR) was normal, skewed slightly towards lower values (i.e., between 0 and 1), with a maximum value of 1.28 for census tracts. In the context of this paper, statistical significance of the distribution is not examined; instead we simply use the tails to focus attention on extreme cases of local or global tag dominance. Values in the upper tail indicate areas where local dominated global, and values in the lower tail of the distribution indicate areas where globally distinctive tags dominated. Globally dominant tags are given as a word cloud in Fig. 3, and exclusively identified place names for large geographies (e.g., state names and large city names) are prominent.

Figure 4 presents tags that are locally distinctive in any of the 14 UAs. Tags that are place references often indicate types of places within an urban region, e.g., ‘park’, ‘city’ and ‘urban’; many non-place references are also included, e.g., ‘instagramapp’, ‘nikon’, ‘2010’. This can be expected since place names were not filtered out a priori.

Fig. 3 Word cloud of tags from lower tail of the local/global *tf-idf* distribution (i.e., globally unique) at the census tract scale

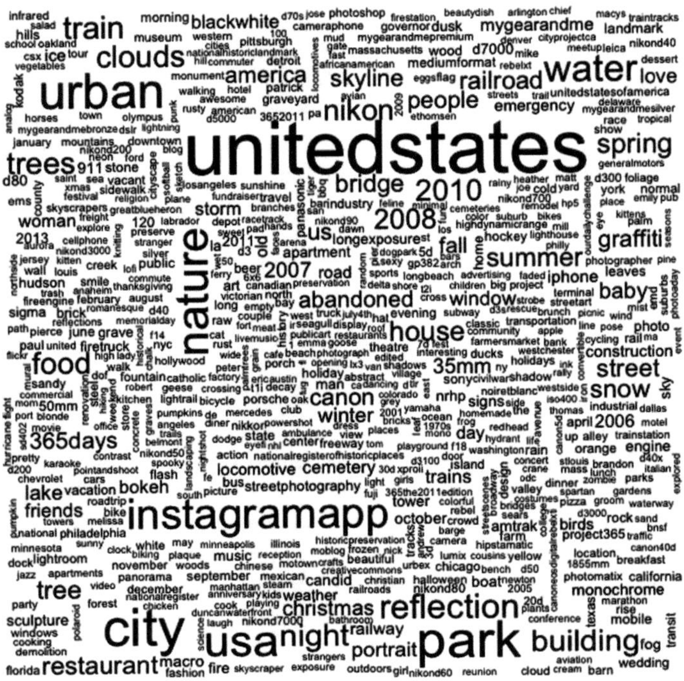

Fig. 4 Word cloud generated from upper tail of local/global *tf-idf* (i.e., locally unique tags) at the census tract scale

Spatial patterns of the LGR across three selected urban regions are presented in Fig. 5. As symbolization is standardized across each map, it can be seen that overall values of the LGR were highest in New York and lowest in Boston indicating more local uniqueness of UCG place descriptors in the former. The degree to which these patterns are spatially structured is visualized in Fig. 6 through the results of an AMOEBA optimization of the local G_i^* statistic. Spatial structure here relates to both the spatial scale of the geographic units (i.e., census tracts) and the place reference counts embedded within the tags.

To explore the combined spatial-semantic characteristics revealed by the LGR further, tags and LGR values are presented for selected sub-areas in Figs. 7, 8 and 9. The symbolization scheme in these maps corresponds to what is used in Figs. 5 and 6, such that higher values of LGR are shades of red, moderate values are yellow, and lower values are green. Therefore we expect to see locally distinctive tags in orange- and red-shaded census tracts and less locally distinctive tags in areas shaded green. While the census tracts in Figs. 7, 8 and 9 are symbolized according to LGR of *tf-idf*, the tag annotation includes only the four tags with highest local *tf-idf* values. To reduce clutter, only census tracts in the top two (i.e. light red and orange) and bottom two classes (light and dark green) of LGR are labelled.

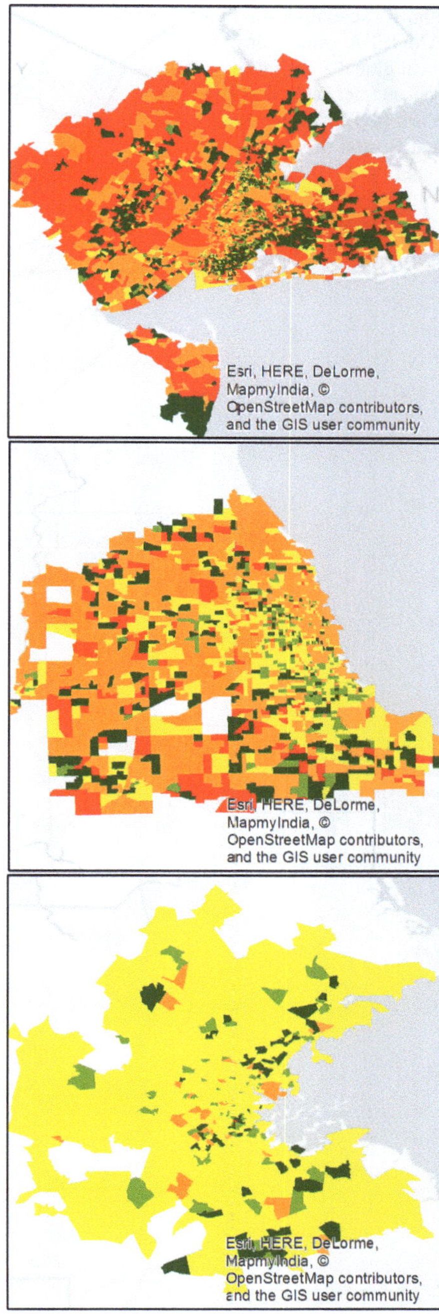

Fig. 5 Local/global *tf-idf* ratios at the census tract scale: New York (*top*), Chicago (*middle*) and Boston (*bottom*) ranging from low (*green*), to moderate (*yellow*) to high (*red*)

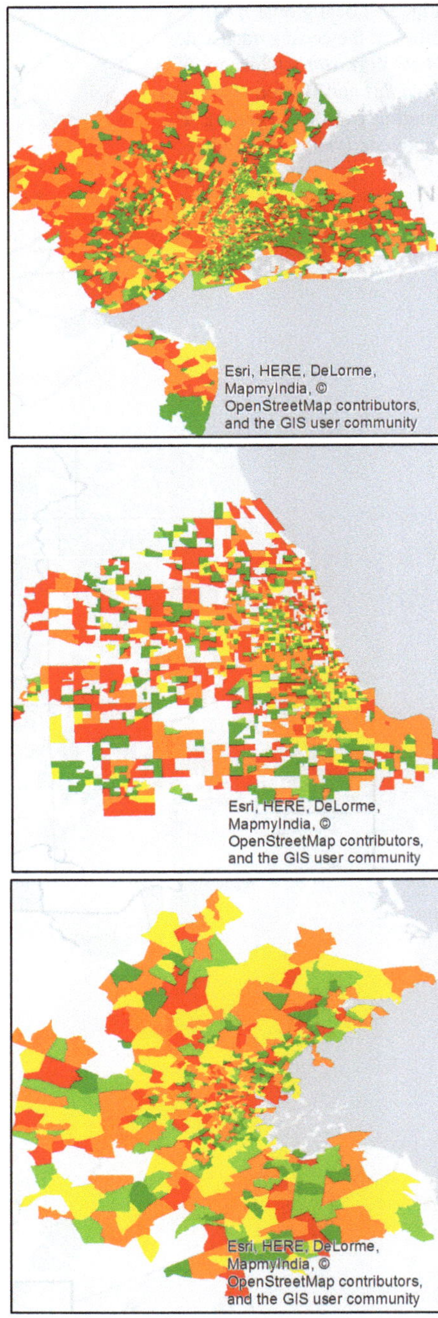

Fig. 6 Spatial clustering of AMOEBA values for local/global *tf-idf* ratios at the census tract scale for New York (*top*), Chicago (*middle*) and Boston (*bottom*)

Fig. 7 New York City census tracts shaded by LGR with high local *tf-idf* tags labelled

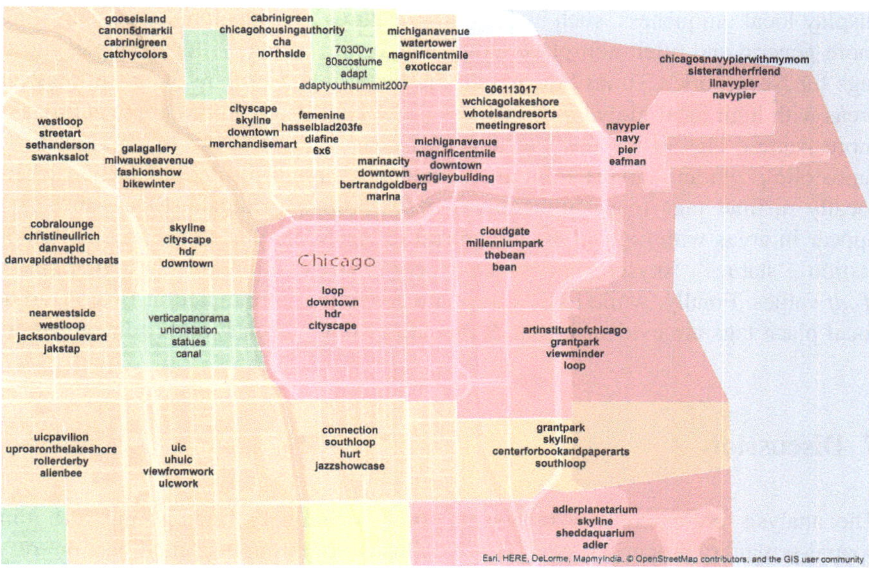

Fig. 8 Chicago census tracts shaded on LGR with high local *tf-idf* tags labelled

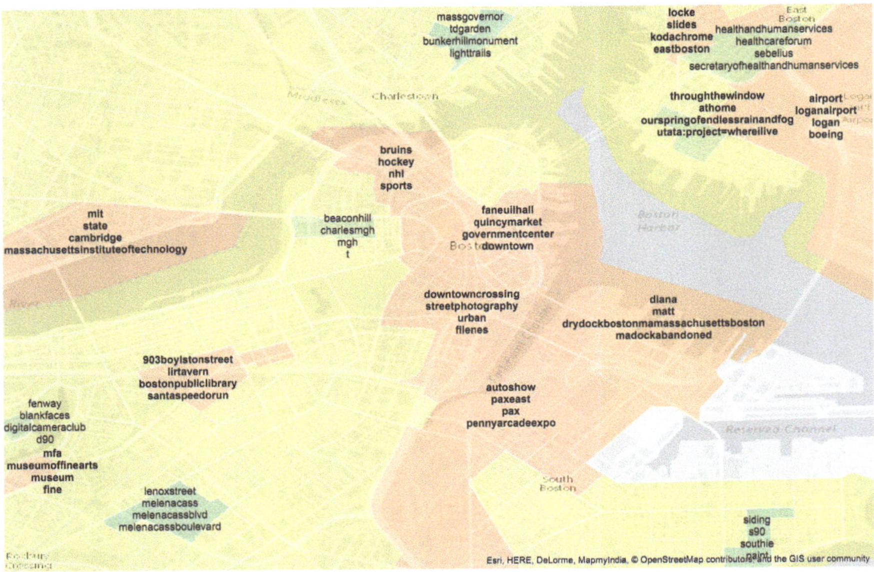

Fig. 9 Boston census tracts shaded on LGR with high local *tf-idf* tags labelled

Through these simple visualizations, several areas and features expected to display local uniqueness, such as landmark features, are apparent along with some more generic and often non-place terms. The centre of Fig. 7, for example, shows tags for New York's Times Square and the Flatiron and Empire State Buildings in areas with high local-global *tf-idf* values. These High-High values are in contrast to more generic tags in the few green-shaded census tracts in the lower right (e.g., 'apostolic', 'garden harvest'). Figure 8 illustrates similar results in Chicago where locally distinct tags (e.g., 'millennium park', 'illinoismedicaldistrict', 'thebean') appear in areas with high LGR, in contrast to candidate place tags such as 'union station', 'statues', 'skyscrapers' which appear in areas with relatively higher *global tf-idf* values. Finally, while LGR values are generally more muted in Boston, some local place tags are evident in Fig. 9 including 'mit', 'bruins' and 'quincymarket'.

7 Discussion

The analysis presented here represents some preliminary findings into the joint spatial-semantic properties of GTPs as illustrated through a modification of a commonly used text analysis measure, the *tf-idf*. By modifying the basis from which the *idf* term is derived, we can compare the degree to which tags are more locally or globally distinctive and start to explore the spatial patterns and relationships embedded within these tagging patterns when visualized in urban areas. While this method is a simple approach, the results indicate some degree of success in capturing locally distinctive tagging through analysis of the LGR measure alone, especially for well-known landmarks. While we envisage uses for this and similar methods in geographic information retrieval tools such as optimizing local queries, we are keenly interested in the degree to which spatial-semantic analysis can shed light on automated detection of place and place-making activities in sources of UGC generally with reduced need for human classification of training tag sets. We are particularly interested in how standard methods for natural language processing might be modified to incorporate elements common to spatial analysis such as local spatial patterns and examining patterns over multiple spatial scales.

References

Aldstadt J, Getis A (2006) Using AMOEBA to create a spatial weights matrix and identify spatial clusters. Geogr Anal 38:327–343

Baeza-Yates R, Ribeiro-Neto R (1999) Modern information retrieval, vol 463. ACM, New York

Blei DM, Ng AY, Jordan MI (2003) Latent dirichlet allocation. J Mach Learn 3:993–1022

Cresswell T (2004) Place. Blackwell, Malden

Feick R, Robertson C (2014) A multi-scale approach to exploring urban places in geotagged photographs. Comput Environ Urban Sys. doi:10.1016/j.compenvurbsys.2013.11.006

Getis A, Ord JK (1992) The analysis of spatial association by use of distance statistics. Geogr Anal 24:189–206

Hollenstein L, Purves R (2010) Exploring place through user-generated content: using Flickr tags to describe city cores. J Spat Info Sci 1:21–48

Jankowski P, Andrienko N, Andrienko G, Kisilevich S (2010) Discovering landmark preferences and movement patterns from photo postings. Trans GIS 14:833–852

Janowicz K, Raubal M, Kuhn W (2011) The semantics of similarity in geographic information retrieval. J Spat Info Sci 2:29–57

Li L, Goodchild MF (2012) Constructing places from spatial footprints. In: Goodchild MF, Pfoser D, Sui D (eds) Proceedings of the 1st ACM SIGSPATIAL international workshop on crowdsourced and volunteered geographic information (GEOCROWD'12). ACM, New York, pp 15–21

Li L, Goodchild MF, Xu B (2013) Spatial, temporal, and socioeconomic patterns in the use of Twitter and Flickr. Cart Geog Info Sci 40:61–77

Mackaness WA, Chaudhry O (2013) Assessing the veracity of methods for extracting place semantics from Flickr tags. Trans GIS 17:544–562

Manzo LC (2005) For better or worse: exploring multiple dimensions of place meaning. J Environ Psychol 25:67–86

Purves R, Edwardes A, Wood J (2011) Describing place through user generated content. First Monday 16 (Sept 5)

Rattenbury T, Naaman M (2009) Methods for extracting place semantics from Flickr tags. ACM Trans Web 3:1–30

Shelton T, Poorthuis A, Graham M, Zook M (2014) Mapping the data shadows of hurricane sandy: uncovering the sociospatial dimensions of 'big data'. Geoforum 52:167–179

Salton G, McGill MJ (1983) Introduction to modern information retrieval. McGraw Hill, New York

Identifying Optimal Study Areas and Spatial Aggregation Units for Point-Based VGI from Multiple Sources

Haydn Lawrence, Colin Robertson, Rob Feick and Trisalyn Nelson

Abstract In this paper, we introduce a new metric for evaluating feasible VGI study areas and the appropriateness of different aggregation unit sizes through three different components of data quality: coverage, density, and user-heterogeneity. Two popular sources of passive VGI are used for initial testing of the metric: Twitter and Flickr. We compare the component and aggregate measures for different simulated point processes and demonstrate the properties of this metric. The three components are assessed iteratively for the point user generated data (tweets and photos) on a local basis by altering grain sizes. We demonstrate the application of this metric with Flickr and Twitter data obtained for three Canadian cities as initial study areas, including Vancouver, Toronto, and Moncton. The utility of the metric for discriminating qualitatively different types of VGI is evaluated for each of these areas based on a relative comparison framework. Finally, we present a use-case for this metric: identifying the optimal spatial grain and extent for a given data set. The results of this analysis will provide a methodology for preliminary evaluation of VGI quality within a given study area, and identify sub-areas with desirable characteristics.

Keywords VGI · Social media · Optimal grain

H. Lawrence (✉)
Department of Geography and Environmental Management,
University of Waterloo, Waterloo, Canada
e-mail: haydnlawrence@gmail.com; hlawrenc@uwaterloo.ca

C. Robertson
Department of Geography and Environmental Studies,
Wilfrid Laurier University, Waterloo, Canada

R. Feick
School of Planning, University of Waterloo, Waterloo, Canada

T. Nelson
Department of Geography, University of Victoria, Victoria, Canada

F. Harvey and Y. Leung (eds.), *Advances in Spatial Data Handling and Analysis*, Advances in Geographic Information Science,
DOI 10.1007/978-3-319-19950-4_5

1 Introduction

In 2007, Goodchild (2007) first coined the subset of web-based user-generated content, including blogs, Facebook posts, or tweets with geographic properties, as volunteered geographic information (VGI). Since then, VGI has figured prominently in the GIScience research agenda, highlighting the geospatial properties of data from applications such as Flickr, Twitter, and OpenStreetMap (OSM). VGI encompasses many types of data, and can be further subdivided into ambient VGI, represented by Twitter and Flickr (Stefanidis et al. 2011), and active VGI, represented by OSM. Ambient VGI is less 'volunteered' (and may even be considered coerced—see McKenzie and Janowicz 2014) as users are likely not creating the data for a specific research purpose. These same sources are now being used to open up new spatial research questions in the social sciences (Goodchild and Janelle 2010). Many have speculated that VGI gives researchers the capability to gauge the sentiment of a geographically defined study population without use of traditional methods of qualitative research, such as questionnaires or direct observation. For example, VGI was recently employed to examine sentiments among community members in the path of a hurricane (Lachlan et al. 2014), allowing people to freely express their emotions, and importantly, enabling timely, and spatially defined assessment of population sentiment. Similarly, VGI may enable access to otherwise difficult to access populations (Stefanidis et al. 2011). While methods for evaluating data quality for new sources of VGI is a burgeoning research area of GIScience (Neis and Zielstra 2014; Jeffery et al. 2014; McKenzie et al. 2014; Hollenstein and Purves 2010), there are few tools available to assess the characteristics of VGI for a given research problem or application area.

The majority of studies into data quality assessment for VGI have been narrowly defined in terms of geographic scale and in the data sources considered (e.g. testing specific sources in isolation of others). Two frequently cited studies of OpenStreetMap that investigated data quality in VGI found OSM data to be very accurate within urban areas and sporadic in remote areas (Zielstra and Zipf 2010; Haklay 2010). However these studies were limited to narrow geographic and temporal scales by choosing areas with expected high OSM data. While initial investigations of data quality of VGI have indicated there is a real potential for VGI as a valuable source of data, these studies have largely focused on areas where large amounts of VGI were expected to be found, such as Tokyo (Stefanidis et al. 2011) and major population centres of Europe (Zielstra and Zipf 2010; Haklay 2010). Others have focused on discrete space-time events such as wildfires (Goodchild and Glennon 2010) and earthquakes (Zook et al. 2012). Other studies such as Hollenstein and Purves' study of Flickr (2010) or Crooks' study of Twitter (2013) are also limited by their source (Flickr or Twitter only). The specific foci found in the studies mentioned above can be seen as particularly problematic when the ephemeral qualities of technology are considered, illustrated by the downfall of the once popular MySpace or the restricted access and terms of service policies for privately held data from social-networking companies such as Facebook. Examples from the literature analyzing multiple sources of VGI simultaneously have begun to emerge, for example Croitoru et al. (2013) using Flickr, Twitter, and YouTube while Li et al. (2013) use both Flickr and Twitter densities for their study of user-demographics in California, though these studies do not look at the data quality characteristics of the data as the central focus. In addition, Mearns et al. (2014) highlights the potential of moving from just Twitter based analysis, which the study focuses on, to data from multiple social media platforms. They also describe a system which works in real time, which would be a very useful tool. Unfortunately, not all sources of VGI currently allow for real-time/streaming access.

Li et al. (2013) examined socioeconomic variables in relation to Twitter and Flickr submissions, finding a narrow subset of the underlying population (mostly rich and educated) were over-represented. However, the Li et al. study (2013) analysis of point densities was temporally limited to a period that was considered optimal and considered only the United States on a geographic scale with the demographic comparison analysis limited to a region of California. While the studies mentioned previously (Hollenstein and Purves 2010; Stefanidis et al. 2011; Li et al. 2013) are instrumental in our understanding of these relatively new forms of data, the isolation of the studies' data to specific scales (geographic, temporal) and individual sources of VGI limits the generality of their results. There is a considerable research need for general data quality assessment tools specific for VGI.

Geographic information has typically been assessed through traditional methods such as the authority or reputation of the data collectors, industry and/or international standards (e.g., ISO 19157:2013), or metrics based on a comparison to reference data (e.g., root mean square error). However, these methods may not be available to researchers attempting to assess VGI quality. A lack of standards, multiple and anonymous data collectors, a lack of comparable reference data, and even multiple and sometimes conflicting contributor motivations all contribute to knowledge gaps in data quality assessment methodology for VGI (Coleman et al. 2009; Foody et al. 2013; Mooney et al. 2013). This constitutes the necessity of a VGI assessment tool stemming from the problem that while traditional data collection methods exert control over sampling plans and study areas, VGI, by definition, allows for virtually no control over the data collection process. The sampling plan when using most types of VGI is necessarily post hoc. For large-scale web-based social media applications, researchers are forced to collect data within the scope of the given APIs, for available time periods and geographic locations. However, four parameters of control available for establishing some research design for VGI include spatial and temporal scales of the study, the thematic focus, and the sources of VGI used in the study (i.e. changing to a core area and using Twitter instead of Flickr). In this paper, we consider whether tools for assessment of VGI data quality can begin to be developed based on explicit consideration of these four parameters.

Few methods exist to offer guidance on whether sufficient VGI is available for a specific research task or, more fundamentally, to what scale VGI can be aggregated and analysed within a specific locale. In lieu of this guidance and given the patchy and heterogeneous nature of VGI, it is difficult for researchers to address questions such as: "What areas have enough VGI of a given type for my analysis?", "If a study area is defined externally, what resolution or spatial units of analysis can be used?", "How representative is the VGI as measured by user-heterogeneity?", or "What correspondence is there between the VGI and pre-defined zones such as census tracts or ecozones?" This paper will explore these issues through the use of a metric designed to evaluate ambient point-based VGI, information with geographic footprints though not actively created as geographic information (Stefanidis et al. 2011). We will examine the metric, computed over various aggregation unit sizes through three components of VGI quality: coverage, density, and user-heterogeneity. The methods described here are generic in that they are designed as a standalone VGI assessment tool, without reference to any authoritative or expert comparative datasets. A key practical outcome of this research will be a set of open source tools which will be directed at VGI evaluation/assessment from a user-oriented perspective, irrespective of scale or source (e.g. Twitter, Foursquare, or Yelp). The last aspect is important as private entities may change their data dissemination policies at any time, potentially restricting access. We start with a definition of the aforementioned metric followed by an examination of its properties for two types of spatial point processes. We follow with a case study of point data obtained from Twitter and Flickr for three different Canadian cities: Toronto, Vancouver, and Moncton, measuring the metric at different grains within two different extents for each city. We conclude with a discussion of current limitations and possibilities for future research for evaluating data quality in VGI point patterns, especially in the context of optimizing initial research study areas.

2 Methods

2.1 Developing a Metric for Evaluating VGI Point Patterns

VGI datasets are highly variable. Data can vary with the particular VGI data source's sharing policies such as Twitter's tiered access model which ranges from 1 % of data being available for free to 100 % of the data (i.e., 'the fire hose') available at a significant cost. Additionally, due to the nature of VGI creation, these data are prone to uncertainty. For example, studies of geotagged photographs often contend with variability introduced by two mechanisms of tagging: in situ tagging while photos are taken, and post hoc bulk-tagging of photos during management and upload to online sharing sites (Hollenstein and Purves 2010). These problems necessitate a method of determining if a chosen study area meets a researcher's needs, preferably before in-depth analysis begins. As the social sciences commonly use auxiliary aggregated datasets, such as census data (Li et al. 2013; Granell et al. 2014), we develop a metric for assessing VGI in aggregated grids of different spatial grains to allow for associations between VGI and other datasets. One of the important aspects of this approach is to mitigate the fact that aggregations, especially with authoritative boundaries, tend to be chosen arbitrarily (Jeffery et al. 2014) and these boundaries can impact analytic results (Openshaw and Taylor 1979). Thus we want a framework for evaluating VGI across aggregated geographies that can be easily quantified in order to test different areal units. Like the K-function, which assesses clustering or dispersion of point patterns at multiple distances (Ripley 1977), the approach here is first explored for multiple grains within the study area for all three metric components.

There are three components to the metric: coverage, density, and user-heterogeneity. Coverage is defined as the ratio of the number of cells that contain data to the total number of cells within the study area grid. While a lower numeric limit on the number of aggregated points within a grid cell could be implemented, this paper considered a cell to contain data on a binary scale—it either contains or doesn't contain data. This component is used as a global indicator of overall VGI coverage, on which Mooney and Corcoran (2013) describe a very discernable contrast between urban and remote areas and even within different urban areas (such as parts of Dublin compared to parts of Paris). This could be used to delineate remote areas or as a socio-economic indicator as Neis et al. (2013) show that lower coverage (completeness) can be attributed to lower socio-economic standing. Values approaching one indicate a greater coverage over the study area (i.e. most cells contain data) while values closer to zero are indicative of the opposite.

Density is assessed by calculating the areas of the four quadrants of a Moran's scatterplot and finding the difference between the largest area and the smallest area, normalized by study area size. This method differs from the local Moran's I value as it accounts for extreme x values instead of being averaged out within the algorithm, helping to find outlier core areas. Experiments with standard measures of spatial autocorrelation revealed that outliers were an important component the metric would need to be sensitive to, as often user-contributed data is patchy. For example, a downtown core or entertainment district may be a severe outlier in terms of number of submissions (high density), but depending on the aggregation scale, this area may or may not be flagged as unusual in a spatial statistic based on adjacent neighbouring areas.

User-heterogeneity is a measure of the number of VGI submissions to the number of unique users averaged over the study area. Values close to one indicate a high number of contributions relative to the number of unique users, which could be considered higher quality as per Linus's Law. However, values nearer to zero could be useful in a study of high value users and their impact on social media (Stefanidis et al. 2011). These three components were chosen for their ability to assess a point dataset irrespective of any thematic content inherent in the data, such as a tweet's text or Flickr photo's tags or image. This helps to maintain the overall goal of this paper to assess VGI point patterns in a general way without creating specific criteria based on the technology or application used. User-heterogeneity was chosen as a data quality component, density was chosen for sensitivity to spatial clustering and extreme outlier detection, and coverage for considering the spatial extent of the data relative to the aggregation unit. The density and user-heterogeneity metrics were designed specifically for assessment of VGI. Whereas most methods for local spatial analysis are tuned for identifying clustering of high or low values, we aim to detect clustering in similar values (not high, not low). Similarly for user-heterogeneity, no existing methods captured the dynamics of the relationship between variance in submissions per user and hypothesized data quality.

2.2 Implementing the Metric

The study area is rasterized into a grid specified by the bounding box size of the point pattern and the current grain size. The raster is intersected with the polygonal study area and all cells outside of the polygonal study area are set to NA. All points within each cell are aggregated by total counts and by counts of unique users. A queen's case neighbourhood matrix is created and the three components built as per the descriptions above. The three components are weighted to sum to a value of 1 and the final metric computed from these weightings.

3 Simulation Study

3.1 Simulation Study Data

Two spatial point processes were chosen to simulate point patterns that could constitute possible VGI distributions under conditions of spatial randomness and spatial clustering. Our objective was to generate point pattern realizations and evaluate the metric under each scenario to examine its sensitivity to different configurations of locations and users. These patterns were created on the unit square, though transformed to a square of 100 by 100 units. The two processes were run 999 times with the metric components computed for each run and averaged for final results. The simulations were done in the R programming language using the spatstat package (Baddeley and Turner 2005). The delineation between a poor quality distribution versus a good quality distribution would be dependent upon the goals of the research in question. For the purposes of this paper, highly clustered areas or areas with extreme outliers are considered poor. Low user-heterogeneity is also considered poor as we consider the idea of community consensus to be the ideal context for many types of analyses of VGI. As such, areas where data is dominated by a one or a small number of users would reflect be less useful than areas with broad participation from many users (Fig. 1).

A uniform marked multitype Poisson process was used to realize random point patterns where points were randomly labelled (marks constitute the user IDs in VGI point data). A randomly distributed point pattern would indicate that each location has an equal probability of a point event, which we consider to be high quality data.

Fig. 1 Matern clustered realization (*left*) and a uniform marked multitype Poisson realization (*right*). Distinct users are shown in different colours and symbols. The purpose of these figures is to outline what is considered a poor quality distribution (*left*) and a good quality distribution (*right*) within the scope of this paper

Similarly, as points are randomly labelled to reflect random spatial allocation of users, each area has equal chance of being visited by each user.

$$X \sim Poisson(\lambda)$$
$$\text{where } \lambda = 10, \text{ types} = 50$$

A stationary Matern cluster process was used to simulate a clustered area of VGI, creating a point pattern mirroring a few users based in a handful of areas creating the majority of data.

$$X \sim Matern(\kappa, r, \mu)$$
$$\text{where } \kappa = 5, \text{ r(radius)} = 0.15, \mu = 100$$

3.2 Simulation Study Results

The simulations match the predicted results for each component of the metric. Coverage for the completely spatial random (CSR) point patterns is very high except at very small grain sizes. This is caused by having too many small cells which contain no points. The clustered pattern yields a low metric value except at high grain sizes, caused by having too few grid cells with only one point needed to be captured by the coverage component. Density findings were consistent with expected spatial autocorrelation values of CSR and clustered point patterns. The CSR point pattern, except at extremely small grains, revealed little spatial autocorrelation or few extreme values of data while the clustered simulations produced values that indicate clustering. User-heterogeneity was found to be very low in a clustered area with few users and high amounts of data. At the smaller grain size (2 units), user-heterogeneity is high in a clustered area as the cell size is so small that very few points are captured by each cell. The opposite can be seen with the CSR point pattern with large grains creating grids of only a few cells capturing a large number of points decreasing user-heterogeneity. However, user-heterogeneity follows the predicted outcomes at most grain sizes for the CSR point pattern by having a high value showing a closer ratio of unique users to data points. These both follow predicted patterns of user-heterogeneity.

4 Empirical Case Study

4.1 Study Areas and Data Collection

Three Canadian cities were chosen as study areas: Toronto, Vancouver, and Moncton. Toronto and Vancouver were chosen as large cities with dense population centres while Moncton was chosen as a contrast to the two larger cities while maintaining a large enough population density to have some VGI (Table 1).

The study areas shown in Fig. 2 were chosen using rectangular bounding boxes required for accessing data from APIs. The second grouping of study areas is

Fig. 2 Moncton, NB (*left*), Toronto, ON (*centre*), Vancouver, BC (*right*)

Table 1 Statistics Canada 2011 data

City	Population	Area (km^2)	Population density/km^2
Toronto	5,583,064	5905	945.4
Vancouver	2,313,328	2882	802.5
Moncton	138,644	2406	57.6

population centres from the 2011 Canadian census, based on minimum populations of 1000 and minimum population densities of 400 people per square kilometre (Fig. 3). The population centres can be found in Fig. 3 for the three cities. These polygons are very different from the previous rectangular bounding boxes, including only the areas with a high density of people allowing for a much higher distribution of VGI. The final study area is a rectangular bounding box of downtown Toronto. It includes the University of Toronto in the top left and the two major sporting event locations and the CN Tower in the bottom.

Data was analysed using the open APIs for Flickr and Twitter, both through Python. In total, 1,541,170 tweets and 63,176 photos were from Toronto, 398,811 tweets" and 44,061 photos for Vancouver, and 25,533 tweets and 1571 photos for Moncton. The tweets were analysed over a four month period from September 2013 while the Flickr photos are all from 2010. This difference in time is an aspect taken into account as one of the purposes of this research is to view the point patterns of different data sets from different sources which might have different collection times or resolutions.

4.2 Case Study Results

Overall, results indicated higher metric values population centre metrics compared to study areas based on bounding boxes (Fig. 4a, b). Coverage and user-heterogeneity of the bounding box study areas show a difference between the two larger cities compared to Moncton. However, density shows similarity between all three cities.

The population centre study areas had very different results. Full coverage was found in the population centres at larger grain sizes while showing similar trends from the lowest grains. As anticipated, density varied between the larger cities compared to the smaller city, Moncton. Finally, user-heterogeneity maintained differentiation between the larger cities compared to Moncton. Exact metric component values can be found in Figs. 5, 6 and 7. User-heterogeneity was never found to be higher than 0.34 (Toronto downtown core). While cursory analyses of only the Rogers Centre and of the University of Toronto did find higher values (0.56 and 0.43 respectively), user-heterogeneity was not found with high values within the Canadian urban city study areas.

The overall trends in all three types of study areas showed decreases in coverage and density as grain size decreased while user-heterogeneity increased as grain size decreased. The downtown Toronto core showed the highest metric values among the three study area types. Coverage is complete at all grain sizes and density and user-heterogeneity is much higher in the Toronto core. Figure 8a, b shows the metric values with all three components combined at different weightings for the chosen study areas and the population centres. When all three components are set at equal weighting, there is little differentiation between the three cities.

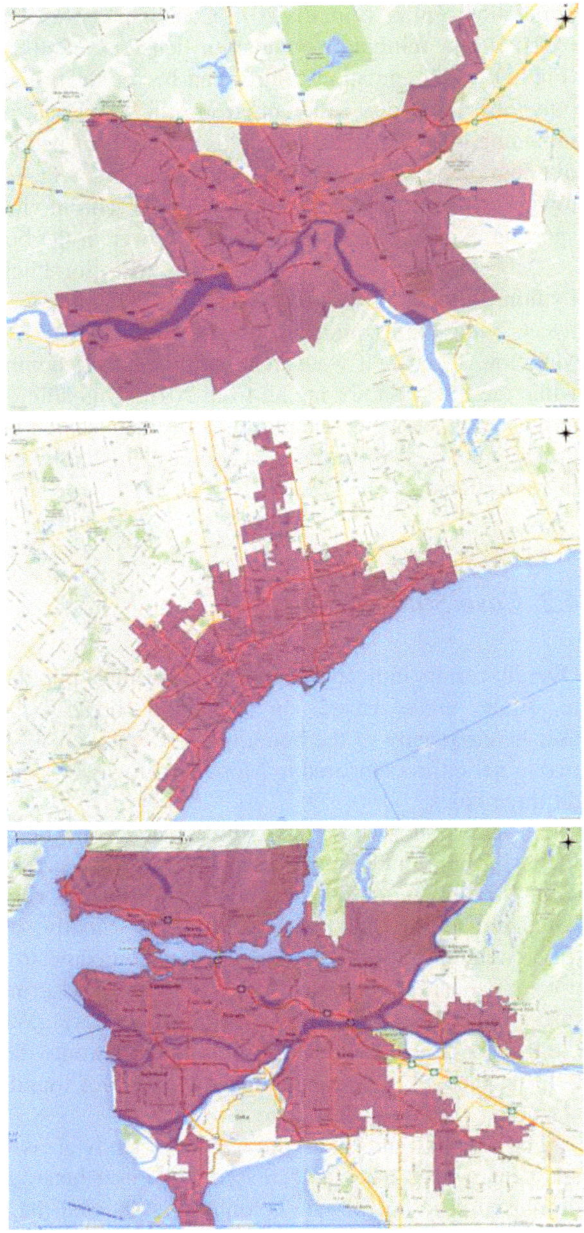

Fig. 3 Moncton (*top*), Toronto (*centre*), and Vancouver (*bottom*) Canadian census population centres

5 Discussion

5.1 Large Versus Small Cities

A distinct difference in metric values was found between Toronto/Vancouver and Moncton. This could be explained by socio-economic factors, as detailed in Li et al.'s conclusions (2013) where tweet density was correlated with well-educated people with advanced degrees, high income, while Flickr photo density was correlated with White and Asian people with advanced degrees. Moncton had an Asian population of less than 1 % and the percent of people with advanced degrees (20.1 %) was less than the national average (22.9 %), Toronto (33.6 %), and Vancouver (30.7 %) as per the 2006 Canadian census. The physical size difference between Vancouver and Moncton is negligible, though population density is very different (Table 1) which may also be a factor in the variation. When using the population centres as opposed to the chosen bounding boxes, similar values can be seen.

The opposite can be seen for density as the population centres show a difference between big/small cities with Moncton having higher density values than the other two. The clustering in Fig. 2 of Moncton follows the three major streets and the downtown and shopping areas. The chosen bounding box created a similar density for all three as it is averaging the water/remote areas. User-heterogeneity shows similar results for all three cities. User-heterogeneity is not reliant on population density nor area with the number of people who tweet or post photos being the more important factor. This is similar to the results found by Mooney and Corcoran (2013) in their study of OSM in London, Paris, and Berlin. Even though Berlin has a much lower population density than Paris, Paris showed less contributed data and only a third of the number of unique users. Further study can be done using different sized cities to see if there is a connection between user-heterogeneity in urban areas in contrast to remote areas or within different urban areas. One important aspect is that the study area did not affect user-heterogeneity, as both the metric values for the bounding box and the population centres were similar. This is due to the fact that while many low user/high data areas would be found in residential/remote areas, the heavily traveled areas such as downtown cores or shopping centres would likely have high user/high data ratios.

5.2 Study Area and Grain Size

The study area had a large effect on coverage and density. It also had a lesser effect on user-heterogeneity, especially with the use of the downtown Toronto core. This is not particularly surprising as correctly choosing a study area is a fairly important aspect of any study. The one point counter to this stems from the use of the population centre. An inset rectangular bounding box for the Toronto core still

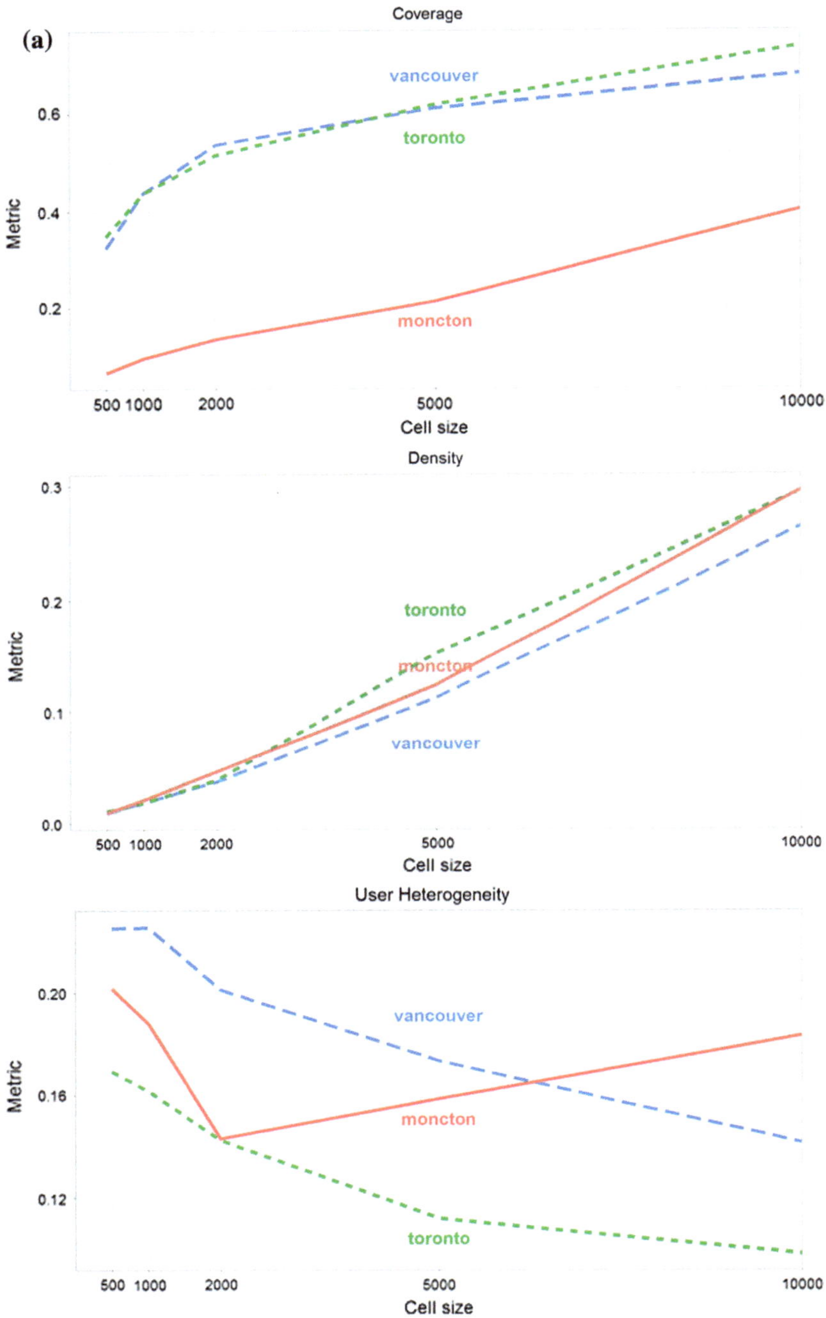

Fig. 4 **a** Metric component values for the chosen bounding boxes at cell sizes 500, 1000, 2000, 5000, and 10,000—coverage (*top*), density (*centre*), and user-heterogeneity (*bottom*). **b** Metric component values for the census population centres at cell sizes 500, 1000, 2000, 5000, and 10,000—coverage (*top*), density (*centre*), and user-heterogeneity (*bottom*)

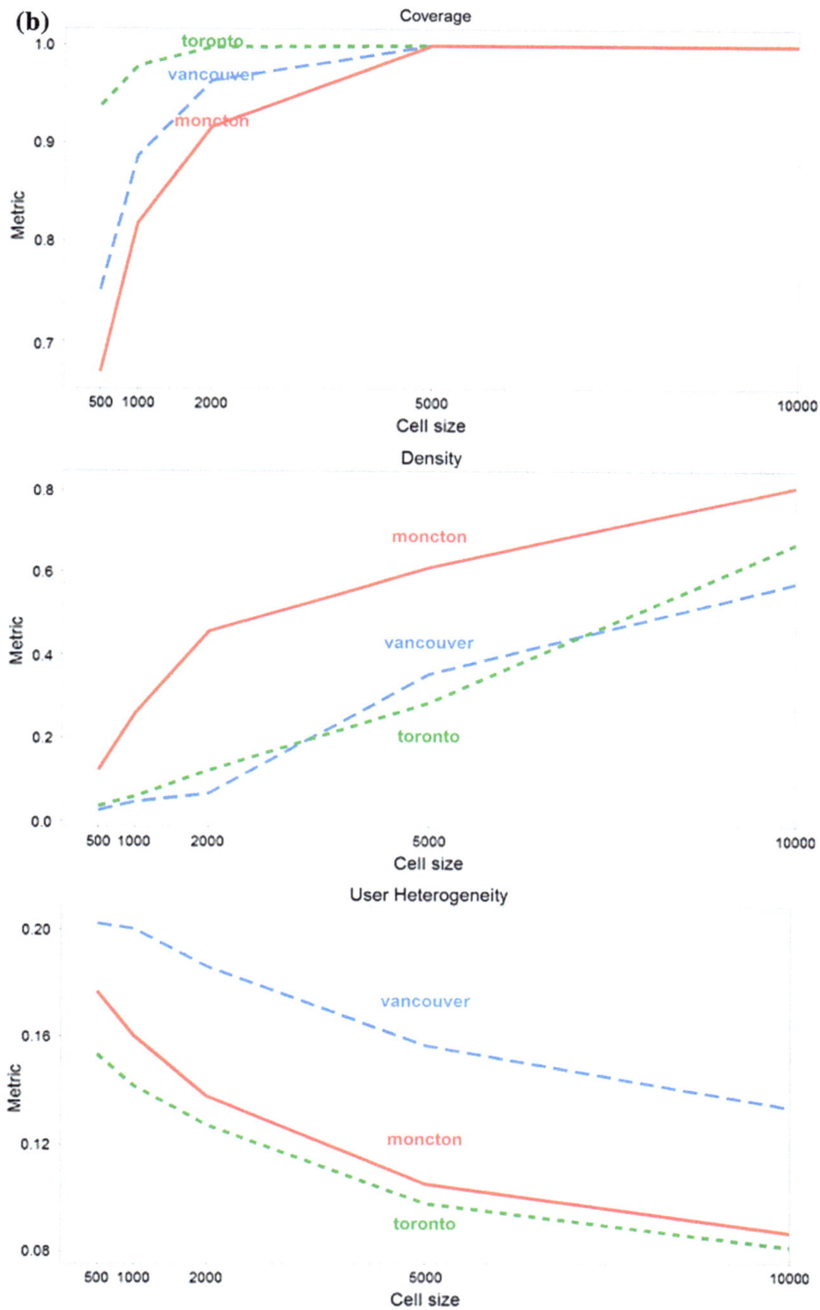

Fig. 4 (continued)

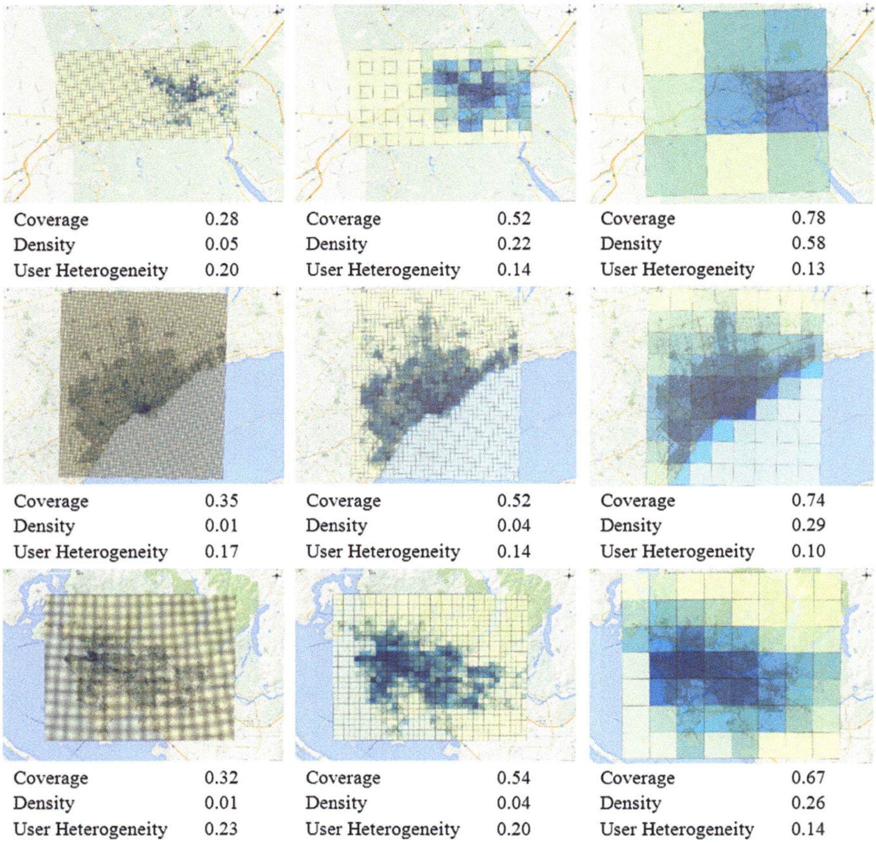

Fig. 5 Metric component values for the chosen bounding boxes at 500 (*left*), 2000 (*centre*), and 10,000 (*right*) grain sizes

produced higher results in all three metric component values. While researchers may be looking for high or low values dependent upon the research at hand, the Toronto core produced results much closer to complete spatial randomness than the population centres derived from the authoritative data sets (i.e. Canadian census). This leads to the idea that a well-chosen study area is superior to using authoritative-driven boundaries such as a census tract, and perhaps suggests data-driven methods for defining study areas is required for the types of VGI analyzed here.

Changes in grain size led to marked differences in the metric values calculated for each study area. When grain size is too large, it creates a grid of very few cells which compares to just using the study area as a whole. For all three cities, a 10,000 metre (m) grid cell size (10 km by 10 km) did not prove to be informative. The 2000 m grid cell size gave a stronger idea of what is happening within each study

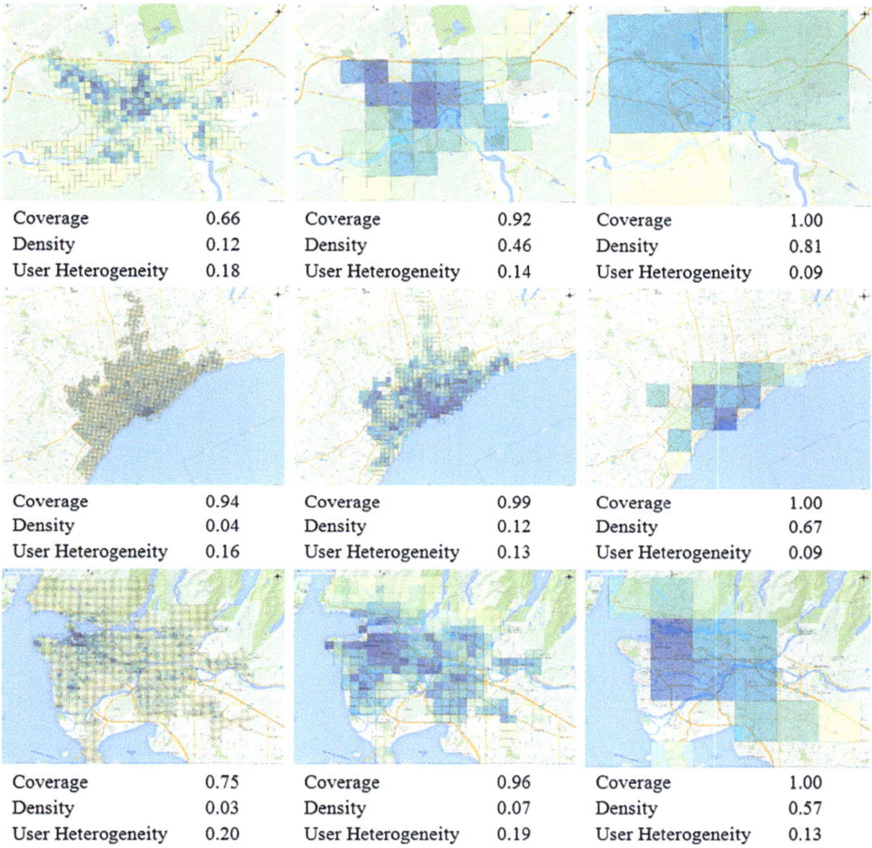

Fig. 6 Metric component values for the census population centres bounding boxes at 500 (*left*), 2000 (*centre*), and 10,000 (*right*) grain sizes

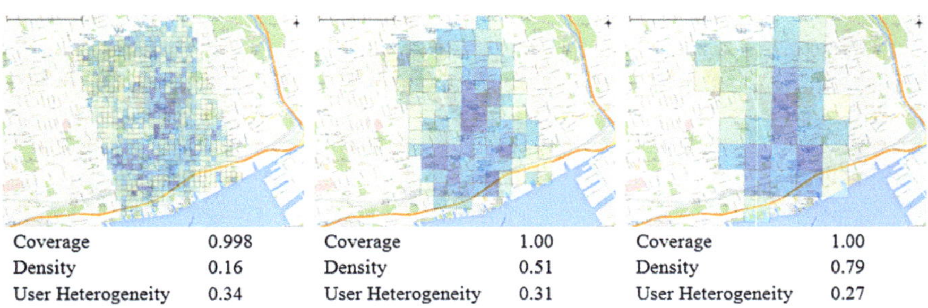

Fig. 7 Metric component values for the Toronto city core bounding boxes at 100 (*left*), 250 (*centre*), and 500 (*right*) grain sizes

(a)

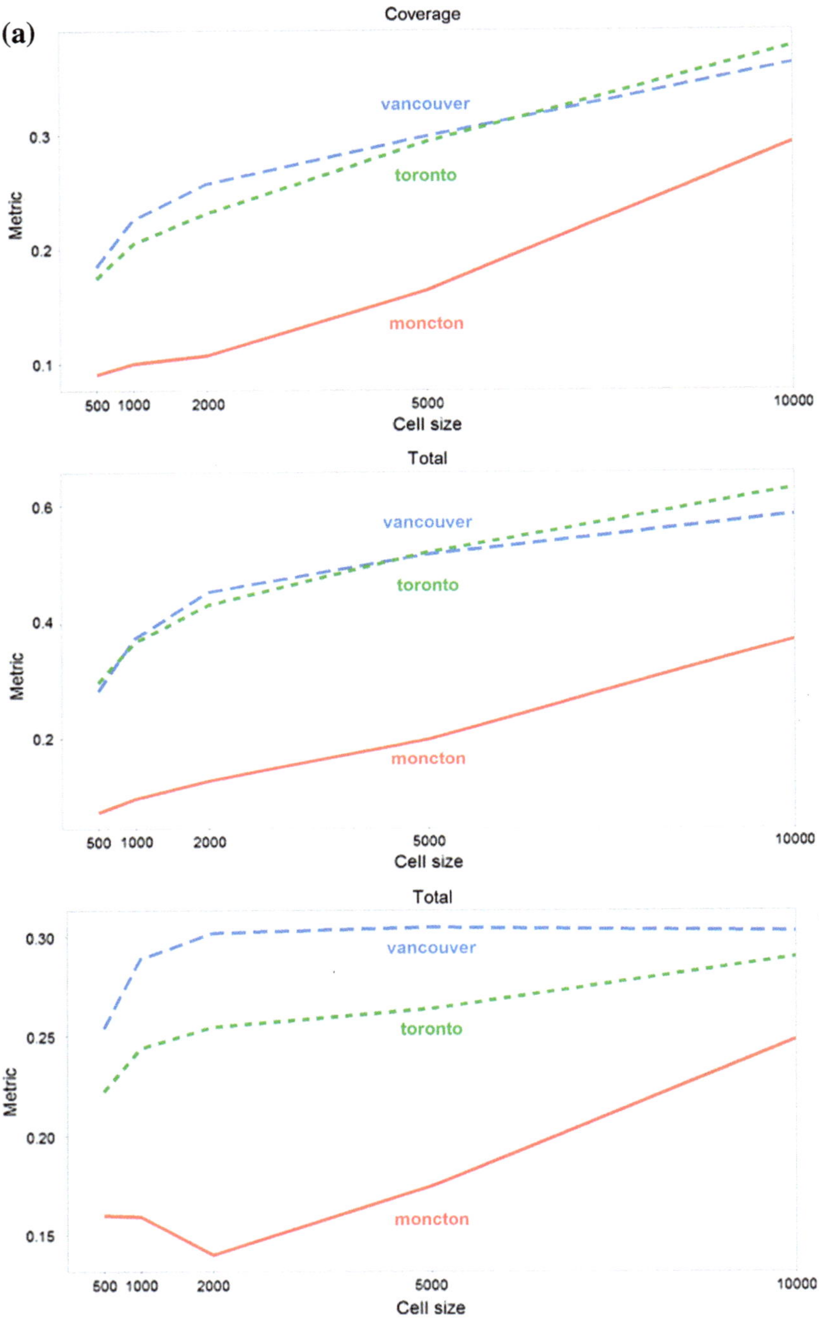

Fig. 8 **a** Metric component values for the chosen bounding by different metric weightings (coverage, density, and user-heterogeneity) at cell sizes 500, 1000, 2000, 5000, and 10,000. **b** Metric component values for the census population centres by different metric weightings (coverage, density, and user-heterogeneity) at cell sizes 500, 1000, 2000, 5000, and 10,000

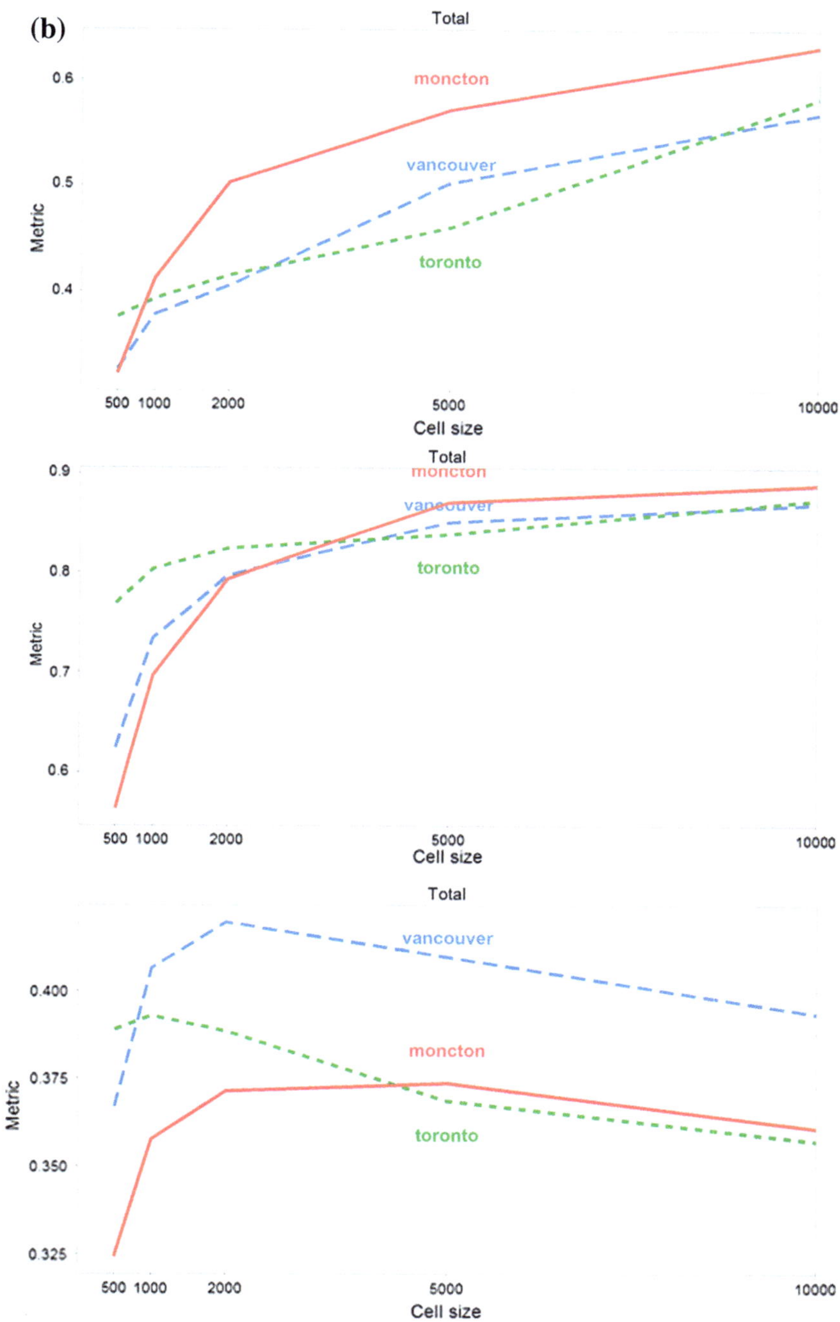

Fig. 8 (continued)

area and while the 500 m grid cell size showed much lower metric values when compared to a complete spatially random point pattern, you can start to see trends within certain sub-areas better. The downtown Toronto core (Fig. 7), the 250 m grid cell size gives a good indication that the data are following either a certain street or possibly a subway line. The 100 m grid cell size doesn't quite give the same overall information, but gives a much more detailed view of the University of Toronto, Rogers Centre, and Air Canada Centre, all considered areas to be relatively high in VGI.

A key point found through both the simulations and the case studies is that there are grains where the metric starts to show less of an increase or decrease (inflection point), where the slope of the values to grain size becomes more level. These inflection points offer the ability to assess diminishing returns in the metric as grain size becomes larger or smaller, allowing for the selection of an optimal grain size. As with traditional data analysis such as identifying priority areas for biodiversity (Jenkins et al. 2013) or in the clustering of disease (Jeffery et al. 2014), finding the optimal grain is critical for analysis of point data. Figure 4a, b demonstrates that both coverage and user heterogeneity don't follow constant rates of increase/decrease, possibly indicating optimal grain sizes at 2000 or 5000 m for coverage and 5000 m for user-heterogeneity. These grains might indicate an optimal spatial grain which could be used by a researcher. While there are strong similarities between Toronto and Vancouver, Moncton tends to differ in overall metric values and sometimes even in trend. Moncton has approximately 140,000 inhabitants compared to roughly 2 million and 5 million found in Vancouver and Toronto respectively. One reason for this could be a limitation on the detectable patterns imposed by data volume of the smaller city. In addition, though Flickr is used for many purposes, there would be much more tourism found within Toronto and Vancouver, which could impact the overall spatial heterogeneity of the dataset, where tourist centres and landmarks in major cities capture a disproportionate share of the data. This is an interesting finding and may allow for the use of metrics such as these to help delineate qualitatively different types of VGI distributions linked to city size and function, though further study with more cities would be required.

6 Conclusions and Future Work

It has always been important to find a proper study area and grain size when using aggregated point data however VGI leaves researchers at the mercy of the data. Without control of the sampling procedures, researchers need to find ways to assess their study areas for data quality based upon the conditions of their studies. Current study areas tend to include authoritative delineations of areas (spatial grain) such as census tracts or census metropolitan areas (CMAs) which are built from authoritative data or historic boundaries that do not capture the transient nature of VGI. Strong increases in the three component values of the current metric are shown to happen as the study area is refined and grain size is modified from finer to coarser resolutions. This provides a basis for the creation of an algorithm using a metric similar to the metric described in this paper to find an optimal study area and spatial grain inherent to a pre-defined criteria. For example, such criteria could be realized by modifying the weights of the three components and optimizing an aggregated measure.

Other properties could potentially be incorporated into the methodology described here. The temporal characteristics of VGI, especially in the sense of user-heterogeneity of a study area, for example, could provide information about the transience of users based on the times they were observed within one area to gauge if they are resident to the area, traveling to the area, or commuting to the area. Similarly, semantic measures of text data could be examined using similar analytical methods. This paper focused solely on the spatial point patterns of VGI though further analysis using the text could prove useful for study area optimizations. Finally, the grids for defining units of analysis were based on a regular lattice, and alternate tessellations could be explored.

By definition, researchers have far less control over VGI compared to traditional data collection methods. However, given the increasing ubiquity of geographically referenced data, it is unlikely for VGI to become less used, and may become normalized into the toolkit of all researchers. Given that traditional research methods were defined on a paradigm of researcher control over the research design, there is cause for greater understanding of how VGI differs from traditional data, and what its value is in a given context. One of VGI's greatest benefits for the social sciences is the potential for a more nuanced, organic and evolving way to sense people (e.g. assessing emotion) and places (e.g., areas people take photos and how they characterize them). To realize this, baseline measures of data quality are required, and the analysis presented here is a first attempt at the creation of a relativistic assessment tool for these new types of data.

References

Baddeley A, Turner R (2005) Spatstat: an R package for analysing spatial point pattern. J Stat Softw 12:1–42

Coleman D, Georgiadou Y, Labonte J (2009) Volunteered geographic information: the nature and motivation of produsers. Int J Spat Data Infrastruct Res 4:332–358

Croitoru A, Crooks A, Radzikowski J, Stefanidis A (2013) Geosocial gauge: a system prototype for knowledge discovery from social media. Int J Geogr Inf Sci 27(12):2483–2508

Crooks A, Croitoru A, Stefanidis A, Radzikowski J (2013) #Earthquake: Twitter as a distributed sensor system. Trans GIS 17(1):124–147

Foody G, See L, Fritz S, Van der Velde M, Perger C, Schill C, Boyd DS (2013) Assessing the accuracy of volunteered geographic information arising from multiple contributors to an internet based collaborative project. Trans GIS 17(6):847–860

Goodchild M (2007) Citizens as sensors: the world of volunteered geography. GeoJournal 69:211–221

Goodchild M, Glennon J (2010) Crowdsourcing geographic information for disaster response: a research frontier. Int J Digital Earth 3:231–241

Goodchild M, Janelle D (2010) Toward critical spatial thinking in the social sciences and humanities. GeoJournal 75:3–13

Granell C, Belmonte O, Diaz L (2014) Geospatial information infrastructures to address spatial needs in health: collaboration, challenges, and opportunities. Future Gener Comput Syst 31:213–222

Haklay M (2010) How good is volunteered geographical information? A comparative study of OpenStreetMap and ordnance survey datasets. Environ Plan 37:682–703

Hollenstein L, Purves R (2010) Exploring place through user-generated content: using Flickr tags to describe city cores. J Spat Inf Sci 1:21–48

Jeffery C, Ozonoff A, Pagano M (2014) The effect of spatial aggregation on performance when mapping a risk of disease. Int J Health Geographics 13(9):1–9

Jenkins C, Pimm S, Joppa L (2013) Global patterns of terrestrial vertebrate diversity and conservation. PNAS 110(28):E2602–E2610

Lacklan K, Spence P, Lin X (2014) Expressions of risk awareness and concern through Twitter: on the utility of using the medium as an indication of audience needs. Comput Hum Behav 35:554–559

Li L, Goodchild M, Xu B (2013) Spatial, temporal, and socioeconomic patterns in the use of Twitter and Flickr. Cartography Geographic Inf Sci 40:61–77

McKenzie G, Janowicz K (2014) Coerced geographic information: the not-so-voluntary Side of User-generated geo-content. In: Extended abstracts of the eighth international conference on geographic information science (GIScience 2014). Vienna, Austria

McKenzie G, Janowicz K, Adams B (2014) A weighted multi-attribute method for matching user-generated points of interest. Cartography Geographic Inf Sci 41(2):125–137

Mearns G, Simmonds R, Richardson R, Turner M, Watson P, Missier P (2014) Tweet my street: a cross-disciplinary collaboration for the analysis of local twitter data. Future Internet 6(2):378–396

Mooney P, Corcoran P (2013) Understanding the roles of communities in volunteered geographic information projects. Progress in location-based services. Springer, Berlin, pp 357–371

Mooney P, Corcoran P, Ciepluch B (2013) The potential for using volunteered geographic information in pervasive health computing applications. J Ambient Intell Humaniz Comput 4 (6):731–745

Neis P, Zielstra D, Zipf A (2013) Comparison of volunteered geographic information data contributions and community development for selected world regions. Future Internet 5:282–300

Neis P, Zielstra D (2014) Generation of a tailored routing network for disabled people based on collaboratively collected geodata. Appl Geogr 47:70–77

Openshaw S, Taylor P (1979) A million or so correlation coefficients: three experiments on the modifiable areal unit problem. In: Wrigley N (ed) Statistical applications in the spatial sciences. Pion, London, pp 127–144

Ripley BD (1977) Modelling spatial patterns. J Roy Stat Soc B Met 172–212

Stefanidis A, Crooks A, Radzikowski J (2011) Harvesting ambient geospatial information from social media feeds. GeoJournal 78(2):319–338

Zielstra D, Zipf A (2010) A comparative study of proprietary geodata and volunteered geographic information for Germany. In: 13th AGILE international conference on geographic information science, 2010

Zook M, Graham M, Shelton T, Gorman S (2012) Volunteered geographic information and crowdsourcing disaster relief: a case study of the Haitian earthquake. World Med Health Policy 2:7–33

A General Framework For Event Detection From Social Media

Khatereh Polous, André Freitag, Jukka Krisp, Liqiu Meng and Smita Singh

Abstract The availability of accurate and/or up-to-date mass data can stimulate the development of innovative approaches for the assessment of spatio-temporal processes. However, extracting meaningful information from these collections of user-generated data is a challenge. Event detection is an interesting concept in the era of Web 2.0 and ubiquitous Internet. Various existing event-detection algorithms share a very simple, yet powerful architecture model; pipes-&-filters. Using this model, the authors in this study developed a generic and extensible programming framework to find meaningful patterns out of heterogeneous and unstructured online data streams. The framework supports researchers with adapters to different social media platforms, optional preprocessing steps. Its graphical user interface supports users with an interactive graphical environment for setting up parameters and evaluating the results through maps, 3D visualization, and various charts. The framework has been successfully tested on Flicker and Instagram platforms for different time periods and locations to detect latent events.

Keywords Event detection · Knowledge discovery · Flicker · Instagram · Social media · Datastream mining

K. Polous (✉) · L. Meng · S. Singh
Department of Cartography, Technical University Munich, Munich, Germany
e-mail: Polous@bv.tum.de

L. Meng
e-mail: Meng@bv.tum.de

S. Singh
e-mail: Singh@tum.de

A. Freitag
Department of Informatics, Technical University Munich, Munich, Germany
e-mail: Andre.Freitag@tum.de

J. Krisp
Department of Geography, Augsburg University, Augsburg, Germany
e-mail: jukka.krisp@geo.uni-augsburg.de

F. Harvey and Y. Leung (eds.), *Advances in Spatial Data Handling and Analysis*, Advances in Geographic Information Science,
DOI 10.1007/978-3-319-19950-4_6

1 Introduction

1.1 Social Media

Over the past decade, the online conversations and discussions have been vastly increased. Different social media platforms (such as Twitter, Facebook, Flickr, YouTube, and Instagram) have become an inseparable part of modern daily life, through which millions of users share their stories, experiences, thoughts, and interests. As a result, these online platforms hold a considerable number of user-generated documents (such as textual and pictorial messages and videos). These conversations are either personal updates related to a user's smaller social circle or responses triggered by events such as natural disasters and political events (Dou et al. 2012). According to Ahlqvist et al. (2008) Social media is a combination of three constituent; namely content, user communities and Web 2.0 technologies. Kaplan and Haenlein (2010) define social media as "a group of Internet-based applications that build on the ideological and technological foundations of Web 2.0 and that allow the creation and exchange of user-generated content." According to Nielsen (2012) the amount of time that internet users spend with social media sites is more than any other type of site. This report also shows that the total time spent on social media sites in the U.S. (using PC and mobile devices) increased by 37 % from 88 billion minutes in July 2011 to 121 billion minutes in July 2012.

1.2 Event Detection

From the information perspective, the social media platforms are great sources of information and a kind of data repository. These information sources provide many research opportunities to analyze and discover latent patterns from big data sets. Although ever growing contents of the social media platforms are like gold mines for social event detection, still many challenges related to the processing of the heterogeneous data (timestamp, location, visual and textual content) should be overcome (Bao et al. 2013). These online text data are not only free-form texts; their spatial and temporal characteristics (Xu 2011), makes it even more difficult to extract and analyze the spatial and temporal information embedded in these large scale data sets.

There are various definitions for event in the literature, however all these definitions share a similar core; an event is something that occurs at a specific time and place. "The specific location and time of an event differentiate it from broader classes of events: for example, "The Eruption of Mt. Pinatubo on June 15th, 1991" is an event whereas "volcanic eruptions" is the more general class of events containing it" (Allan et al. 1998). "Events serve as a succinct summary of social media streams" (Dou et al. 2012), which show the evolution of a specific social phenomena over a period of time. Furthermore, investigating the relationships between events and people's responses triggered by these events can be illuminative in understanding the impact of public policies (Dou et al. 2012). This clearly shows the importance and practicality of event detection from available social media big data.

1.3 Clustering

Clustering is one of the most broadly accepted methods for detecting events from document sets and social media sets (Bao et al. 2013). Clustering is a sort of grouping a set of things, items or objects in a way that the similar items or objects—considering a criterion or some criteria—are gathered in the same group or cluster. There are various clustering algorithms in the literature that can be used for very different purposes. Even similar algorithms with different configurations act very different; therefore clustering is an "iterative process of knowledge discovery or interactive multi-objective optimization that involves trial and failure" (Mourya and Prasad 2013). Normally the data pre-processing and model parameters should be repetitively modified until the desired results are achieved. This multi-objective optimization problem is used in many fields such as data mining, machine learning, pattern recognition, image analysis, statistical data analysis, information retrieval, and bioinformatics (Geetha and Shyla 2014). However for social media analysis, due to heterogeneous characteristic of the metadata which come from multiple modalities, the most challenging part of event detection is integrating the associated metadata together appropriately (Bao et al. 2013). However, exploring the effectiveness and efficiency of various clustering algorithms for event detection from social media is not in the scope of this paper and will be discussed in another study separately.

1.4 Problem Statement

Although social media sites are like gold mines for social event detection, there are many challenges associated with the heterogeneous metadata to be overcome. Furthermore, each event detection case requires a different clustering algorithm that should be defined using special techniques. There are possibly hundreds published approaches with different clustering algorithms towards the problematic of event detection in social media. Unfortunately, there is no clear common basis to be found in these works. Most of the works are targeted at a single specific platform. They are created independently from each other leading to the lack of a common interface and process, which makes it hard to evaluate and compare different works against each other. This lack also hinders further research based on the existing works since researchers are forced to re-implement the works in order to evaluate and extend them. In addition, researchers face a huge overhead when dealing with the topic and are forced to replicate already existing implementations including the adaptation to different social media platforms and the presentation of the outputs.

Whereas a closer study of the most of the social media platforms shows that they share many basic features, making it possible to use the same algorithms and detection processes for all of these platforms. Furthermore various existing event-detection algorithms share a very simple architecture. Hence, to circumvent the aforementioned challenges, in this study a framework is developed that divides the general task of event detection in social media into independent subtasks and allows the development of platform independent models. This enables the complete separation of data retrieval, event detection and presentation layers. The framework is built on a modular architecture, which makes the integration of any generic event-detection algorithm possible through a simple python adapter. The Modular structure of the framework also offers the possibility to integrate and adopt any clustering methodology easily. This general framework provides the possibility of event detection from any type of social media and has the potential to be used for event detection from multi social streaming sources. These abilities help the researchers to easily test, evaluate and compare different event-detection algorithms for different sources of information; here social media. Currently two data adapters for two different social media platforms (Flickr and Instagram) have been developed. Furthermore, the framework includes a web-based graphical user interface (GUI) that can be run even on restricted devices such as tablets or smartphones. The GUI has many interactive capabilities such as maps (Openstreetmap) and geo-spatial 3D plots for better visualization of the detected patterns. The outputs are also saved as comma-separated values (CSV) files, which makes possible further statistical analysis of the results in any statistical package. The framework requires various parameters for any online data stream such as the location where events are searched (a bounding box), the time of the events (a period of time), the parameters of clustering algorithm like number of contributions and common word usage. Upon running the tool, available data for the defined space and time are downloaded through the social media API and stored in a local event database. These data are then clustered in space and time based on initial parameters. The framework has been successfully tested on Flicker and Instagram platforms for different periods of time in different locations to detect many latent events. Due to the importance of spatial characteristics of the data in this work, only geo-tagged photos are stored.

This paper is organized as follows. Section 2 provides a literature review about event detection from different social media platforms. Section 3 describes in detail the proposed approach in this paper. Section 4 presents achieved results and discusses the implications of the results. Finally, the last section contains conclusions and discussions.

2 Literature Review

The rapid spread of the concept that uses web as participatory platforms and consequently a huge amount of information which is constantly uploaded on internet stimulates the development of innovative approaches for the assessment of spatio-temporal processes and event detection.

Bao et al. (2013), proposed a social event detection approach, named SDE-RHOCC, to detect the real world events from uploaded photos on online photo sharing platforms like Flickr. The advantages of the proposed SDE-RHOCC approach can be listed as; implementation of a star-structured K-partite graph for heterogeneous metadata integration during co-clustering process, consideration of intra-relationships within time spaces for clustering performance improvement, and adaptation of information-theoretic co-clustering framework with no limitation on the numbers of clusters in each metadata set. The experimental experiences on Mediaeval Social Event Detection Dataset showed the effectiveness of the proposed approach in social media datasets.

Sakaki et al. (2010) applied a semantic analysis on the tweets on Twitter to test the real time nature of Twitter, especially in the case of event detection. In the study the tweets were classified into two different domains; positive and negative. They considered each tweet as a sensor observing and reporting about happening or not happening of an event. The location of events was estimated by using Kalman-filtering and particle filtering approach. The obtained results revealed that the particle filter worked better than other compared methods in estimating the centers of the event (here the center of earthquakes and typhoons). In addition an earthquakes reporting system was developed to quickly make the people aware of the event which is happening.

Gao et al. (2013) analyzed spatio-temporal distribution of the tweets and analyzed their content in Sina Weibo on realistic data sets. They used an adaptive K-means clustering algorithm for the tweets published in the geographical area and counted the number of the tweets in each cluster. Their experiments approved the benefit of their tool in location related social event detections.

Becker et al. (2010) implemented a weighted clustering algorithm considering multiple features listed as title, description, tags, location, and time. They continued their work by presenting a framework in 2010 to achieve high quality clustering results. They examined ensemble based and classification-based techniques for combining a set of similarity metrics. This offers the possibility of finding similarity among detected events. Their experiments revealed that the similarity metric learning techniques produce better performance.

Weng and Lee (2011) focused on event detection in Twitter by analyzing the contents of the tweets published in the platform. They introduced a framework named EDCoW (Event Detection with Clustering of Wavelet-based Signals). In EDCoW, the signal of each individual word is computed by applying wavelet analysis on the frequency based raw signals of the words. By considering corresponding signal autocorrelations unimportant words are deleted. The remaining words were then clustered to form events with a modularity-based graph partitioning technique. Based on their experimental experiences, the authors claimed EDCoW achieved a fairly good performance in the study.

Parikh and Karlapalem (2013) developed a scalable system, called ET, for detecting real world events from a set of micro blogs (tweets). The key feature of their system was clustering the related keywords based on content similarity and appearance similarity among keywords. ET used a hierarchical clustering process for determining the events. It was tested on two different datasets from two different domains. The results for both of these domains were precise.

Petkos et al. (2012) introduced a supervised multimodal clustering algorithm. The algorithm was tested on the challenge data of MediaEval social event detection and is compared to an approach using multimodal spectral clustering and early fusion. Using the explicit supervisory signal, the algorithm was able to achieve higher clustering accuracy and at the same time it required the specification of a much smaller number of parameters. The authors claim that their algorithm can be applied not only to the task of social event detection, but to a wider scope for other multimodal clustering problems as well.

Rabbath et al. (2011) introduced an event-based photo clustering approach in social media for automatically detecting media elements, which match a specific query in a user's social network and arranging them into a photo book. Through combining the content analysis of texts and images the photos of a specific story are selected. An Expectation-Maximization algorithm was used to calculate the probabilities of each two photos to belong to the same event. For selecting important photos to create a photo book, different semantic information such as people's tags, the interaction between the users, and between the photos were exploited.

Tamura and Ichimura (2013) proposed a density-based spatio-temporal clustering algorithm for extracting bursty areas from Georeferenced Documents. The proposed clustering algorithm was able to recognize the temporally- and spatially-separated clusters. The clustering algorithm separated coordinate space from time space.

Zhou and Chen (2013) proposed a framework for monitoring online social events from tweet streams for real applications such as crisis management. To represent the tweets a graphical model called location-time constrained topic (LTT) was proposed to fuse various information such as social content, location and time of tweets. The similarity of messages was caught using a complementary distance, which considered the differences between two messages over four attributes; content, location, time, and link. To prove the effectiveness and efficiency of their proposed approach, they conducted two experiments over long tweet streams during two occurred crisis in Australia.

Li et al. (2012) proposed a Twitter-based Event Detection and Analysis System (TEDAS) to; (1) detect new events, (2) analyze the spatial and temporal pattern of events, and (3) identify the importance of events.

Yang et al. (1998) proposed an agglomerative clustering algorithm (GAC: augmented Group Average Clustering) to extract retrospective events from news story. To analysis the correlation between cluster quality and efficiency of computations an iterative bucketing and re-clustering model was applied. Hierarchical and non-hierarchical document clustering algorithms were applied to 15,836 stories to exploit their content and temporal information. The cluster hierarchies were the key to detect previously unidentified events retrospectively, supporting both query-free and query-driven retrievals. Furthermore, the temporal distribution of document clusters provided useful information to improve both retrospective detection and online detection of events.

Reuter et al. (2011) addressed the complex issue of finding appropriate clusters composed of photos, which demonstrate the same event in Flicker. They presented a novel approach which relied on state-of-the-art techniques from the area of record linkage. Their results showed that the performance of their clustering methods and the obtained parameters can be transferred to similar datasets without a major degradation.

A probabilistic model for Retrospective Event Detection in news (RED) was proposed by Li et al. (2005). In their model they used Expectation Maximization (EM) algorithm to maximize log-likelihood of the distributions and learn the model parameters. It is required to provide the number of events in a model, which is considered as the difficult part in practice.

Ilina et al. (2012) proposed a semi-automatic approach for detecting event related tweets. They aimed to extract event related information from tweets content to use them in web applications that list the events like concerts, with their relevant information as dates, locations or performers. They used a classification approach based on Naive Bayes and n-gram features to extract the event related content from broadcasters and their followers.

Papadopoulos et al. (2012) developed a framework called Social Event Detection (SED) to provide an opportunity for examining and evaluating different approaches to the problem of social event detection in multimedia collections. In their study they explored how social events can be detected automatically by analyzing social multimedia content. They evaluated their results using two different method; Harmonic mean (F-score) of Precision and Recall for the retrieved images and Normalized Mutual Information (NMI). The value of both evaluation varied between [0; 1], while higher values indicating better agreements with the real world event.

In this paper after a comprehensive literature review, authors decided to adopt density based criteria to implement their own DBSCAN clustering algorithm. Main motivations to opt density based clustering algorithm are that: (a) it is an unsupervised clustering algorithm which has potential to detect arbitrary shapes in noisy dataset; (b) it can be used for real time data; (c) we can use it at local as well at global level: this is indeed the main reason for its popularity compared to other algorithms. It does not require prior knowledge about number of clusters such as k-mean which is a tedious job when facing large datasets.

3 Approach

When reviewing different works, it turns out that event detection is a quite challenging task in the field of data analysis in social media. The basic process comprises the detection of meaningful patterns in the available data, which can be used to infer social events which take place. In addition, researchers have to cope with a large overhead of processes not related to the actual problem but necessary for the overall task. Indeed before starting the actual core process several questions about the data acquisition and preparation arise. The programming interfaces of the platforms have to be examined and data retrieval and buffering have to be considered. Once the data has been made available it still often requires additional preprocessing steps. It is only after these steps that the actual event detection algorithm can be considered, however there are still open tasks left in order to finalize the whole process. The implicit information inhered in the detected content clusters need to be extracted in order to describe the events; furthermore the results still should be presented and evaluated. Indeed aside from the actual basic detection process researchers face a huge overhead of additional tasks dragging away the attention from the actual problem and pacing down the process of research. Most of these additional tasks are shared across different implementations and are just replicated in new studies. That is why the authors in this study decided to develop and support the research community with a framework in doing overhead tasks allowing them to concentrate on their actual task: the detection of events.

3.1 Methodology

Several existing event detection algorithms and solutions have been examined and the most basic shared structure has been identified. These various event-detection algorithms share a very simple, yet powerful architecture. The generic structure of such algorithms can be seen in Fig. 1. It is a linear processing pipe where the data

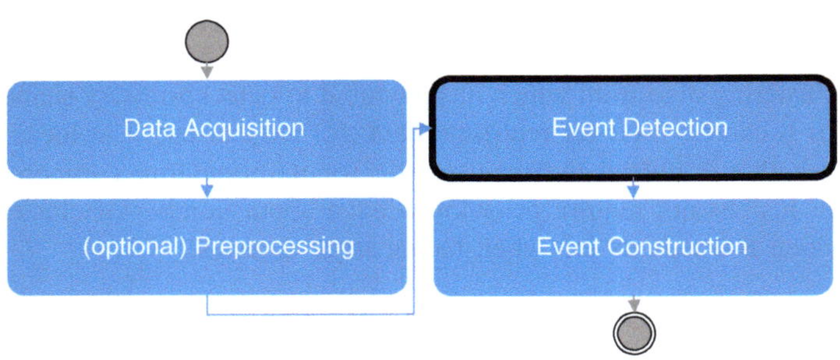

Fig. 1 Event-Basic Structure shared across common detection algorithms

undergoes successive transformation from raw data to a descriptive representation of the detected events.

The actual event detection algorithm represents the core of this processing pipe and is surrounded by the data acquisition, the optional preprocessing, and the event construction. Event construction includes the extraction of inherent information from detected events (content clusters) as well as their representation. This structure emphasizes the clear separation of different parts of an event detection approach and encourages the creation of a modular framework, the independent development of different parts, as well as the reuse of already existing implementations. Using this model, the authors in this study developed a generic and extensible programming framework to find meaningful patterns out of heterogeneous and unstructured online data streams. The separation of different processing stages comes with the benefit that each stage can be exchanged and replaced by an alternative implementation, allowing a dynamic adaptation to different objectives such as choosing a target social media platform or adapting the output of the process. However the development of such a generic framework requires some modifications to the previously shown processing pipe.

Indeed the event detection stage is not the only stage that requires preprocessing of the data acquired through data acquisition step; all stages in the process including any possible stage following the output may require preprocessing of the data from the previous stage. Because the characteristics of data in different social media platforms differ from each other. Therefore preprocessing must not only be applied before the event detection step but an optional preprocessing/filtering step must be available for all stages. Figure 2 shows this generalized processing model. The basic processing steps stay the same as in Fig. 1, but each of them is succeeded by an optional filtering step.

As can be seen, the new processing pipe is a linear combination of transformation and filtering steps and therefore an instance of the well-known pipes and filters pattern (Buschmann et al. 1996). The single transformation steps have no

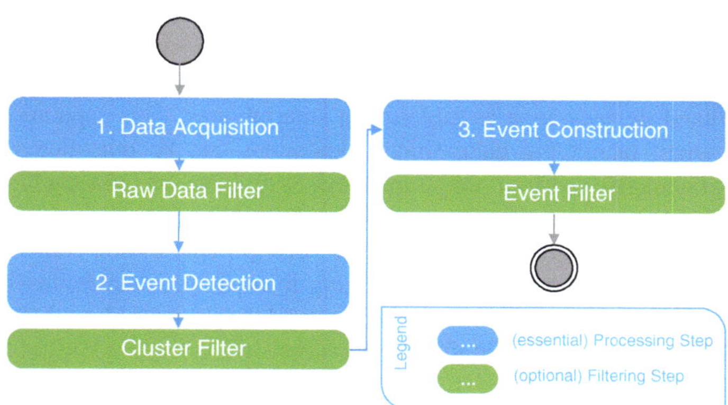

Fig. 2 Generalized Event-Detection processing pipe

cross-references on one another but solely rely on the data produced by the preceding step, granting the single components complete independence from each other as long as the data-contracts between them are met. This not only allows single components to be developed independently but also allows the construction of modular applications through a dynamic composition of data acquisition, event detection and event construction components.

As an additional enhancement, the framework comes with a graphical user interface easing the composition of different process stages and providing an easy way of adapting its parameterization. This facilitates testing, evaluating and plain usage of the algorithms and also provides users with the possibility to get deeper into the matter by literally grasping the algorithms, comprehending the impact of its parameters and getting visual feedback. The framework also ships with an already implemented selection of all of its components, including data adapters to Flickr and Instagram, predefined filtering steps, information extraction components and visualization features like maps, plots and charts. The concrete implementation of the framework and the graphical user interface is discussed in the following sub-sections.

3.2 Framework

The framework is built upon the Python language which has recently received much attention in the academia; especially due to the clear structure of its code and many available standard toolkits. Python and similar languages such as R, have even become a de facto standard in the field of data mining and machine learning. It is not only supported by field specific frameworks like NumPy (NumPy Developers 2014) and SciPy (SciPy Developers 2014), which enable the language for mining big data; it is also equipped with numerous general third party tools (Travis 2007). Furthermore, a graphical user interface based on a state-of-the-art combination of HTML and JavaScript is added to the framework. This separates the basic framework and the user interface in a server-client manner (Buschmann et al. 1996). It allows the basic framework to be the part that mines for events—using non trivial calculations on powerful machines—whereas the user interface can still be run on more restricted devices such as tablets and smartphones. This separation is more important when facing big data, since calculations can be outsourced to cloud, while the interactive part of the framework will have no constraint for client devices. In addition, since the uprising of Web 2.0, HTML/JavaScript is prevalence in the field of light-weight graphical interface design. Due to this combination and numerous available third party tools—ranging from simple data frameworks up to 3d graphic engines—developers are able to extend the framework without requiring knowledge in a Python specific graphical framework.

3.3 Model

The most essential part of the framework is the model representing the data flow in the event detection process. This model specifies the data contract between different stages, therefore is the key in assuring a clean separation of the stages and providing the possibility for independent development of the different components (Fig. 3). Like the event detection process, it is also separated into three parts.

The contract for the data acquisition component is specified by the Basic Content model. It models the content acquired from an arbitrary social media platform and is kept very general only including parts which are shared by the great majority of existing platforms. Its representation contains a timestamp specifying the date of creation, a geo location reference and a text field. This model covers the ground case and is compatible to the majority of platforms, including Facebook, Twitter, Flickr and Instagram. But in order to support the generic nature of the framework the BasicContent model is supposed to represent the top most element of a hierarchy of content models. As can be seen in Fig. 3, it can be extended to support any kind of content by a simple inheritance; such as AudioContent and VideoContent. The hierarchical model—using the concept of information hiding—allows the components to decide on the degree of information, which they want to extract from the models. Hence, it ensures the compatibility of existing components with future extensions while placing no restrictions on the extensions itself. So, components designed for BasicContent will also be able to handle extended versions such as Audio Content and newly introduced components can still benefit from the additional information that the extended model provides.

The data contract for the second processing step (the event detection) is represented by the Content Cluster model. Just as with the content models, ContentCluster only covers the ground case and can be extended freely if the component needs to communicate any additional information.

The last data model belongs to the event construction stage and represents the final output of the whole process; the BasicEvent. This model represents a detected event and contains an identifier, usually representing the name of the event, as well

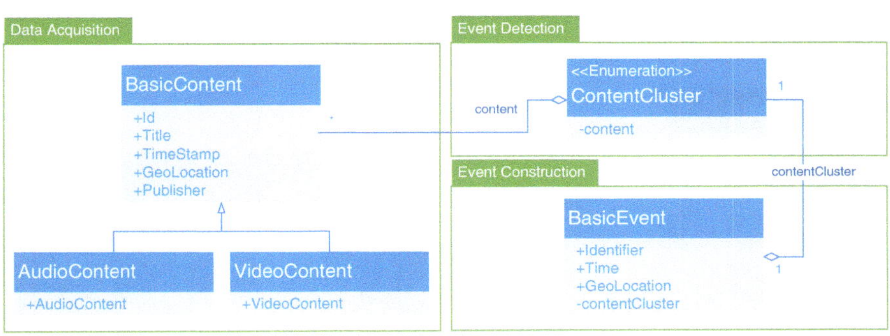

Fig. 3 UML class diagram: data models

as fields for the date and location of the event. It additionally holds a reference to the ContentCluster to which it belongs to grant access to the actual content related to that event. Again, the model can be seen as the most top element of a hierarchy of models, covering only the ground case and being open for any extension.

The discussed data models define contracts between different processing stages and represent the first step towards a modular system. But in order to ensure the independence of the sub-modules and in order to make the components substitutable, the exact interfaces of the single processing steps is still required to be defined. The model for these processing steps is displayed in Fig. 4. Here again, the diagram is split into three parts representing the three basic stages of the previously introduced processing model. The interface for the data acquisition stage is defined by SocialMediaContentProvider which serves the sole purpose of retrieving all contents available on the platform for a queried timespan and a geo bounding box.

As previously discussed, the separation and the intended independence of single stages require an optional available filtering step following each stage. This filtering serves the purpose of postprocessing each stage's output, smoothing the data transition such as filtering outliers. The framework allows the application of multiple successive filters and the filter interfaces presented in Fig. 4; RawDataFilter,

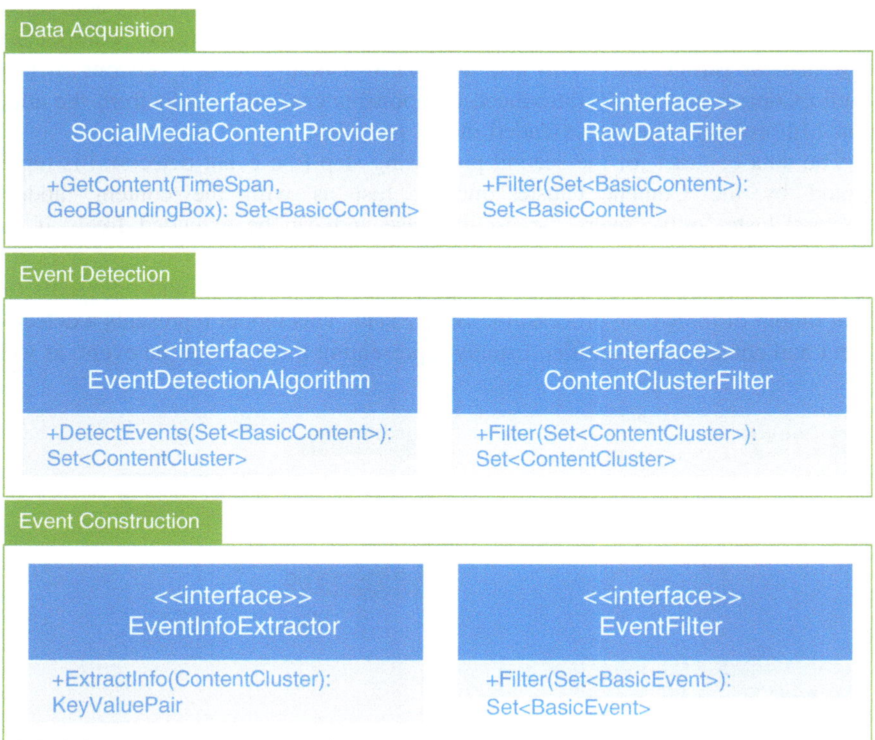

Fig. 4 UML class diagram: processes interfaces

ContentClusterFilter and EventFilter. The event detection stage is represented by the EventDetectionAlgorithm interface to extract clusters from a set of contents to form events.

The first two stages are represented with a similar implementation in the framework process. The process requires one data provider and one event detection algorithm. In contrast, the final stage (the event construction) is of a more dynamic nature. The construction transforms the raw content cluster identified as forming an event into a more descriptive representation embodied by the BasicEvent model. This model holds basic information about time, location and identification of the event and is ought to be extended depending on the purpose and objective of the actual application. Therefore, the construction stage is designed to allow a dynamic assembling of this representation, which allows defining a set of EventInfoExtractors to extract the desired information and dynamically extend the BasicEvent model. EventInfoExtractors may be used to infer time and location of an event but may also extract more abstract information such as statistics and graphical plots.

The framework process representing the actual interplay of the defined data models and process interfaces is shown in Fig. 5. As can be seen it reassembles the original process model of the Fig. 2. Using the introduced models and interfaces, different parts of event detection algorithms can now be developed independently.

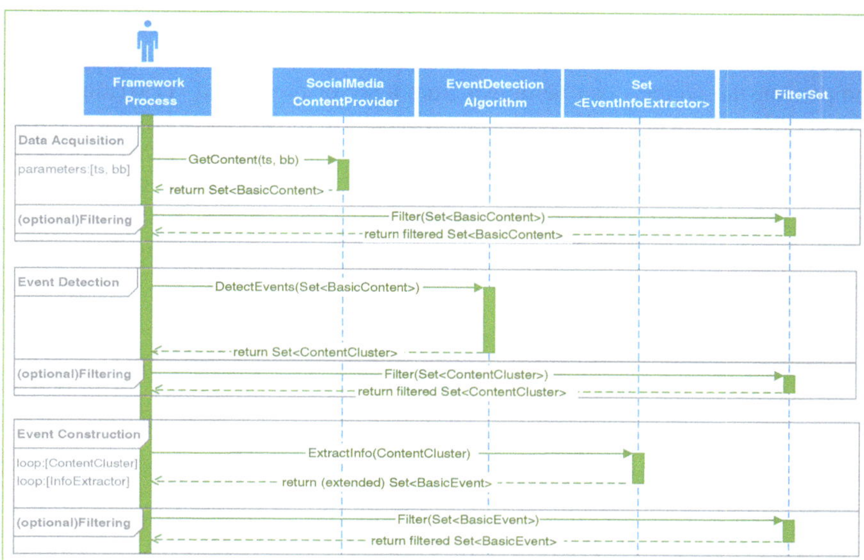

Fig. 5 UML sequence diagram: general framework process

3.4 Graphical User Interface

The graphical user interface enhances the framework by a graphical component and is targeted at the evaluation and production stage. The interface is built upon a combination of HTML/JavaScript and is to be run in a common web browser. It therefore is an optional component to the framework, separated from the core in terms of a client-server pattern (Buschmann 1996). This not only allows the core to be outsourced to a powerful machine, but due to the high distribution of web browsers even allows restricted devices such as tablets or smartphones to be used as clients of the framework. The graphical user interface was developed in the manner of a drag-n-drop user interaction concept which aside from its user friendly interaction abilities targets at touch enabled devices like tablets. The basic idea of the interface is that users can easily adapt their applications and evaluate different parameterizations by using a graphical interface on one of their favorite devices while the non-trivial and time consuming mining processes happen on a powerful background machine.

A screenshot of a basic and exemplary configuration of the interface can be seen in Fig. 6. The view is split up into three main parts, the toolbox, the workbench and the result panel. The toolbox displays all available components as discussed in the previous sections. These components can be dragged into the workbench area in order to fill the drop-areas representing the different stages of the framework process. When being dropped, the components unfold and offer the possibility to configure their parameters. Any component compatible to the interfaces discussed in the previous sections can be integrated into the UI. It only requires the definition of a constructor and its parameters. For the unfolded workbench that represents a component, the widgets can be drawn from a set of predefined HTML/JavaScript widgets. The screenshot in Fig. 6 contains a small exemplary subset of them; a map, a date-time chooser as well a slider widget. Due to the nature of HTML/JavaScript there is no restriction on extending this set. A great majority of the widgets is provided by jQuery-UI (jQuery Foundation 2014) but the user is free to use any

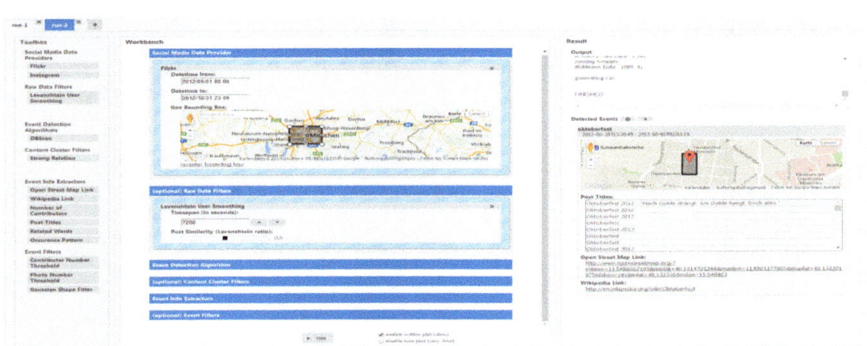

Fig. 6 A screenshot of an exemplary configuration of the framework's graphical user interface

additional third party libraries. With this concept the user can integrate its components into the user interface by simply providing a constructor, its parameters and its visual representation in a simple configuration file.

After composing the components and adapting the parameters, the result panel shows a live output of the original code while the algorithm is running, allowing easy debugging during runtime. After the algorithm is finished, the detected events also appear in the result panel. The basic data of the events is displayed and a miniature map visualizes the approximate location as well as the approximate size of the events. Other results can be dynamically composed based on the actual BasicEvent implementation as well as their extensions from the user's choice of EventInfoExtractors.

For visualizing the results—like the workbench—it is possible to draw widgets from a set of predefined HTML/JavaScript elements. This enables developers to completely adapt the application's visual output. The predefined components include different visualization elements such as maps, simple hypertext links and lists. In addition, the graphical interface offers a selection of plots, including a 3d scatter plot as well as a state-of-the-art t-Distributed Stochastic Neighbor Embedding (t-SNE) plot (van der Maaten and Hinton 2008) for visualizing the achieved data as a whole. The output plots are used to evaluate the impact of different parameter settings on the whole data set (Fig. 7). The graphical user interface also exports the result data in a comma separated value (CSV) format reflecting the BasicEvent model.

The framework is publicly available and can be downloaded from (https://polous@bitbucket.org/polous/smm.git).

4 Results and Discussions

A clustering based method was used as an exemplary event detection algorithm in order to evaluate the flexibility of the developed framework. The assumption is that during the happening of an event people publish an unusual concentration of content related to this event. With this assumption detecting an event comes down to separating the content belonging to such a concentrated cluster from the rest of the data. This is done by detecting clusters of dense data regarding temporal, locational and textual features in the social media content. We use here a density-based spatial clustering of applications with noise (DBSCAN) (Ester et al. 1996) as an exemplary event detection algorithm in the framework. Due to the simple interfaces of the framework, the DBSCAN algorithm was integrated using a simple Python adapter. After this integration, the framework is ready for new setups; choosing a target social media platform, applying preprocessing/filters, and defining the information which should be extracted from the events.

As a first step, the algorithm was targeted at the Flickr and Instagram adapters to provide access to the content of these platforms with a temporal SQLite (SQLite 2014) database as a cache in between. The results for the application of the

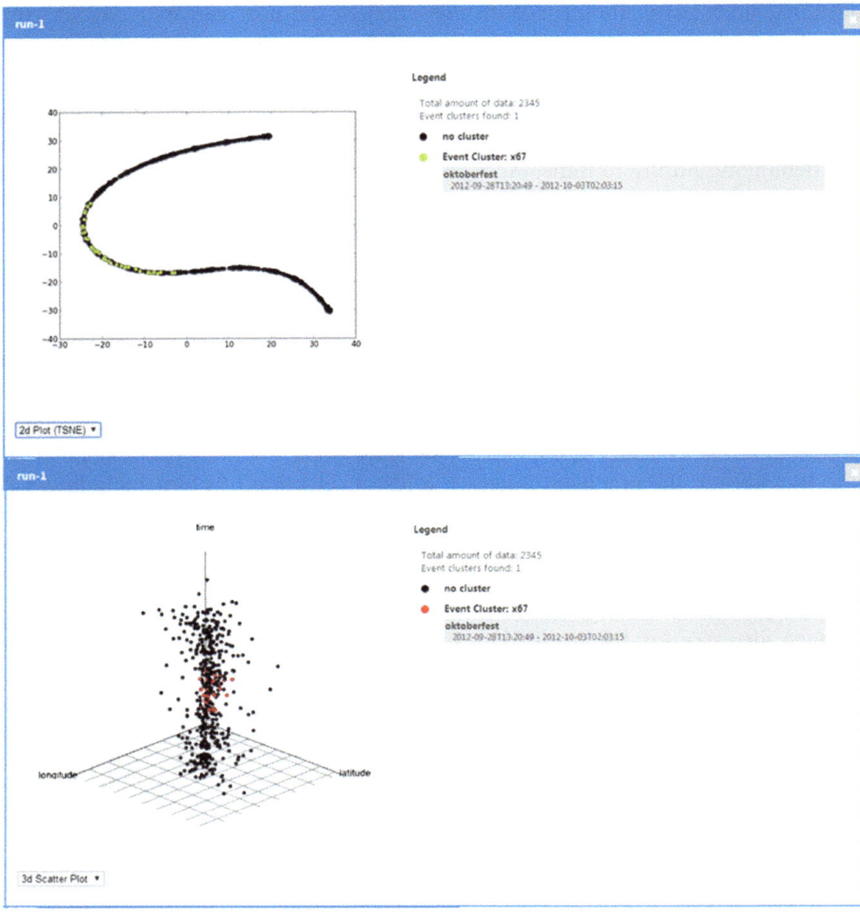

Fig. 7 Two screenshots of two different integrated plots

algorithm to data published at Flickr around October 2013 in Munich are shown in Fig. 8 (left).

As can be seen in Fig. 8 (left), some events (17) were detected, one of which is "Oktoberfest". Oktoberfest is an annual Bavarian fair held around the beginning of October and located at the Theresienwiese in Munich. The algorithm detected its existence solely based on the content that users published around this date. This information was then used by the framework in order to extract knowledge from the content of the detected clusters such as a name, the presumable time and the location of the event. The extracted knowledge used for the purpose of evaluation includes; the basic event description features and an additional list of words likely related to that event. The visualization of the event and the extracted information can be seen at the bottom of Fig. 8 (left) where the map widget indicates the event's

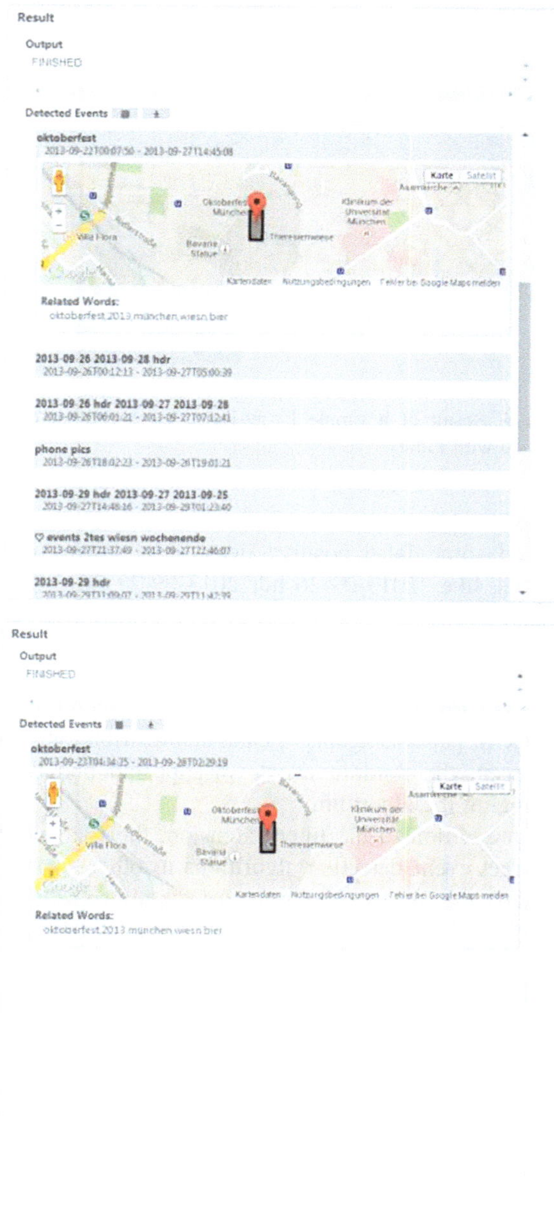

Fig. 8 Result of a simple Event-Detection-Algorithm on Flickr, October 2013, Munich (*left*). Result of a simple Event-Detection-Algorithm on Flickr, October 2013, Munich, refined with the filters (*right*)

presumed location. It approximately matches the actual location, which is the Theresienwiese in Munich. The related words list does also seem to match quite well, including words like “munich”, “2013” and “beer”.

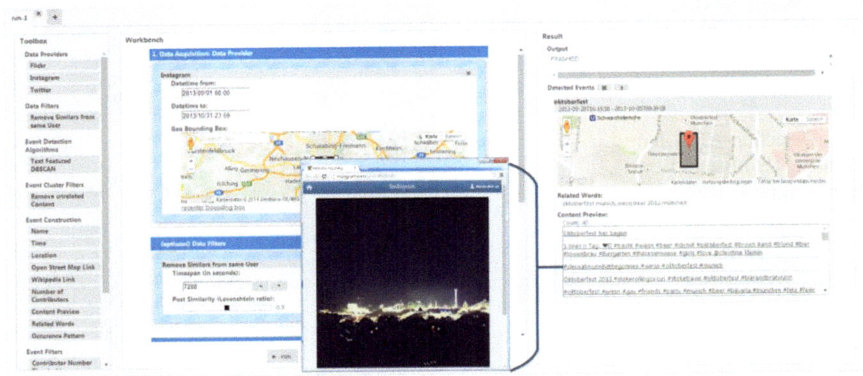

Fig. 9 Result of a simple Event-Detection-Algorithm on Instagram, October 2013, Munich, refined with Filters

Although the algorithm was accurate with the "Oktoberfest" event, it also produced some false positive detections, suggesting events consisting of random words, like "2013-09-26 hdr 2013-09-27 2013-09-28" event. This probably happened due to the structure of the clustering algorithm, which only applies a density based method. The result is refined and improved without adapting the algorithm itself, by simply using additional filtering and preprocessing steps. The results of an adapted run using filters such as user smoothing, thresholds, and matching the shape of patterns against a Gaussian distribution, is shown in Fig. 8 (right). As can be seen, the random events disappeared and the result was improved without changing the algorithm.

In addition to the filter extension, the framework also offers the possibility to retarget event detection algorithms to other social media platforms. For instance, exactly the same configuration of the above setup was retargeted at the Instagram platform. This was done by simply selecting the Instagram provider from toolbox in the user interface and dragging it onto the social media stage in the workbench. The result of a run targeted at the same area and timespan is demonstrated in Fig. 9. The "Oktoberfest" event was detected again, but this time from content published at Instagram.

5 Conclusion

Detecting events from social media is a known challenge for researchers in the field. A new framework for event detection in different social media is presented. The major characteristic of the presented platform that distinguishes it from all other implementations is its generality. To reach this generality, different event detection stages are separated from each other into a module-based approach, which offered the possibility to compose different independent components into a powerful event detection application. The framework can be used in a very wide range of social media platforms like flicker, Instagram, Twitter, Facebook and also a combination of these platforms to find complicated patterns from heterogonous big data. Unlike other similar implementations, its object-oriented and module based implementation provide a unique possibility for researchers to integrate any clustering algorithm and social media platform in the framework in a few simple steps. The platform works like a high level API and prevents the direct interaction of researchers with many complexities of the work. This is further enhanced by a graphical user interface (GUI) with a simple drag-and-drop interaction concept, which allows the composition of a powerful event detection algorithm in just a few clicks. The visual output could be adapted by choosing from the set of predefined knowledge extractors. The GUI also provides visual aid for the evaluation of an optimal parameter setting. Finally its open source character provides the researchers the possibility to extend, modify and customize the framework to their needs.

Ongoing research in this study is focused on adding new clustering algorithms to the framework. Implementation of this step is important in order to have a better understanding of different clustering algorithms and evaluating their usability for diverse areas. Another possible field of research that can increase the generality of the approach is adding more adaptors to the framework for other social media platforms such as Facebook and Twitter. The reason for this investigation is examining the possibility of event detection from multiple sources of data at the same time. An interesting field of research might be to use the output results of the approach as input for a higher level decision support system such as crisis management tools or prediction algorithms.

Acknowledgments The authors gratefully acknowledge the support by the International Graduate School of Science and Engineering (IGSSE), Technische Universität München, under project 7.07.

References

Ahlqvist T, Bäck A, Halonen M, Heinonen S (2008) Social media road maps: exploring the futures triggered by social media. Technical report, VTT Technical Research Center of Finland, Finland

Allan J, Papka R, Lavrenko V (1998) On-line new event detection and tracking. In: 21st annual international ACM SIGIR conference on research and development on information retrieval. ACM, New York, pp 37–45

Bao BK, Min W, Lu K., Xu C (2013) Social event detection with robust high-order co-clustering. In: 3rd ACM international conference on multimedia retrieval. ACM, New York, pp 135–142

Becker H, Naaman M, Gravano L (2010) Learning similarity metrics for event identification in social media. In: 3rd ACM international conference on web search and data mining. ACM, New York, pp 291–300

Buschmann F, Meunier R, Rohnert H, Sommerlad P, Stal M (1996) Pattern-oriented software architecture: a system of patterns. Wiley, New York

Dou W, Wang K, Ribarsky W, Zhou M (2012) Event detection in social media data. In: 2nd IEEE workshop on interactive visual text analytics: task-driven analysis of social media content. IEEE Press, New York, pp 971–980
Ester M, Kriegel HP, Sander J, Xu X (1996) A density-based algorithm for discovering clusters in large spatial databases with noise. In: 2nd international conference on knowledge discovery and data mining. AAAI Press, Palo Alto, pp 226–231
Gao X, Cao J, He Q, Li J (2013) A novel method for geographical social event detection in social media. In: 5th ACM international conference on internet multimedia computing and service. ACM, New York, pp 305–308
Geetha S, Shyla M (2014) An efficient divergence and distribution based similarity measure for clustering of uncertain data. Int J Sci Res 3(3):333–339
Ilina E, Hauff C, Celik I, Abel F, Houben GJ (2012) Social event detection on twitter. In: 12th international conference on web engineering. Springer, Berlin, pp 169–176
jQuery (2014) jQuery Foundation HTTP, www.jqueryui.com. Accessed 30 January 2014
Kaplan AM, Haenlein M (2010) Users of the world, unite! the challenges and opportunities of social media. Bus Horiz 53:59–68
Li R, Lei KH, Khadiwala R, Chang KC (2012) Tedas: a twitter-based event detection and analysis system. In: 28th IEEE international conference on data engineering. IEEE Press, New York, pp 1273–1276
Li Z, Wang B, Li M, Ma WY (2005) A probabilistic model for retrospective news event detection. In: 28th annual international ACM SIGIR conference on research and development on information retrieval. ACM, New York, pp 106–113
Mourya M, Prasad P (2013) An effective execution of diabetes dataset using WEKA. Inter J Comput Sci Inf Technol 4(5):681–682
Nielsen (2012) State of the media: The Social Media Report 2012. Technical report, Nielsen Holdings N.V
NumPy (2014) NumPy Developers HTTP, www.numpy.org. Accessed 30 January 2014
Papadopoulos S, Schinas E, Mezaris V, Troncy R, Kompatsiaris I (2012) Social event detection at mediaeval 2012: challenges, dataset and evaluation. In: Multimedia benchmark workshop 2012. CEUR-WS, Aachen
Parikh R, Karlapalem K (2013) ET: events from tweets. In: 22nd international conference on world wide web companion. IW3C2, Geneva, pp 613–620
Petkos G, Papadopoulos S, Kompatsiaris Y (2012) Social event detection using multimodal clustering and integrating supervisory signals. In: 2nd ACM international conference on multimedia retrieval. ACM, New York, p 23
Rabbath M, Sandhaus P, Boll S (2011) Automatic creation of photo books from stories in social media. In: TOMCCAP, pp 27–27
Reuter T, Cimiano P, Drumond L, Buza K, Schmidt-Thieme L (2011) Scalable event-based clustering of social media via record linkage techniques. In: 5th international AAAI conference on weblogs and social medi. AAAI Press, Palo Alto, pp 172–202
Sakaki T, Okazaki M, Matsuo Y (2010) Earthquake shakes twitter users: real-time event detection by social sensors. In: 19th ACM international conference on World Wide Web. ACM, New York, pp 951–860
SciPy (2014) SciPy Developers HTTP, www. scipy.org. Accessed 30 January 2014
SQLite (2014) SQLite HTTP, www.sqlite.org. Accessed 30 January 2014
Tamura K, Ichimura T (2013) Density-based spatiotemporal clustering algorithm for extracting bursty areas from georeferenced documents. In: 2013 IEEE international conference on systems, man, and cybernetics. IEEE Press, New York, pp 2079–2084
Travis E (2007) Python for scientific computing. Comput Sci Eng 9(3):10
van der Maaten LJP, Hinton GE (2008) Visualizing high-dimensional data using t-SNE. J Mach Learn Res 9:2579–2605

Weng J, Lee BS (2011) Event detection in twitter. In: 5th international AAAI conference on weblogs and social media. AAAI Press, Palo Alto, pp 401–408
Xu S (2011) Discovering and tracking events from news, blogs and microblogs on the web. In: 10th international conference on spatial information theory. Springer, Berlin, pp 20–23
Yang Y, Pierce T, Carbonell J (1998) A study of retrospective and on-line event detection. In: 21st annual international ACM SIGIR conference on research and development on information retrieval. ACM, New York, pp 28–36
Zhou X, Chen L (2013) Event detection over twitter social media streams. VLDB J 23(3):381–400

The Uncertain Geographic Context Problem in Identifying Activity Centers Using Mobile Phone Positioning Data and Point of Interest Data

Xingang Zhou, Jianzheng Liu, Anthony Gar On Yeh, Yang Yue and Weifeng Li

Abstract People aggregate at different areas in different times of the day, thus forming different activity centers. The identification of activity centers faces the uncertain geographic context problem (UGCoP) because people go to different places to conduct different activities, and also go to the same place for carrying out different activities in different times of the day. In this paper, we employ two kinds of novel dynamic data, namely mobile phone positioning data and Point of Interest (POI) data to identify the activity centers in a city in China. Then mobile phone positioning data is utilized to identify the activity centers in different times of a working day, and POI data are used to show the activity density variations at these activity centers to explain the temporal dynamics of geographic context. We find that mobile phone positioning data and POI data as two kinds of spatial-temporal data demonstrate people's activity patterns from different perspectives. Mobile phone

X. Zhou · J. Liu · A.G.O. Yeh (✉) · W. Li
Department of Urban Planning and Design, The University of Hong Kong, Hong Kong, China
e-mail: anthony.yeh@hku.hk

X. Zhou
e-mail: xingang_zhou@hku.hk

J. Liu
e-mail: jzliu@hku.hk

W. Li
e-mail: wfli@hku.hk

Y. Yue
Department of Transportation Engineering, College of Civil Engineering, Shenzhen University, Shenzhen, China
e-mail: yueyang@szu.edu.cn

Y. Yue
Shenzhen Key Laboratory of Spatial Smart Sensing and Services, Shenzhen University, Shenzhen, China

F. Harvey and Y. Leung (eds.), *Advances in Spatial Data Handling and Analysis*, Advances in Geographic Information Science,
DOI 10.1007/978-3-319-19950-4_7

positioning data provide a proxy to represent the activity density variations. POI data can be used to identify activity centers of different categories. These two kinds of data can be integrated to identify the activity centers and clarify the UGCoP.

Keywords Activity center · UGCoP · Mobile phone positioning data · Point of interest

1 Introduction

In a city, different areas provide people with the opportunity to conduct various activities, thus forming different activity centers at different times of the day (Cervero 1991). Activity centers help us understand how people use the urban space, which is of great importance to urban planning and transport policy making (Ratti et al. 2006). The identification of activity centers faces the uncertain geographic context problem (UGCoP) (Kwan 2012b), because the geographic context of people's activities is not only related to land-use, but also what people are doing at a particular time and space for particular reasons (Erickson and Schultz 1997). The UGCoP is related to but different from the modifiable area unit problem (MAUP) (Openshaw 1984). These two problems are common in studies that examine the influence of area-based variables on individual behavior variables. The MAUP is about the zoning scheme and geographic unit scale. While the UGCoP is about spatial contextual uncertainty, and the temporal dynamics of geographic context (Kwan 2012b). The essence of the UGCoP is that the authenticity of research might be compromised due to the inaccurate description of the true geographic context (Kwan 2012b). In the study of people's activity centers, true geographic context refers to the interactions of land uses and activities in it that exert contextual influences on the individuals being studied. It not only represents the geographic environment, but also includes geographic variables related to people's activities. People's activities are not only influenced by land uses, but also their characteristics. Different people go to different places for different purposes, and also go to the same place for different activities in different times of the day.

Mobile phone data and Point of Interest (POI)) data have given a new momentum to explore people's activities in a smart way (Batty et al. 2012; Yue et al. 2014). Mobile phone data reflect phone users' activity patterns (Calabrese et al. 2013), and can be used to explore functional regions (Gonzalez et al. 2008). Studies on a large number of mobile phone data have found that people's activities are rather regular and could be predicted at a staggering 93 % potential predictability (Song et al. 2010). Mobile phone positioning data reflect the mobility of the people, which is closely related to land-use patterns (Phithakkitnukoon et al. 2010). POI data are obtained from social networks, such as Twitter and Facebook. They identify the business and recreation establishments throughout the city, reflecting land-use intensity and diversity (Yuan et al. 2012).

This paper utilizes mobile phone positioning data and POI) data to identify activity centers respectively, and to illustrate the UGCoP by elucidating the uncertainty caused by geographic context. With this study, we can enhance the understanding of the UGCoP, especially in the fields of human mobility.

2 Data Description and Processing

We use anonymous mobile phone positioning data and POI data in a city in China. The mobile phone positioning data in this study is tower-based periodic location update data. Mobile phone datasets include phone user anonymized ID, date, time, and phone tower locations. The location of the mobile phone user is approximated by the Voronoi service area of the nearest mobile phone tower. The mobile phone towers are more concentrated in the central city than in the peripheral areas, as shown in Fig. 1.

POI data are the other type of data employed in this study to identify activity centers. It is collected from Sina Weibo operating since August 2009, by using Sina Weibo's API (application programming interface, http://open.weibo.com/). The dataset contains 34,876 POIs with latitude, longitude, name, address, category, and number of check-in times. The category of one POI indicates the utility and function of the POI (e.g., restaurant, shopping mall and hospital). It in total contains about 207 kinds of different categories. The number of check-in for one POI) is the accumulated number of users' check-in in the POIs by September 2, 2013, which shows the active status and hotness of POIs.

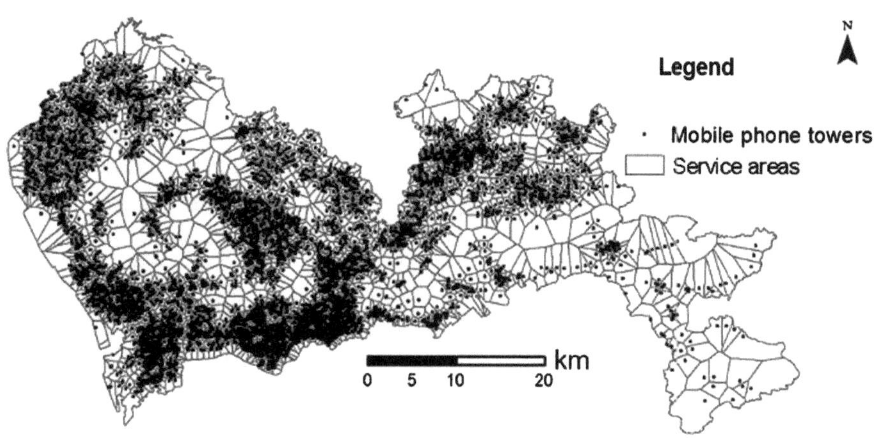

Fig. 1 Voronoi service areas of mobile phone towers

3 Identification of Activity Centers Using Mobile Phone Positioning Data

Two methods are employed to identify the activity centers from mobile phone positioning data respectively, as shown in Fig. 2.

First, the mobile phone positioning data are queried over an hour interval to facilitate comparisons among different hours of the day. Then, time-cumulative activity density is used to understand the spatial intensity of various activities at different time of the day in different places. The mobile phone positioning records between 9–11 am and 3–5 pm are selected as most typical working hours to infer the employment activity centers. The mobile phone positioning records between 4 and 5 am, which is sleeping time at night for most people are used to infer the residential activity centers. The mobile phone positioning records from 8 to 9 pm, which is leisure time for most people are used to infer the non-employment or living activity centers.

Then, inverse distance weighted method is employed to generate a continuous activity density distribution surface. Spatial autocorrelation is utilized to find the appropriate distance band where the activity density distribution is most spatially correlated. The kernel density estimation model is used to represent the density

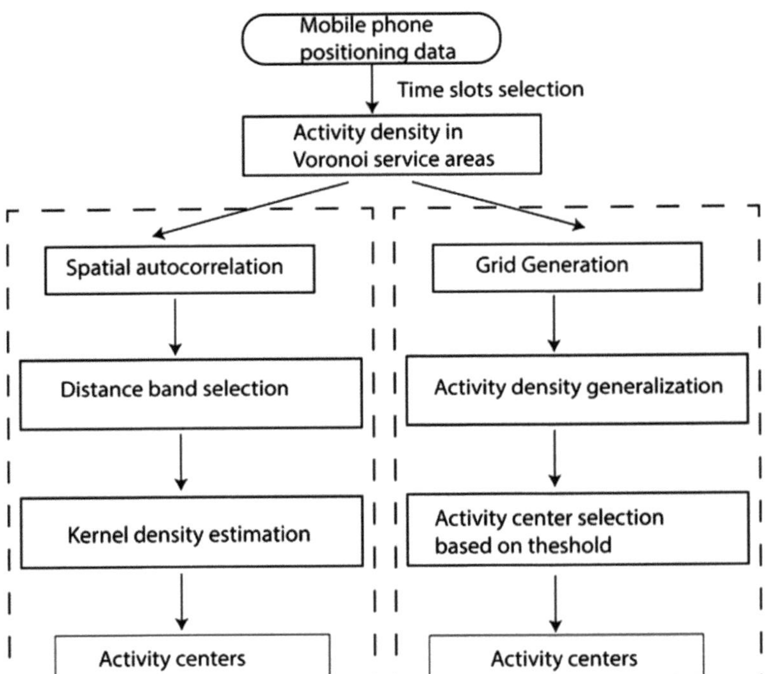

Fig. 2 A framework for identification of activity centers using mobile phone positioning data

distribution of different activities in real geographical space at certain time of the day from a macro perspective. Figure 3 shows several activity centers and sub-centers during sleeping time and working time. During working time, the employment activity density at the city centers are rather high. The activity density distribution during sleeping time at the city centers is relatively lower than that during working time.

In order to compare the activity density distribution at different parts of the city, a lattice-based framework is employed to identify the activity centers from a micro perspective. At first, the activity density in Voronoi service areas are transformed to grid-based density distribution. The grid size is 500 m, which is coarse enough to

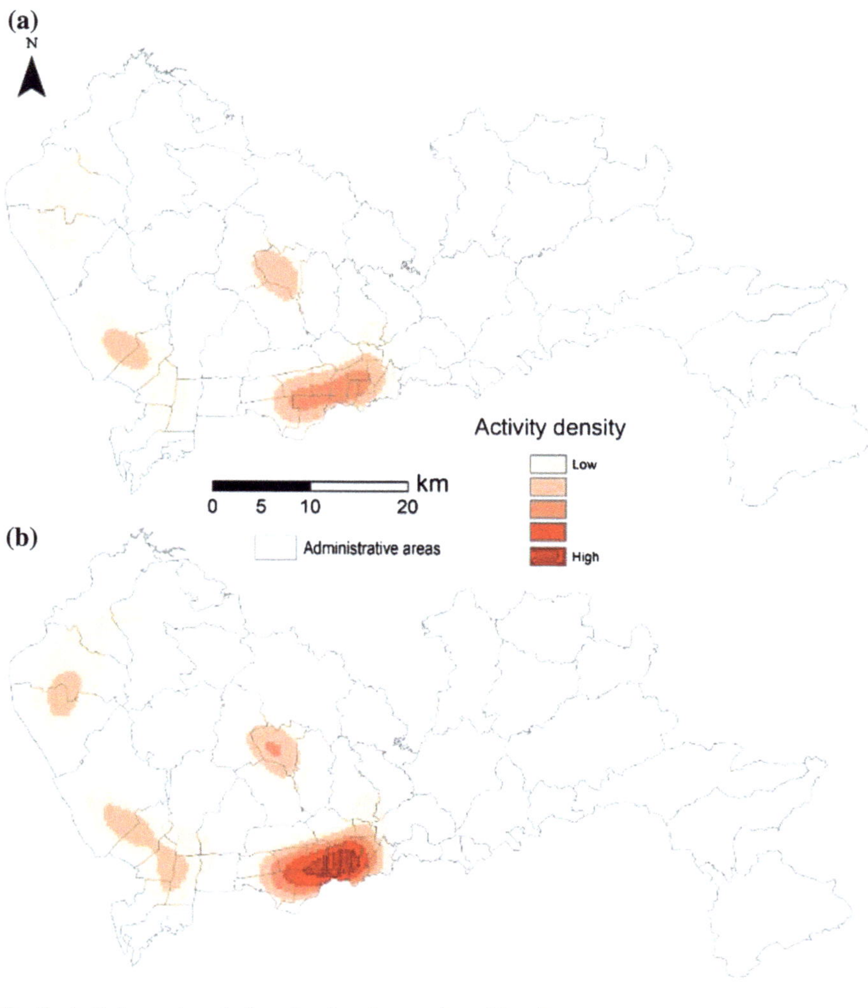

Fig. 3 Activity centers during sleeping time and working time

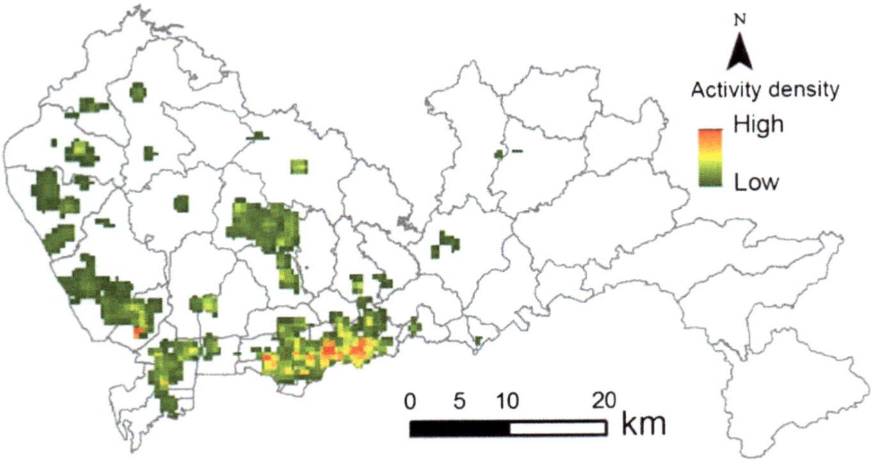

Fig. 4 Employment activity centers

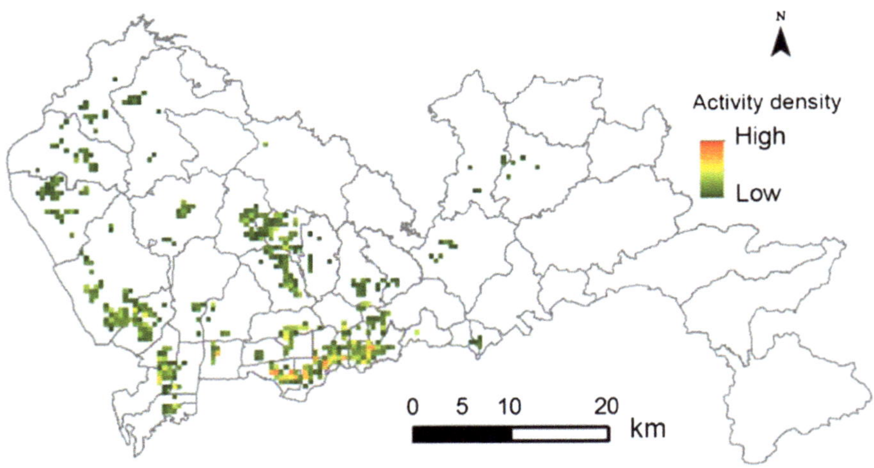

Fig. 5 Residential activity centers

reduce the spatial inaccuracy. The number of people within each grid is used to represent the activity density. Based on the sample activity centers, we select a threshold to identify the activity centers. The employment activity centers and the residential activity centers are shown in Figs. 4 and 5.

Comparing Figs. 4 and 5, we find that the employment centers are generally more scattered than the residential centers. The activity density at these employment centers are much higher than those at the residential centers.

4 Identification of Activity Centers Using POI Data

Identification of activity centers using POI) data draws on a few inferences and characteristics of activity centers: (1) activity centers receive more visits from people than surrounding areas; (2) activity centers have higher density of POIs than surrounding areas; (3) activity centers have higher degree of functional integrity than surrounding areas. Functional integrity here refers to the quality and condition of a place being a whole in terms of function and service. Higher functional integrity means that the place has enough POIs with different functions that can support itself on its own, without much reliance on other places. The rationale behind these assumptions is that a typical activity center usually has many visits from a lot of people. While doing their business in the activity center, people also need to dine, reside, shop, etc., which means that there would be more POIs related to restaurants, hotels and shopping malls around.

For one grid, we use the sum of check-in number of all POIs within the grid to denote people's visits. Number of POIs within the grid is used to proxy the POI density of the grid. Number of function categories is used to proxy the functional integrity of the grid.

Based on these inferences and assumptions, this study employs a lattice-based framework to identify activity centers, as shown in Fig. 6. At first we generalize the activity distribution with a uniform grid. Grid size is determined as 500 m, which is coarse enough to reduce the noise and detailed enough not to mix different areas. Then characterization within each grid is done by calculating POI check-in number, density and POI functional integrity based on POI attributes. After characterization,

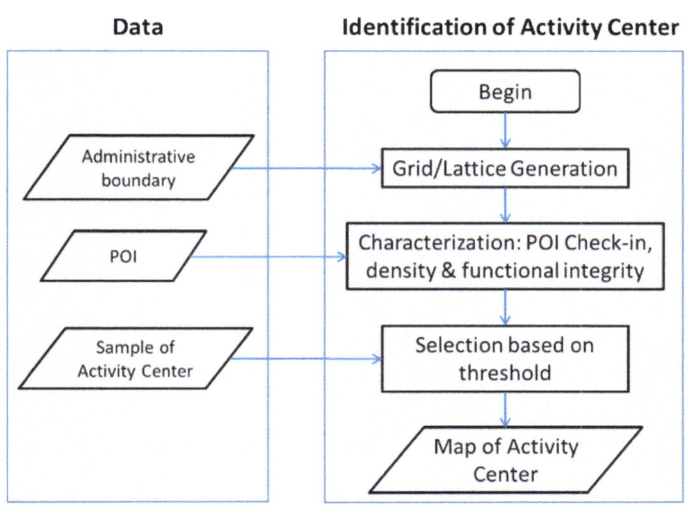

Fig. 6 A framework for identification of activity centers using POI data

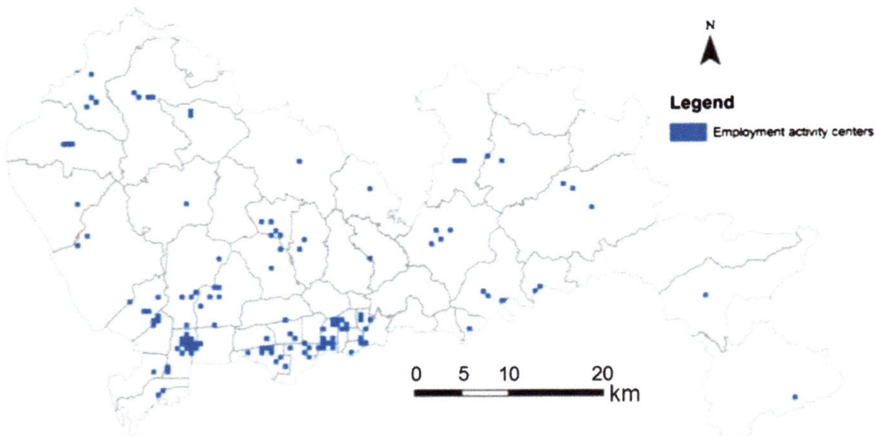

Fig. 7 Employment activity centers

we determine a threshold for selecting activity centers according to sample activity centers. In the end, activity centers are identified using the threshold.

We conducted the identification of activity center using POI data by implementing the framework in Fig. 6. The activity centers of different activity types such as working, residential and recreation, are presented as Figs. 7, 8 and 9.

In order to obtain a macro perspective of the activity centers, we employ kernel density estimation model to estimate activity centers using POI) data. Activity centers identified using kernel density estimation based on POI data is presented in Fig. 10.

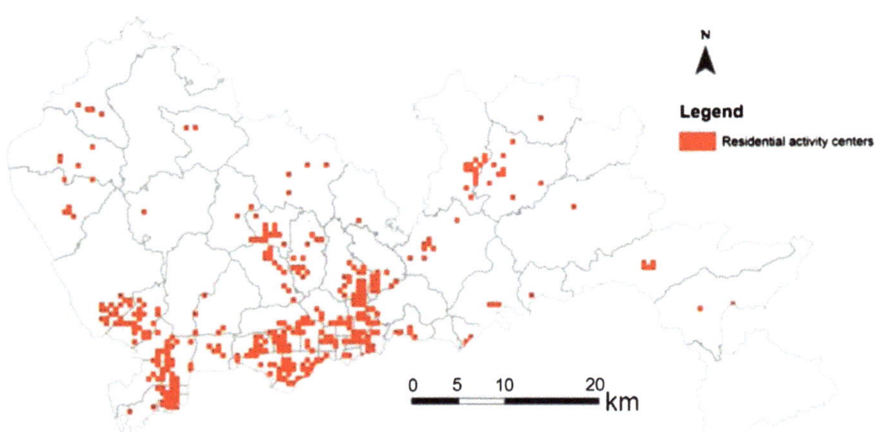

Fig. 8 Residential activity centers

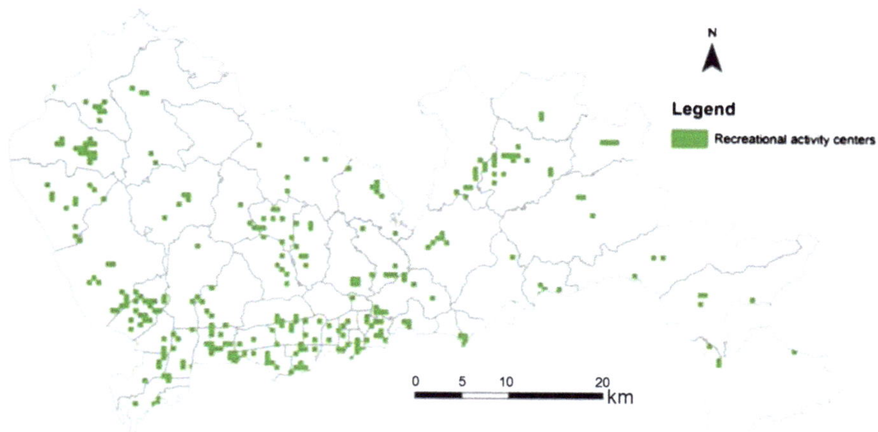

Fig. 9 Recreational activity centers

Figure 10 shows several activity centers and sub-centers of employment, residential and recreation activities. The recreational activity density is much higher than that of residential and employment activities. POI data mainly reflect people's recreational activities, as people are more likely to check-in during recreational activities.

5 The Uncertain Geographic Context Problem in Identifying Activity Centers

People's activities take place in different times of the day and in disparate areas (Kwan 2012a). In the same geographic environment, there are different activities at different periods. In order to compare the activity density variations during different times of the day, the original irregular mobile phone data have been converted into hourly intervals for analysis. We select Huaqiangbei as a typical employment activity center and High-tech park as a typical employment activity sub-center for analysis, to examine the activity density variations at these activity centers.

The activity density distribution at these two activity centers are calculated and compared, as shown in Fig. 11. At the activity centers, the activity density distribution in different times of the day show great differences. This figure illustrates how people use the urban space for different activities over 24 h of a typical working day. The activity density at Huaqiangbei is much higher than that at High-tech Park. The activity density distribution at Huaqiangbei is like "M", which increases in the morning from about 8 am, and decreases a little during lunch time, with peak in the afternoon at around 5 pm.

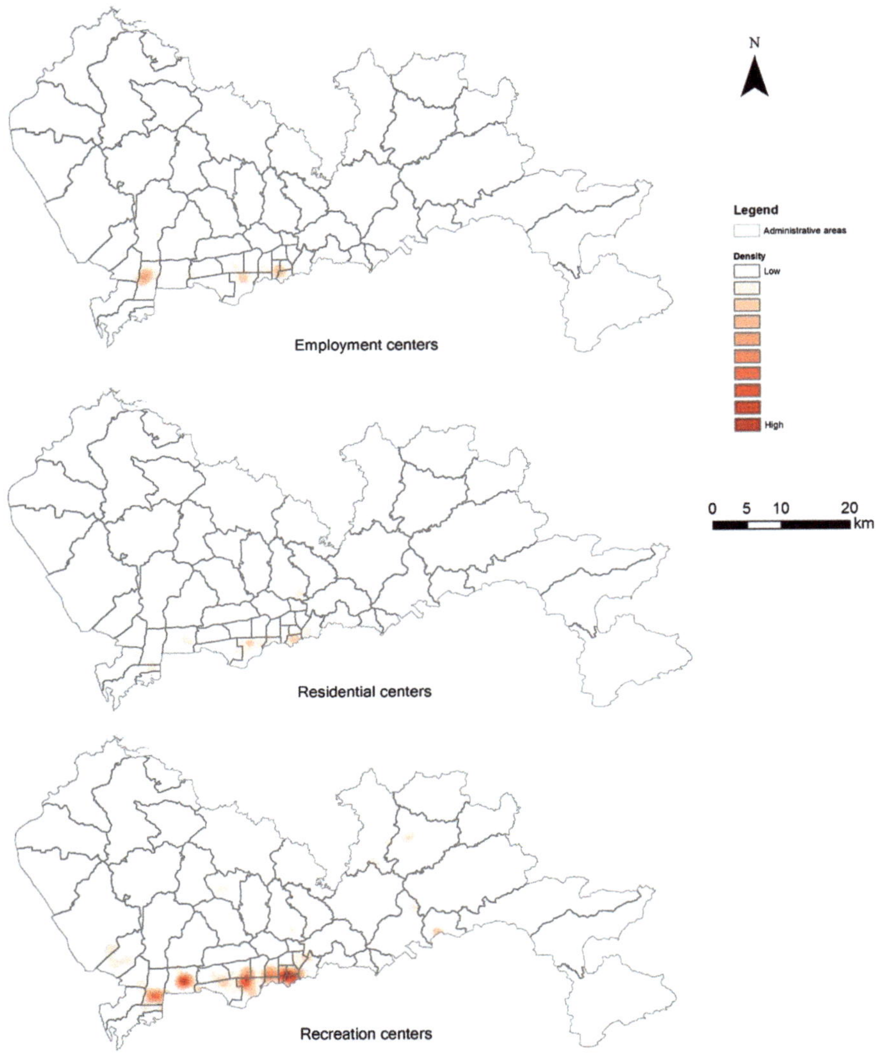

Fig. 10 Activity centers identified using kernel density estimation based on POI data

POI data are employed to explain the activity density variations. As we can see from Fig. 12 and Table 1, the percentage of POI check-in for recreation activities and employment activities are rather high. During working hours, people aggregate there to work, and many people stay there for recreation and shopping activities after work. The activity density at High-tech Park is much lower than that at Huaqiangbei, and the fluctuation is much smaller, because High-tech Park is mainly for office and residential development. As we can see from Fig. 13 and Table 1, the percentage of POI check-in for employment activities is rather high.

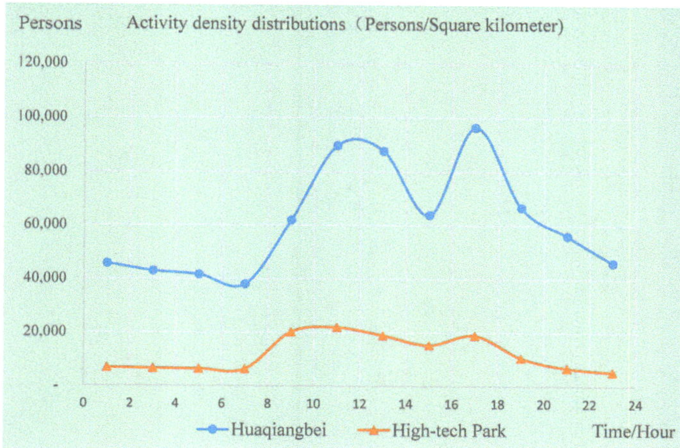

Fig. 11 Activity density variations with time (Zhou et al. 2014)

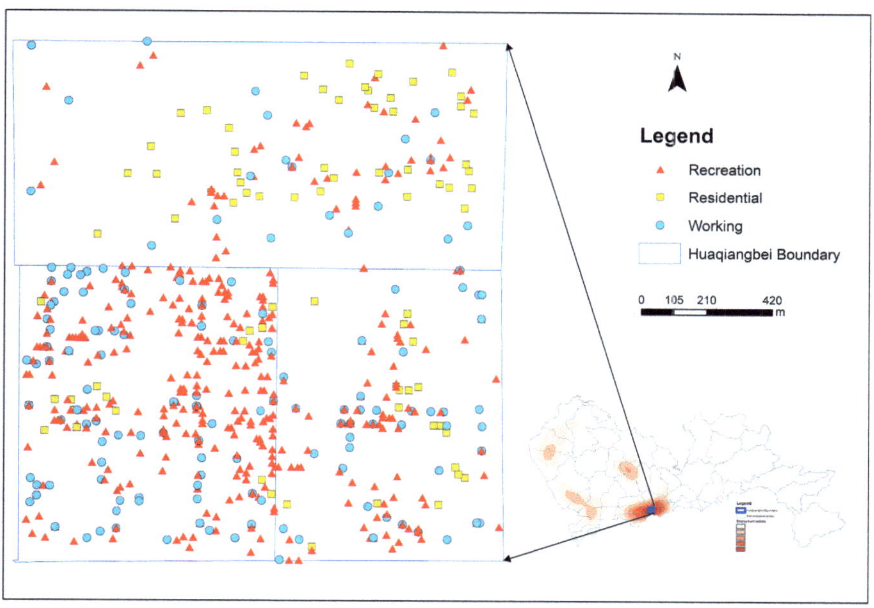

Fig. 12 Activity category distribution at Huaqiangbei from POI data

Table 1 POI check-in distribution

	Percentage	
Activity categories	Huaqiangbei (%)	High-tech Park (%)
Recreation	72.9	8.1
Residential	7.8	12.4
Employment	19.3	79.5

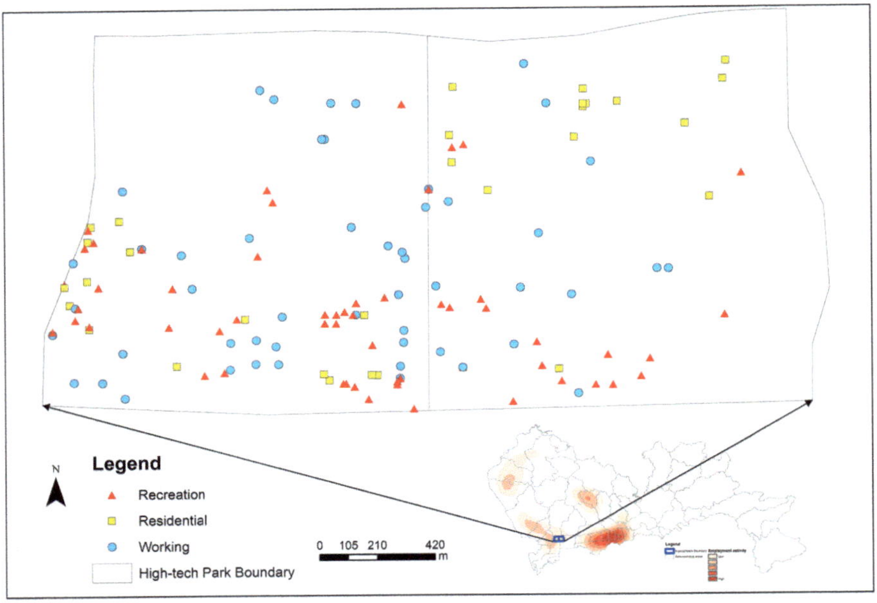

Fig. 13 Activity category distribution at High-tech Park from POI data

6 Conclusion

We find that the mobile phone positioning data and POI data show distinct spatial and temporal characteristics, and demonstrate people's activity centers from different perspectives. Mobile phone positioning data provide a proxy to estimate the frequency of people aggregating in a place. It can be used to identify the activity centers, and how intensively the urban space is used by people in different times of day. POI data represent the popularity or intensity of people's use of certain functions in a place. It can be used to identify activities centers of different categories. The identification of activity centers faces the uncertain geographic context problem. This study adopts a place-based (using POI data) to people-based (using mobile phone positioning data) approach to explain the UGCoP in the activity centers. The mobile phone positioning data are collected from the largest mobile phone operator in the city. It is assumed that mobile phone users are randomly distributed in the city. So the mobile phone users are a large sample of the whole people. Mobile phone positioning data can be used to identify activity centers in different times of the day. Whereas, POI) data can be used to explain the temporal dynamics of activity density in these activity centers in different times of the day. The spatial and temporal variations of the activity centers are not only influenced by the land-use, but also the spatial contextual uncertainty and temporal dynamics of geographic context. The study shows that mobile phone positioning data and POI data can be used to identify people's activity centers from different perspectives, and can be integrated to clarify the UGCoP. POI data have some limitations in check-in times because they are collected from young and middle-aged urban residents aged 10–60. The findings of the study can only be generalized to urban areas where the majority of the population are young and middle-aged people.

Acknowledgments This research was supported by the National Science Foundation of China (No. 41471378, 41231171, 41171348), and Shenzhen Scientific Research and Development Funding Program (JCYJ20121019111128765, JCYJ20130329144141856). Weifeng Li would like to thank the support from the Francis SK Lau Research Fund.

References

Batty M, Axhausen KW, Giannotti F, Pozdnoukhov A, Bazzani A, Wachowicz M, Ouzounis G, Portugali Y (2012) Smart cities of the future. Eur Phys J Spec Top 214(1):481–518. doi:10.1140/epjst/e2012-01703-3

Calabrese F, Diao M, Di Lorenzo G, Ferreira J Jr, Ratti C (2013) Understanding individual mobility patterns from urban sensing data: a mobile phone trace example. Transp Res Part C 26:301–313. doi:10.1016/j.trc.2012.09.009

Cervero R (1991) Land uses and travel at suburban activity centers. Transp Quaterly 45 (4):479–491

Erickson F, Schultz J (1997) When is a context? Some issues and methods in the analysis of social competence. In: Cole M, Engestrom Y, Vasquez O (eds) Mind, culture, and activity: seminal papers from the laboratory of comparative human cognition. Cambridge, pp 22–31

Gonzalez MC, Hidalgo CA, Barabasi A-L (2008) Understanding individual human mobility patterns. Nature 453(7196):779–782

Kwan M-P (2012a) How GIS can help address the uncertain geographic context problem in social science research. Ann GIS 18(4):245–255

Kwan M-P (2012b) The uncertain geographic context problem. Ann Assoc Am Geogr 102 (5):958–968. doi:10.1080/00045608.2012.687349

Openshaw S (1984) Concepts and techniques in modern geography number 38: the modifiable areal unit problem. Geo Books, Norwick

Phithakkitnukoon S, Horanont T, Di Lorenzo G, Shibasaki R, Ratti C (2010) Activity-aware map: identifying human daily activity pattern using mobile phone data. In: Human behavior understanding. Springer, pp 14–25

Ratti C, Williams S, Frenchman D, Pulselli R (2006) Mobile landscapes: using location data from cell phones for urban analysis. Environ Plann B 33(5):727–748

Song C, Qu Z, Blumm N, Barabási A-L (2010) Limits of predictability in human mobility. Science 327(5968):1018–1021

Yuan J, Zheng Y, Xie X (2012) Discovering regions of different functions in a city using human mobility and POIs. In: Proceedings of the 18th ACM SIGKDD international conference on knowledge discovery and data mining. ACM, pp 186–194

Yue Y, Lan T, Yeh AGO, Li Q-Q (2014) Zooming into individuals to understand the collective: a review of trajectory-based travel behaviour studies. Travel Behav Soc 1(2):69–78. doi:10.1016/j.tbs.2013.12.002

Zhou X, Yue Y, Yeh AGO, Wang H, Zhong T (2014) Uncertainty in spatial analysis of dynamic data—identifying city center. Geomatics Inform Sci Wuhan Univ 39(6):701–705 (in Chinese)

Part II
Network Analysis

Mining Co-location Patterns Between Network Spatial Phenomena

Jing Tian, Fu-quan Xiong and Fen Yan

Abstract The mining of co-location patterns is a popular issue in the field of spatial data mining. However, little attention has been paid to the co-location patterns between network spatial phenomena. This paper addresses this issue by extending an existing method to mining the co-location patterns between network spatial phenomena. The approach consists of two stages: (1) defining a co-location model on a network space based on skeleton partitioning of a road network to have co-occurrence relationships; (2) computing statistical diagnostics for these co-occurrence relationships. Our method was then applied to a case study regarding the mining of co-location patterns of manufacturing firms in Shenzhen City, China. These co-location patterns were also analyzed qualitatively according to the three mechanisms derived from agglomeration economies. Our method was compared with the existing method and the differences were verified by the network cross K-function.

Keywords Network spatial phenomena · Co-location patterns · Manufacturing firms · Agglomeration · Network cross K-function

J. Tian
Key Laboratory of Geographic Information System, Ministry of Education, Wuhan University, 129 Luoyu Road, 430079 Wuhan, China
e-mail: yutaka-2010@163.com

J. Tian · Fu-quanXiong · F. Yan (✉)
School of Resource and Environment Science, Wuhan University, 129 Luoyu Road, 430079 Wuhan, China
e-mail: 2011301130025@whu.edu.cn

F. Harvey and Y. Leung (eds.), *Advances in Spatial Data Handling and Analysis*, Advances in Geographic Information Science,
DOI 10.1007/978-3-319-19950-4_8

1 Introduction

Many phenomena and events take place on or in close proximity to networks. Real-world examples include traffic accidents, street crimes and facility location. These phenomena can be referred to as network spatial phenomena, represented as points lying on or close to networks geometrically (Okabe and Yamada 2001; Okabe et al. 2006). The analysis of network spatial phenomena has drawn much attention in the field of spatial analysis (Batty 2005; Okabe et al. 2006), and its noticeable feature is the transformation of distance metric from the Euclidean distance on a plane to the shortest-path distance on a network. This leads the continuous and homogenous plane to become discrete and non-homogeneous. In many urban areas, the Euclidean distance between two points differs significantly from the shortest path distance between these two points on a network. Therefore, we often reach false conclusions when we apply the methods of planar Euclidean space to network spatial phenomena (Yamada and Thill 2004; Lu and Chen 2007). One of the main approaches for analyzing of network spatial phenomena is to extend typical methods implemented on a planar Euclidean space to a network space, such as the network K-function and network kernel density estimation (Okabe and Yamada 2001; Borruso 2008; Xie and Yan 2008).

A co-location pattern is a set of spatial features or events that are frequently located together, and it represents a subset of Boolean spatial features, the instances of which are often located in spatial proximity (Huang et al. 2004, 2006), for example, the Nile Crocodile and the Egyptian Plover in ecology, the supplier and the producers in economics (Hu 2008). Mining co-location patterns is the process of identifying co-location patterns from spatial datasets with a large number of Boolean spatial features. The process of mining co-location patterns has become an important data mining task exhibiting the distribution of phenomena and events along with supporting policy decision-making (Huang et al. 2006; Hu 2008). The two most common approaches to mining co-location patterns are spatial statistics and data mining approaches (Huang et al. 2004). Spatial statistics-based approaches incorporate measures of spatial correlation to characterize relationships between different types of spatial features. These measures of spatial correlation include the cross-K function with Monte Carlo simulation, mean nearest-neighbor distance, spatial regression models (Huang et al. 2004), Q-statistics (Ruiz et al. 2010), co-location coefficient (Leslie and Kronenfeld 2011), indicators of association between points and areas (Guo et al. 2011) and so forth (Li et al. 2006; Okabe et al. 2006; Arnia et al. 2008). Data mining approaches mainly create transactions by choosing co-location patterns, and then design Apriori-like algorithms to mine co-location patterns. The frequency of these co-location patterns are decided by a participation ratio and a participation index. The methodologies expounded by Koperski and Han (1995), Zhang et al. (2004), Bembenik and Rybinski (2009), Shekhar and Huang (2001) are representatives of this type of approaches. In general, there are three major co-location models of discovering co-location patterns in spatial datasets: the reference feature centric model, the window centric model and the event centric model (Shekhar and Huang 2001). (1) The reference feature centric model (Koperski and Han 1995) focuses on one specific type of feature while defining neighborhood relationships to create transactions; (2) The window centric model, also known as the data partitioning model, defines windows with an appropriate size first, and then creates transactions based on these windows; (3) The event centric model (Huang et al. 2004) is often incorporated to find all subsets of spatial features that are likely to co-locate. As with the reference feature centric model, it creates transactions by defining neighborhood relationships. Few researches have focused on mining co-location patterns between network spatial phenomena. Major tools applied to discovering the co-location patterns between network spatial phenomena are the network cross K-function method (Okabe and Yamada 2001) and the network conditional nearest-neighbor distance method (Okabe et al. 2006).

An approach (Sierra and Stephens 2012) based on Natural Language Processing (NLP) has also been proposed. Its main steps are to first set out a co-location model in order to extract subsets of Boolean features that have co-occurrence relationships, and then estimate statistical independence between different kinds of Boolean feature classes to determine whether co-occurrence relationships are statistically significant. In the co-location model of this approach, the plane was partitioned by grids; points located in the same grid cell were defined as co-occurrences. In this paper, we extend this approach to a network space. Firstly, we define a co-location model on a network space based on skeleton partitioning to have co-occurrence relationships and then, we estimate statistical independence between feature classes to determine whether co-occurrence relationships are frequent enough to statistically identify co-location patterns.

The rest of this paper is organized as follows. Section 2 proposes a method of mining co-location patterns between network spatial phenomena. Section 3 applies our method to mining co-location patterns between manufacturing firms in Shenzhen city, China, and these co-location patterns are analyzed qualitatively according to the three mechanisms derived from agglomeration economies. Also Sect. 3 compares our method with the method proposed by Sierra and Stephens, and the differences are verified by the network cross K-function. Finally, this paper ends with perspectives of further research.

2 Methodology

2.1 Defining the Co-location Model Based on Skeleton Partitioning

In the method proposed by Sierra and Stephens, the plane is partitioned by uniform grids. If different kinds of Boolean feature classes have instances in the same grid cell, they are defined to have a co-occurrence. In the same grid cell, our concern is simply whether instances of different feature classes co-occur, as this study does not investigate the times of their instances. Okabe's book (Okabe et al. 2000) introduces many methods for tessellating spaces. We would like to point it out that this study aims to discover co-location patterns between network spatial phenomena, which reveals how Boolean feature classes co-locate on or close to networks. Therefore, how to partition the network space affects our results and becomes a major concern.

There exist four ways of partitioning network spaces. The first is Planar graph, a road network consists of road segments and nodes, and the road segments intersect at the nodes. The second is proposed by Xie and Yan and called lixel partitioning method in this paper, each segment is divided into linear units of a defined network length l. We notice that how to define a reasonable network length l is a hard issue, because it is tough to identify the connections between the network length l and the co-location patterns. Moreover, the final partition results using the lixel partitioning method is indefinite even when the network length l is fixed. The third way, proposed by Shiode (2008), is called the network-based Quadrat method. The basic idea of this method is to decompose a road network into several subnetworks with equal total length, and the constituent road segments of the subnetworks are connected to each other. Shiode has admitted that the quadrats change with the starting node in most cases. It means that if we start from different nodes, the results are likely to be different. Besides, it's a difficult task to determine a proper total length of a subnetwork since a great deal of time and efforts are needed to understand how the total length affects the co-location patterns. The fourth way is the network Voronoi diagram (Okabe et al. 2000). A road network is decomposed into several Voronoi subnetworks based on the shortest-path distance between facility location and any point on the network. Each facility location has its Voronoi subnetwork, which goes against the goal of this paper, discovering which Boolean feature classes co-locate on or close to networks.

Given such considerations, we tend to adopt a skeleton partitioning method to partition networks. Our co-location model is based on skeleton partitioning of a road network. Firstly, constrained delaunay triangulation is applied to the road segments, then skeletons are generated. Following this, approximate Voronoi polygons can be formed based on these skeletons (Liu et al. 2009, 2010). If different kinds of Boolean feature classes have instances in the same approximate Voronoi polygon, they are defined to have a co-occurrence.

As show in Fig. 8.1a, b, the feature classes A, B and C have 3, 2, 5 instances, respectively. Through grid partitioning, as shown in Fig. 8.1a, the co-occurrences are {A2, C3}, {B1, C4, C5}, {B2, A3}. B1, C4 and C5 are in the same grid cell, so they are defined to be a co-occurrence. Through skeleton partitioning, as shown in Fig. 8.1b, the co-occurrences are {A2, B2}, {B1, C4, C5}. B1, C4 and C5 are in the same approximate Voronio polygon, so they are defined to be a co-occurrence. A2 and C3 are in the same grid cell but in different approximate Voronoi polygons. A2 and B2 are in different grid cells but in the same approximate Voronoi polygon.

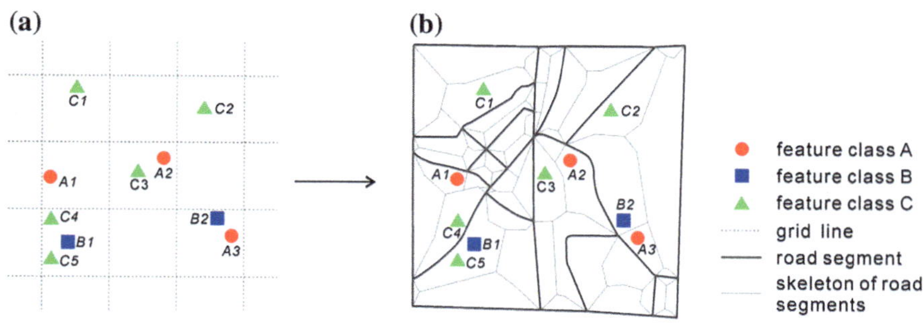

Fig. 8.1 **a** Grid partitioning in a Euclidean space **b** skeleton partitioning in a network space

2.2 Statistical Diagnostics of Co-occurrence Relationships

After identifying co-occurrence relationships, the next step is to determine whether feature classes X and Y are independently distributed by using statistical diagnostics (Sierra and Stephens 2012). This is in order to evaluate whether the corresponding co-occurrences are frequent enough that these two feature classes can be statistically identified as co-location patterns.

Suppose $P(X, Y)$ represents the probability of a co-occurrence of feature classes X and Y, with N representing the total amount of approximate Voronoi polygons and N_{xy} representing the number of approximate Voronoi polygons where X and Y have a co-occurrence. Then $P(X, Y)$ can be approximated by

$$\hat{P}(X, Y) = \frac{N_{xy}}{N} \tag{1}$$

Next, when N_y is the number of approximate Voronoi polygons where instances of Yoccur, then

$$\hat{P}(X|Y) = \frac{N_{xy}}{N_y} \tag{2}$$

estimates the probability of X's occurrence in the presence of Y.

Then, we define S_x and $S_{x|y}$ to be the number of occurrences of X in N_y approximate Voronoi polygons with probability $P(X)$ and the number of occurrences with probability $P(X|Y)$, respectively. It then needs to be determined that whether S_x and $S_{x|y}$ are statistically different. If they deviate a lot from each other, then this suggests with a high confidence level that the distributions of X and Y are different.

The following procedure is used to measure the deviation of S_x and $S_{x|y}$. Since the occurrences of X and Y in Eq. 1 only take the number of approximate Voronoi polygons, where X and Y have a co-occurrence, into account. In this case, S_x is a binomial random variable whose expected value and variance is $N_yP(X)$ and

$N_y P(X)(1 - P(X))$, respectively. Thus, $E(S_x)$ and σ_{S_x}, the expected value and the standard deviation of S_x, can be approximated by

$$E(S_x) = N_y \hat{P}(X) \tag{3}$$

$$\sigma_{S_x} = \sqrt{N_y \hat{P}(X)(1 - \hat{P}(X))} \tag{4}$$

Moreover, we can approximate the expected value of $S_{x|y}$:

$$E(S_{x/y}) = N_y \hat{P}(X|Y) \tag{5}$$

Afterwards, a binomial test can be used to measure the influence of Y on the distribution of X:

$$z = \frac{E(S_{x/y}) - E(S_x)}{\sigma_{S_x}} = \frac{N_y \hat{P}(X|Y) - N_y \hat{P}(X)}{\sqrt{N_y \hat{P}(X)(1 - \hat{P}(X))}} \tag{6}$$

According to the central limit theorem, with a sufficient sample size, the binomial distribution can be approximated by the normal distribution. And 1.96 and −1.96 correspond to the critical value of z under the confidence level of 95 %. If $z > 1.96$, this suggests that the probability of co-occurrences between X and Y is much larger than the expected one if X and Y are independently distributed. In other words, X and Y frequently co-occur and can be statistically identified as a co-location pattern. By contrast, the case where $z < -1.96$ suggests that the probability of co-occurrences between X and Y is much smaller than the expected one if X and Y are independently distributed. In other words, X and Y have repulsive relationship.

3 Case Study

To demonstrate our methodology, it is applied to mining co-location patterns among different industries in agglomeration economics, and determine which industrial pairs among two-digit manufacturing industries can be identified as co-location patterns. Moreover, The comparison with the method proposed by Sierra and Stephens has been done, and the differences have been verified by the network cross K-function.

Shenzhen is a leader in manufacturing communication equipment, electronics, chemical products, metals, transport equipment etc. and has been influential in the economical growth of Guangdong province, attracting labors from across the country (Zhou 2010). Therefore, this research on the distribution of manufacturing firms in Shenzhen provides implications for the selection of locations by firms in the area.

3.1 Data

The 2009 data of manufacturing firms in Shenzhen was provided by National Geomatics Center of China in Shenzhen, and is a subset of a point of interest (POI)) dataset on firms in Shenzhen. Using the two-digit level of the Industrial classification identifier for national economic activities,[1] the firms were classified based on their industry and main products. Additional information was obtained from Internet and phone consultations for firms with missing data. Finally, Table 8.1 shows the 50,784 manufacturing firms.[2] The 1:10,000 scale road network data and manufacturing firms of Shenzhen are presented in Fig. 8.2.

3.2 Result and Analysis

As show in Table 8.2, 311 among 351 industrial pairs can be statistically identified as co-location patterns with a confidence level of 95 %, no pairs are subject to repulsion and 40 industrial pairs are neither co-located nor repulsive. Noticeably, C26 and all other manufacturing industries can be identified as co-location patterns. This result is not unique to C26 as C36, C39, C40 and C42 also possess the same feature.

According to three significant mechanisms of agglomeration in agglomeration economics, the results can be analyzed qualitatively. The three mechanisms are: Labor market pooling, Knowledge spillovers and Input sharing (Monseny et al. 2011). These mechanisms and co-location patterns have positive influences on each other. Moreover, these mechanisms are not isolated, and thus, a co-location pattern is possibly related to all of them. To clarify, the results can be analyzed on the basis of major mechanism.

Labor market pooling A labor market is formed when laborers gather at a certain area, which leads to agglomeration effects promoting labor transfer and labor sharing between industries. This process then decreases firms' costs on laborers as well as reduces unemployment (Wu 2010). Being restricted to fierce competition on products and skills, firms of an industry may moderately cluster. To reap benefits from the agglomeration effects, firms may choose the locations that are close to firms of other industries. Hence, co-location patterns are likely to form. This spatial proximity brings great opportunities for labor sharing and enhances the

[1]The classification is based on Industrial classification for national economic activities, GB/T 4754-2002, which is published by General Administration of Quality Supervision, Inspection and Quarantine of The People's Republic of China in 2007.

[2]Due to the absence of firms of Manufacture of coke, refined petroleum products and nuclear fuel (C25), Manufacture of chemical fiber (C28) and Recycling (C43) in the original data, these three industries are omitted because the number of dimensions-above firms of the three industries in 2009 was only 12, 5, 1, respectively, according to the 2010 statistical yearbook of Shenzhen.

Table 8.1 Manufacturing firms of Shenzhen city

Classification of industries	Number of firms	Classification of industries	Number of firms
C13 manufacture of farm and sideline products	132	C29 manufacture of rubber products	77
C14 manufacture of food products	57	C30 manufacture of plastic products	2206
C15 manufacture of beverages	131	C31 manufacture of non-metallic mineral products	639
C16 manufacture of tobacco products	8	C32 smelting and pressing of ferrous metals	43
C17 manufacture of textiles	138	C33 smelting and pressing of nonferrous metals	17
C18 manufacture of wearing apparel	2933	C34 manufacture of basic metals	3287
C19 manufacture of leather and related products	774	C35 manufacture of general-purpose machinery	346
C20 sawmilling and planing of wood	507	C36 manufacture of special-purpose machinery	3655
C21 manufacture of furniture	845	C37 manufacture of transport equipment	326
C22 manufacture of paper and paper products	1112	C39 manufacture of electrical equipment	1652
C23 printing and reproduction of recorded media	1875	C40 manufacture of communication equipment, computer and other electronic apparatus	18,550
C24 manufacture of stationery and sports goods	823	C41 manufacture of instrumentation and office machinery	514
C26 manufacture of chemicals and chemical products	1708	C42 manufacture of arts and crafts and other manufacturing	7978
C27 manufacture of basic pharmaceutical products and pharmaceutical preparations	451		

agglomeration effects. As labor plays a more important role in labor-intensive industries than other industries, and laborers are more likely to flow between industries with similar skills or with low technology, typical labor-intensive industries tend to form co-location patterns. In our study, C18 and C19, which are low-tech industries, need a large amount of laborers, in reality, labor sharing is quite common between them. Occupations in C29 and C30 are of high similarity, and these industries also experience a great deal of labor transfer and labor sharing. Due to benefits from the agglomeration effects, C18 and C19, C29 and C30 form two typical co-location patterns.

Knowledge spillovers Workers' movement, communications between firms etc. promote knowledge sharing between industries and increase the knowledge contents of industries. Explicit knowledge spills over by products, language and other

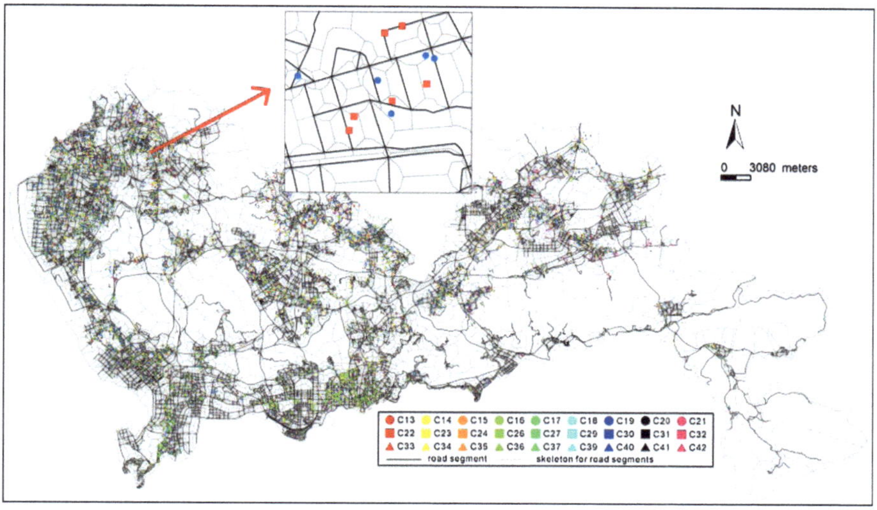

Fig. 8.2 Skeleton partitioning and distribution of manufacturing firms in Shenzhen city

forms, while highly personal implicit knowledge spills over only by face-to-face communication. Knowledge spillovers mainly promote the implicit knowledge spillovers which are subject to distance constraints and similarity of occupations between industries (Wang 2010). Hence, industries form co-location patterns in order to obtain knowledge spillovers as much as possible. Understandably, industries that benefit much from knowledge spillovers tend to form co-location patterns. Spillovers of skills, technology and experience of producing chemical products largely promote complementarities and increase knowledge contents between industries which explains the co-location of C26 and C27 in this study. Manufacture of electric and electronic components belongs to C39, Manufacture of Optoelectronic devices and other electronic device belongs to C40. Spillovers of designing and improving electronic components significantly raise knowledge level and innovation level of those two industries. Benefiting from the agglomeration effects, C26 and C27, C39 and C40 form two typical co-location patterns.

Input sharing For two industries that have forward linkages or backward linkages, firms of one industry are probably located close to firms of another industry in order to save transport costs of inputs and reduce the loss of product in transit. These two industries form a co-location pattern accordingly. One empirical example (Amiti 2005) is rubber (C29) and transport equipment (C37), C37 relies on the output from C29 to manufacture transport equipment. This leads these two industries to form a co-location pattern. Industries, which have forward linkages or backward linkages tend to form co-location patterns, particularly those producing semi finished products and those producing consumer products. In our study, C20 supplies C21 with processed timber and C34 supplies C35 with metal products. The transports of processed timber and metal products cost a large amount of human and material resources and unavoidably suffer from loss of products due to accidents.

Table 8.2 Results of statistical diagnostics of our method

	C13	C14	C15	C16	C17	C18	C19	C20	C21	C22	C23	C24	C26
C13		**6.56**	**7.35**	*−0.25*	*0.90*	**7.77**	**3.56**	**6.37**	**5.42**	**4.87**	**3.69**	**5.34**	**7.82**
C14	**6.54**		**5.42**	*−0.17*	**3.55**	**4.74**	*1.08*	*0.51*	**3.74**	**5.90**	**3.31**	**4.64**	**7.93**
C15	**7.35**	**5.43**		*−0.24*	**5.12**	**18.27**	**9.13**	**2.60**	**8.75**	**6.57**	**13.74**	**12.57**	**18.93**
C16	*−0.25*	*−0.17*	*−0.24*		*−0.25*	*1.64*	*1.28*	*−0.42*	*−0.54*	*−0.65*	*0.50*	**2.93**	**2.22**
C17	*0.90*	**3.55**	**5.12**	*−0.25*		**16.31**	**8.66**	**3.33**	**8.75**	**7.99**	**12.70**	**9.32**	**11.37**
C18	**8.13**	**4.98**	**19.12**	*1.73*	**17.06**		**27.34**	**14.23**	**23.20**	**28.65**	**34.15**	**27.32**	**33.54**
C19	**3.63**	*1.10*	**9.29**	*1.31*	**8.81**	**26.58**		**10.19**	**16.30**	**23.82**	**22.48**	**22.22**	**15.50**
C20	**6.42**	*0.51*	**2.62**	*−0.43*	**3.36**	**13.72**	**10.10**		**19.27**	**13.20**	**12.78**	**12.33**	**13.13**
C21	**5.52**	**3.82**	**8.91**	*−0.55*	**8.90**	**22.55**	**16.30**	**19.43**		**20.40**	**20.84**	**20.46**	**20.13**
C22	**5.01**	**6.07**	**6.75**	*−0.67*	**8.21**	**28.12**	**24.04**	**13.44**	**20.59**		**30.49**	**22.42**	**18.90**
C23	**3.84**	**3.46**	**14.32**	*0.53*	**13.23**	**34.01**	**23.01**	**13.20**	**21.35**	**30.94**		**25.69**	**26.90**
C24	**5.44**	**4.75**	**12.82**	**3.00**	**9.50**	**26.62**	**22.26**	**12.47**	**20.51**	**22.26**	**25.15**		**23.18**
C26	**8.06**	**8.20**	**19.53**	**2.30**	**11.73**	**33.06**	**15.70**	**13.42**	**20.41**	**18.98**	**26.62**	**23.44**	
C27	**6.64**	**5.41**	**27.14**	**2.21**	**3.95**	**23.41**	**6.48**	**4.67**	**8.88**	**11.74**	**16.44**	**16.28**	**24.66**
C29	*1.76*	*−0.54*	*−0.75*	*−0.19*	**4.19**	**4.37**	**6.89**	**9.09**	**8.13**	**10.07**	**7.61**	**6.43**	**3.29**
C30	**4.42**	**6.73**	**6.27**	*−0.90*	**7.72**	**26.26**	**26.63**	**14.79**	**22.00**	**31.69**	**30.17**	**26.65**	**21.96**
C31	**8.60**	**4.98**	**16.41**	*1.48*	**8.81**	**25.85**	**13.58**	**14.04**	**16.85**	**15.22**	**22.02**	**15.71**	**27.83**
C32	*−0.58*	**2.12**	*1.25*	*−0.14*	**4.45**	**3.29**	**4.92**	**3.02**	**6.53**	**6.32**	**5.78**	**4.58**	**4.54**
C33	*−0.35*	*−0.24*	*−0.33*	*−0.09*	**2.45**	**3.20**	*1.82*	**2.75**	**5.75**	**3.45**	**2.53**	*1.66*	**5.19**
C34	**7.54**	**8.26**	**9.74**	*0.02*	**10.83**	**30.58**	**25.01**	**17.29**	**25.56**	**32.09**	**32.57**	**26.68**	**27.50**
C35	**5.22**	**5.87**	**14.12**	*−0.37*	**7.03**	**23.63**	**10.65**	**5.29**	**12.30**	**13.22**	**17.83**	**15.97**	**22.59**
C36	**6.96**	**7.67**	**15.06**	**3.36**	**11.37**	**35.71**	**22.46**	**13.65**	**21.29**	**29.47**	**33.73**	**24.54**	**35.72**
C37	**5.73**	**5.72**	**12.41**	*−0.38*	**6.19**	**16.18**	**7.34**	**7.02**	**9.84**	**7.28**	**15.86**	**9.55**	**19.17**
C39	**5.49**	**5.95**	**18.77**	**3.49**	**9.31**	**31.89**	**18.33**	**13.62**	**17.14**	**24.78**	**31.88**	**22.10**	**31.14**

(continued)

Table 8.2 (continued)

	C13	C14	C15	C16	C17	C18	C19	C20	C21	C22	C23	C24	C26
C40	**6.47**	**6.67**	**15.61**	**2.31**	**13.15**	**38.97**	**23.82**	**13.05**	**20.78**	**29.91**	**35.15**	**24.82**	**34.51**
C41	*1.91*	**5.08**	**6.76**	*−0.46*	**9.78**	**22.64**	**19.39**	**8.53**	**14.88**	**17.81**	**21.06**	**18.15**	**16.76**
C42	**8.38**	**8.02**	**16.85**	**3.94**	**10.45**	**38.85**	**21.75**	**16.82**	**24.54**	**26.83**	**33.86**	**23.68**	**35.17**

	C27	C29	C30	C31	C32	C33	C34	C35	C36	C37	C39	C40	C41	C42
C13	**6.60**	*1.76*	**4.20**	**8.48**	*−0.59*	*−0.35*	**7.02**	**5.19**	**6.49**	**5.70**	**5.30**	**5.77**	*1.89*	**7.70**
C14	**5.36**	*−0.54*	**6.38**	**4.89**	**2.12**	*−0.24*	**7.67**	**5.83**	**7.14**	**5.68**	**5.73**	**5.92**	**5.01**	**7.35**
C15	**26.95**	*−0.75*	**5.96**	**16.17**	*1.25*	*−0.33*	**9.06**	**14.04**	**14.05**	**12.34**	**18.10**	**13.90**	**6.69**	**15.47**
C16	**2.19**	*−0.19*	*−0.85*	*1.46*	*−0.14*	*−0.09*	*0.02*	*−0.37*	**3.12**	*−0.37*	**3.35**	**2.05**	*−0.45*	**3.60**
C17	**3.93**	**4.20**	**7.34**	**8.69**	**4.47**	**2.46**	**10.09**	**6.99**	**10.61**	**6.16**	**8.98**	**11.71**	**9.68**	**9.60**
C18	**24.33**	**4.58**	**26.12**	**26.66**	**3.45**	**3.36**	**29.80**	**24.59**	**34.88**	**16.83**	**32.20**	**36.32**	**23.44**	**37.34**
C19	**6.55**	**7.03**	**25.76**	**13.62**	**5.02**	*1.85*	**23.70**	**10.78**	**21.34**	**7.42**	**18.00**	**21.59**	**19.52**	**20.33**
C20	**4.68**	**9.18**	**14.18**	**13.96**	**3.06**	**2.79**	**16.24**	**5.31**	**12.85**	**7.05**	**13.25**	**11.73**	**8.51**	**15.58**
C21	**8.98**	**8.28**	**21.27**	**16.89**	**6.67**	**5.87**	**24.21**	**12.45**	**20.22**	**9.95**	**16.82**	**18.83**	**14.98**	**22.93**
C22	**11.98**	**10.36**	**30.93**	**15.41**	**6.51**	**3.55**	**30.69**	**13.50**	**28.25**	**7.43**	**24.56**	**27.36**	**18.10**	**25.31**
C23	**17.01**	**7.94**	**29.88**	**22.61**	**6.04**	**2.64**	**31.60**	**18.48**	**32.81**	**16.43**	**32.05**	**32.62**	**21.71**	**32.41**
C24	**16.50**	**6.57**	**25.83**	**15.79**	**4.69**	*1.70*	**25.33**	**16.20**	**23.36**	**9.69**	**21.75**	**22.55**	**18.31**	**22.18**
C26	**25.27**	**3.40**	**21.53**	**28.29**	**4.69**	**5.37**	**26.41**	**23.18**	**34.38**	**19.66**	**30.98**	**31.70**	**17.10**	**33.31**
C27		*0.39*	**6.72**	**18.87**	*−0.93*	*1.29*	**11.57**	**16.08**	**18.20**	**15.27**	**21.05**	**18.45**	**10.14**	**20.57**
C29	*0.39*		**7.95**	**3.94**	**6.20**	**7.13**	**10.81**	**4.01**	**7.84**	**3.04**	**4.09**	**7.43**	**2.70**	**6.29**
C30	**7.03**	**8.38**		**16.04**	**6.25**	**6.59**	**36.29**	**12.44**	**33.47**	**8.00**	**27.77**	**32.02**	**19.51**	**29.73**
C31	**19.02**	**4.01**	**15.47**		**5.56**	**3.51**	**18.92**	**22.25**	**22.66**	**15.24**	**23.04**	**18.83**	**13.56**	**21.72**
C32	*−0.92*	**6.19**	**5.92**	**5.47**		**4.79**	**9.64**	*1.46*	**7.68**	**2.54**	**5.78**	**8.22**	**5.46**	**6.19**
C33	*1.28*	**7.12**	**6.24**	**3.45**	**4.78**		**4.31**	*1.42*	**5.86**	*1.37*	**3.78**	**4.06**	*0.91*	**5.10**
C34	**12.35**	**11.63**	**37.04**	**20.02**	**10.38**	**4.64**		**17.17**	**38.40**	**13.08**	**31.88**	**34.97**	**21.71**	**34.30**

(continued)

Table 8.2 (continued)

	C27	C29	C30	C31	C32	C33	C34	C35	C36	C37	C39	C40	C41	C42
C35	**16.06**	**4.03**	**11.89**	**22.05**	*1.47*	*1.43*	**16.07**		**22.63**	**14.70**	**22.68**	**17.81**	**14.40**	**20.17**
C36	**19.37**	**8.42**	**34.08**	**23.93**	**8.26**	**6.31**	**38.31**	**24.11**		**19.14**	**38.08**	**40.40**	**20.63**	**37.78**
C37	**15.25**	**3.06**	**7.65**	**15.10**	**2.56**	*1.39*	**12.25**	**14.70**	**17.97**		**17.20**	**15.73**	**6.13**	**16.04**
C39	**21.68**	**4.24**	**27.36**	**23.53**	**6.00**	**3.93**	**30.76**	**23.38**	**36.84**	**17.72**		**38.18**	**19.73**	**32.46**
C40	**20.58**	**8.35**	**34.17**	**20.83**	**9.25**	**4.57**	**36.56**	**19.89**	**42.34**	**17.56**	**41.36**		**22.04**	**44.27**
C41	**10.18**	**2.74**	**18.74**	**13.51**	**5.53**	*0.93*	**20.44**	**14.48**	**19.46**	**6.16**	**19.24**	**19.84**		**18.39**
C42	**22.25**	**6.86**	**30.76**	**23.31**	**6.76**	**5.57**	**34.78**	**21.84**	**38.40**	**17.36**	**34.09**	**42.93**	**19.81**	

In the table, the numbers measure the influences of industries in the first row on the industry in the first column, the bold numbers are bigger than 1.96, italic numbers are between −1.96 and 1.96, no number is smaller than −1.96

Driven by saving costs and reducing loss, C20 and C21, C34 and C35 form two typical co-location patterns.

As discussed above, it is reasonable to use the agglomeration economics theory to explain the existences of co-location patterns.

3.3 Comparison

To validate our method, we apply both our method and the method proposed by Sierra and Stephens to the mining of co-location patterns between network spatial phenomena, and test the differences between the mining results by the network cross K-function method.

According to the method proposed by Sierra and Stephens, the space is partitioned by uniform grids, and the number of the cells should be bigger than the number of the greater point set when analyzing two point sets. In our experiment, the number of the point sets vary a lot, too many cells may cause false conclusions for point sets with few points. Therefore, we think the number of cells should be quite close to the number of approximate Voronio polygons (13,543 in our case). Finally, we make the resolution 385 m × 385 m and the number of the cells is 13,644, as shown in Fig. 8.3. Different feature classes that have instances in the same cell are defined to have a co-occurrence. The same statistical diagnostics were incorporated to analyze this relationship as in Sect. 2.2 and the results are shown in Table 8.3. Interestingly, 327 among 351 industrial pairs can be statistically identified as co-location patterns with a confidence level of 95 %, no pairs are subject to repulsion and 24 industrial pairs are neither co-located nor repulsive.

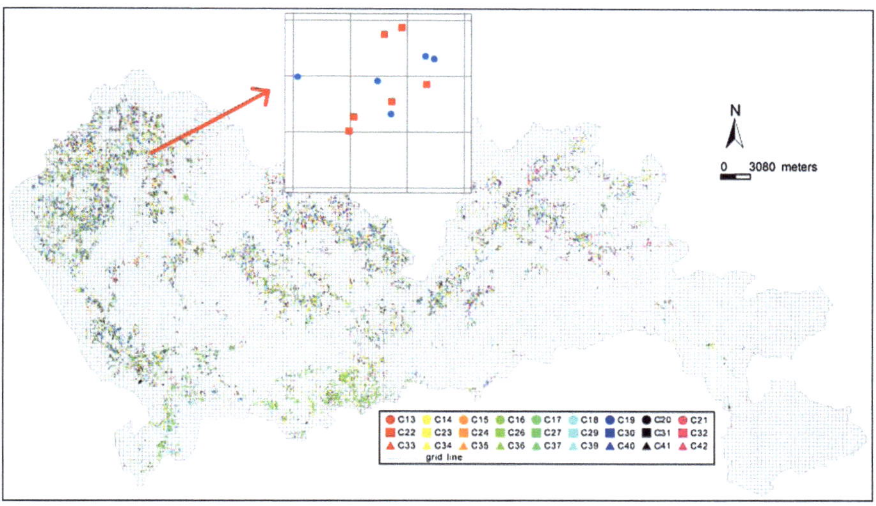

Fig. 8.3 Grid partition and distribution of manufacturing firms in Shenzhen city

Table 8.3 Results of statistical diagnostics of the method proposed by Sierra and Stephens

	C13	C14	C15	C16	C17	C18	C19	C20	C21	C22	C23	C24	C26
C13		**8.21**	**11.48**	*−0.24*	**5.07**	**11.45**	**7.83**	**7.84**	**7.30**	**6.47**	**8.95**	**10.06**	**11.38**
C14	**8.19**		**10.48**	**5.74**	**5.01**	**6.22**	**4.52**	**6.37**	**7.39**	**5.40**	**6.08**	**8.70**	**8.84**
C15	**11.48**	**10.50**		**8.77**	**8.85**	**18.40**	**10.25**	**3.55**	**10.11**	**7.32**	**17.35**	**16.01**	**21.17**
C16	*−0.24*	**5.73**	**8.74**		*−0.25*	**3.18**	**3.25**	*−0.42*	*−0.55*	*−0.64*	**1.98**	**3.07**	**3.87**
C17	**5.08**	**5.02**	**8.86**	*−0.25*		**18.82**	**12.47**	**7.46**	**8.66**	**12.91**	**17.33**	**15.28**	**11.55**
C18	**11.90**	**6.48**	**19.14**	**3.32**	**19.56**		**35.06**	**15.28**	**24.88**	**32.55**	**42.59**	**35.85**	**36.44**
C19	**7.96**	**4.61**	**10.42**	**3.32**	**12.67**	**34.27**		**9.09**	**18.61**	**27.33**	**29.87**	**25.38**	**19.37**
C20	**7.91**	**6.44**	**3.58**	*−0.43*	**7.52**	**14.82**	**9.03**		**22.87**	**13.94**	**14.87**	**13.29**	**14.02**
C21	**7.44**	**7.55**	**10.30**	*−0.57*	**8.81**	**24.37**	**18.64**	**23.09**		**23.08**	**23.52**	**22.87**	**19.20**
C22	**6.64**	**5.55**	**7.51**	*−0.66*	**13.24**	**32.12**	**27.60**	**14.18**	**23.25**		**35.40**	**26.67**	**22.82**
C23	**9.28**	**6.31**	**18.00**	**2.06**	**17.96**	**42.48**	**30.47**	**15.29**	**23.95**	**35.77**		**30.17**	**29.26**
C24	**10.24**	**8.88**	**16.31**	**3.14**	**15.56**	**35.11**	**25.42**	**13.42**	**22.87**	**26.46**	**29.63**		**23.97**
C26	**11.71**	**9.11**	**21.78**	**4.00**	**11.88**	**36.05**	**19.60**	**14.29**	**19.39**	**22.87**	**29.02**	**24.21**	
C27	**9.58**	**8.73**	**31.30**	**2.40**	**12.49**	**27.95**	**13.33**	**7.63**	**11.26**	**14.47**	**23.09**	**20.01**	**30.99**
C29	*1.84*	**3.14**	*−0.72*	*−0.19*	**4.24**	**7.82**	**6.57**	**9.03**	**9.52**	**10.66**	**8.23**	**10.12**	**5.91**
C30	**7.27**	**7.38**	**9.34**	*0.39*	**13.52**	**35.77**	**32.17**	**17.64**	**26.78**	**38.81**	**39.70**	**33.08**	**28.57**
C31	**12.34**	**7.27**	**18.51**	*−0.51*	**13.76**	**27.97**	**15.52**	**13.08**	**18.88**	**17.31**	**24.46**	**19.62**	**28.43**
C32	*−0.55*	**4.75**	*1.44*	*−0.14*	**6.45**	**6.83**	**2.07**	**4.20**	**4.23**	**6.03**	**7.16**	**7.41**	**4.40**
C33	*−0.31*	*−0.22*	**3.13**	*−0.08*	**2.75**	**3.02**	**2.19**	**3.08**	**6.18**	**3.95**	**5.25**	**4.81**	**6.06**
C34	**11.87**	**6.21**	**10.94**	*1.22*	**14.43**	**40.88**	**32.45**	**21.17**	**28.89**	**37.73**	**41.49**	**33.02**	**31.23**
C35	**11.55**	**5.11**	**22.68**	**5.41**	**9.67**	**27.26**	**16.13**	**6.94**	**12.96**	**15.69**	**23.75**	**17.39**	**25.85**
C36	**10.63**	**7.10**	**16.36**	**3.67**	**16.74**	**46.34**	**29.95**	**18.27**	**25.48**	**35.60**	**45.05**	**31.52**	**40.41**
C37	**10.61**	**7.00**	**20.75**	*−0.36*	**8.10**	**19.98**	**10.63**	**9.57**	**11.11**	**8.66**	**19.31**	**14.85**	**21.46**
C39	**9.29**	**7.35**	**19.14**	**5.19**	**12.70**	**37.89**	**23.21**	**14.44**	**18.42**	**27.74**	**36.35**	**26.20**	**32.29**

(continued)

Table 8.3 (continued)

	C13	C14	C15	C16	C17	C18	C19	C20	C21	C22	C23	C24	C26
C40	**11.42**	**7.90**	**18.07**	**3.79**	**16.74**	**49.35**	**31.45**	**17.11**	**26.53**	**37.48**	**44.95**	**33.60**	**38.58**
C41	**7.50**	**5.30**	**10.18**	*−0.44*	**15.52**	**27.57**	**21.01**	**12.05**	**13.97**	**19.61**	**24.49**	**21.33**	**18.62**
C42	**12.58**	**6.59**	**18.81**	**4.37**	**13.70**	**44.89**	**28.99**	**19.25**	**25.35**	**33.59**	**42.62**	**32.31**	**38.85**

	C27	C29	C30	C31	C32	C33	C34	C35	C36	C37	C39	C40	C41	C42
C13	**9.53**	*1.84*	**6.93**	**12.17**	*−0.55*	*−0.32*	**11.11**	**11.50**	**10.01**	**10.56**	**8.99**	**10.43**	**7.42**	**11.72**
C14	**8.67**	**3.14**	**7.02**	**7.16**	**4.75**	*−0.22*	**5.81**	**5.08**	**6.68**	**6.95**	**7.10**	**7.20**	**5.24**	**6.12**
C15	**31.11**	*−0.72*	**8.90**	**18.25**	*1.44*	**3.14**	**10.23**	**22.56**	**15.40**	**20.63**	**18.52**	**16.50**	**10.08**	**17.51**
C16	**2.37**	*−0.19*	*0.37*	*−0.50*	*−0.14*	*−0.08*	*1.13*	**5.36**	**3.45**	*−0.36*	**5.00**	**3.45**	*−0.44*	**4.05**
C17	**12.43**	**4.25**	**12.89**	**13.58**	**6.47**	**2.77**	**13.52**	**9.63**	**15.77**	**8.06**	**12.30**	**15.29**	**15.37**	**12.76**
C18	**28.91**	**8.14**	**35.46**	**28.69**	**7.12**	**3.15**	**39.79**	**28.21**	**45.39**	**20.67**	**38.15**	**46.88**	**28.39**	**43.46**
C19	**13.48**	**6.69**	**31.17**	**15.55**	**2.11**	**2.24**	**30.88**	**16.31**	**28.67**	**10.75**	**22.84**	**29.20**	**21.15**	**27.43**
C20	**7.66**	**9.13**	**16.97**	**13.01**	**4.25**	**3.11**	**19.99**	**6.97**	**17.36**	**9.61**	**14.11**	**15.77**	**12.03**	**18.08**
C21	**11.41**	**9.71**	**26.00**	**18.97**	**4.32**	**6.32**	**27.54**	**13.14**	**24.44**	**11.26**	**18.16**	**24.68**	**14.09**	**24.03**
C22	**14.77**	**10.96**	**37.97**	**17.53**	**6.20**	**4.06**	**36.26**	**16.02**	**34.41**	**8.85**	**27.56**	**35.13**	**19.93**	**32.09**
C23	**23.82**	**8.54**	**39.25**	**25.02**	**7.45**	**5.47**	**40.28**	**24.51**	**44.01**	**19.92**	**36.49**	**42.57**	**25.15**	**41.15**
C24	**20.27**	**10.32**	**32.11**	**19.71**	**7.56**	**4.92**	**31.48**	**17.62**	**30.23**	**15.04**	**25.83**	**31.25**	**21.51**	**30.63**
C26	**31.70**	**6.08**	**28.02**	**28.84**	**4.54**	**6.26**	**30.07**	**26.47**	**39.15**	**21.96**	**32.15**	**36.24**	**18.97**	**37.20**
C27		0.54	**12.23**	**22.32**	**2.76**	*1.63*	**16.62**	**23.39**	**24.00**	**22.01**	**26.86**	**25.49**	**16.62**	**25.15**
C29	*0.54*		**9.74**	**6.59**	*1.83*	**11.70**	**13.26**	**6.91**	**10.35**	**5.00**	**4.43**	**10.97**	**2.17**	**7.22**
C30	**12.76**	**10.23**		**17.40**	**9.01**	**8.53**	**47.48**	**16.36**	**44.70**	**12.94**	**35.35**	**45.93**	**20.70**	**40.37**
C31	**22.50**	**6.70**	**16.81**		**2.41**	**2.47**	**26.12**	**23.03**	**28.62**	**17.03**	**24.76**	**27.08**	**17.67**	**25.99**
C32	**2.74**	*1.82*	**8.57**	**2.37**		**10.92**	**6.45**	**4.17**	**9.29**	**5.30**	**4.59**	**8.17**	**4.97**	**6.94**
C33	*1.61*	**11.68**	**8.11**	**2.43**	**10.91**		**7.53**	*1.72*	**6.30**	*1.68*	**5.62**	**4.82**	**2.92**	**5.64**
C34	**17.66**	**14.19**	**48.35**	**27.52**	**6.91**	**8.08**		**20.48**	**51.44**	**16.74**	**38.56**	**50.25**	**26.30**	**45.98**

(continued)

Table 8.3 (continued)

	C27	C29	C30	C31	C32	C33	C34	C35	C36	C37	C39	C40	C41	C42
C35	**23.37**	**6.95**	**15.67**	**22.82**	**4.20**	*1.74*	**19.27**		**26.61**	**21.18**	**23.43**	**22.27**	**17.95**	**23.93**
C36	**25.34**	**11.01**	**45.24**	**29.96**	**9.89**	**6.71**	**51.12**	**28.12**		**23.95**	**45.99**	**55.18**	**27.59**	**49.06**
C37	**22.00**	**5.04**	**12.40**	**16.88**	**5.34**	*1.69*	**15.75**	**21.19**	**22.67**		**20.50**	**19.69**	**10.97**	**20.96**
C39	**27.59**	**4.58**	**34.80**	**25.22**	**4.76**	**5.82**	**37.29**	**24.09**	**44.74**	**21.07**		**46.39**	**21.43**	**38.82**
C40	**27.75**	**12.03**	**47.94**	**29.25**	**8.97**	**5.29**	**51.50**	**24.26**	**56.90**	**21.45**	**49.17**		**29.54**	**56.66**
C41	**16.69**	**2.19**	**19.93**	**17.60**	**5.03**	**2.96**	**24.86**	**18.04**	**26.25**	**11.02**	**20.95**	**27.25**		**24.74**
C42	**26.86**	**7.76**	**41.34**	**27.54**	**7.48**	**6.08**	**46.23**	**25.59**	**49.64**	**22.40**	**40.37**	**55.59**	**26.32**	

In the table, the numbers measure the influences of industries in the first row on the industry in the first column, the bold numbers are bigger than 1.96, italic numbers are between −1.96 and 1.96, no number is smaller than −1.96

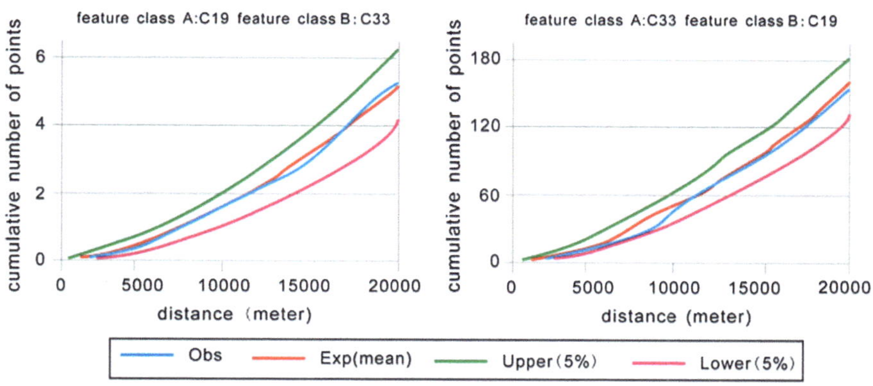

Fig. 8.4 Test for C19 and C33 using the network cross K-function method

When comparing the mining results using our method and the method proposed by Sierra and Stephens, we see differences on 18 industrial pairs: 17 industrial pairs form co-location patterns using the method proposed by Sierra and Stephens, but do not form co-location patterns using our method, they are {C13, C17}, {C13, C41}, {C14, C16}, {C14, C19}, {C14, C20}, {C14, C29}, {C15, C16}, {C15, C33}, {C16, C18}, {C16, C19}, {C16, C23}, {C16, C35}, {C19, C33}, {C23, C33}, {C27, C32}, {C32, C35}, {C33, C41}; 1 industrial pair form co-location patterns using our method but do not form co-location pattern using the method proposed by Sierra and Stephens, it is {C29, C32}.

Given two sets of points, a set of type A points and that of type B points, placed on a given network, the network cross K-function method[3] test whether the configuration of type A points affect the distribution of type B points, if there is a relationship, this means that point set pair {A, B} can form a co-location pattern. This is the common method applied to the test for co-location patterns between network spatial phenomena (Okabe and Yamada 2001; Huang et al. 2004; Okabe et al. 2006). This study found that the industrial pairs, which do not form co-location patterns using our method, are proved by the network cross K-function to be randomly distributed. For example, the test for industrial pair {C19, C33} are shown in Fig. 8.4, the distribution of C19 and C33 do not affect each other. The industrial pair {C29, C32} is proved by the network cross K-function to be aggregated distribution. The test, as presented in Fig. 8.5, shows that C29 tend to cluster around C32 and vice versa. In summary, the differences are verified by the network cross K-function method.

[3]The network cross K-function method is included in SANET. SANET stands for "Spatial Analysis along Networks". It is a software package developed by Atsu Okabe and his group. Details about SANET can be found at http://sanet.csis.u-tokyo.ac.jp/.

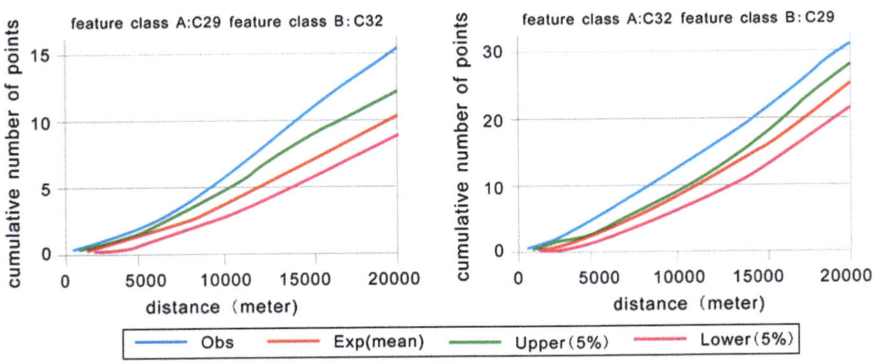

Fig. 8.5 Test for C29 and C32 using the network cross K-function method

4 Conclusion

This paper extends current method proposed by Sierra and Stephens to a network space. We define a co-location model to determine co-occurrence relationships between different Boolean feature classes based on skeleton partitioning of a road network. Statistical indicator was then used to determine whether the co-occurrence relationships were frequent to be statistically significant. Our method was validated by mining co-location patterns among manufacturing industries in Shenzhen and the comparison with the method proposed by Sierra and Stephens.

The industrial pairs, which form different patterns using our method and the method proposed by Sierra and Stephens, were tested by the network cross K-function method, and our next step is to test all industrial pairs and compare the results of our method with the network cross K-function method. Recently, with the help of big data and cloud computing, conditions about acquisition of data and complex computing have improved. Possibilities for further research include statistical data in the field of economics, including types and numbers of employees, supply relations between industries, similarity of technology in products etc. This would allow us quantitatively validate the co-location patterns identified in this paper.

Skeleton partitioning is chosen as the way of tessellating network spaces in this paper because the skeleton partitioning of a road network is unique if the road network is defined. Trying more ways of tessellating network spaces will also be included in our further study. The network-based Quadrat method and the lixel partitioning method are likely to be applied in mining co-location patterns between network spatial phenomena and compared with the skeleton partitioning method. The influences of different partitioning methods on co-location patterns are to be evaluated in the future work.

Acknowledgments The authors greatly appreciate the helpful comments of two anonymous reviewers. The authors also want to thank Dr. Okabe and his group for providing the program package, SANET, which allowed the calculation of the network cross K-function in this research. The work presented in this paper was supported by National Science Foundation for Fostering Talents in Basic Research of the National Natural Science Foundation of China (Grant No. J1103409) and by Innovation and Entrepreneurship Training Project for College Students of Wuhan University (Grant No.S2014438).

References

Amiti M (2005) Location of vertically linked industries: agglomeration versus comparative advantage. Eur Econ Rev 49(4):809–832

Arnia G, Espa G, Quah D (2008) A class of spatial econometric methods in the empirical analysis of clusters of firms in the space. Empirical Economics 34(1):81–103

Batty M (2005) Network geography: relations, interactions, scaling and spatial processes in GIS. In: Unwin D, Fisher P(eds) Re-presenting geographical information systems. Wiley, Chichester, pp 149–170

Bembenik R, Rybinski H (2009) FARICS: a method of mining spatial association rules and collocations using clustering and Delaunay diagrams. J Intell Inf Syst 33(1):41–64

Borruso G (2008) Network density estimation: A GIS approach for analysing point patterns in a network space. Trans GIS 12(3):377–402

General Administration of Quality Supervision, Inspection and Quarantine of The People's Republic of China (2007) Industrial classification for national economic activities, GB/T 4754-2002. China Zhijian Publishing House, Beijing

Guo L, Du SH, Haining R, Zhang LJ (2011) Global and local indicator of spatial association between points and polygons: a study of land use change. Int J Appl Earth Obs Geoinf. doi:10.1016/j.jag.2011.11.003

Hu W (2008) Co-location pattern discovery. Encyclopedia of GIS. Springer, Berlin, p 2008

Huang Y, Shekhar S, Xiong H (2004) Discovering co-location patterns from spatial data sets: a general approach. IEEE Trans Knowl Data Eng 16(12):1472–1485

Huang Y, Pei J, Xiong H (2006) Mining co-location patterns with rare events from spatial data sets. Geoinformatica 10(2):239–260

Koperski K, Han JW (1995) Discovery of spatial association rules in geographic information databases. In: 4th International symposium on large spatial databases. Maine, USA, pp 47–66

Leslie TF, Kronenfeld BJ (2011) The co-location quotient: a new measure of spatial association between categorical subsets of points. Geogr Anal 43(3):306–326

Li DR, Li DY, Wang SL (2006) Spatial data mining theories and applications. Science Press, Beijing

Liu XJ, Ai TH, Liu YL (2009) Road density analysis based on skeleton partitioning for road generalization. Geo-spatial Inf Sci 12(2):110–116

Liu XJ, Zhan FB, Ai TH (2010) Road selection based on Voronoi diagrams and "strokes" in map generalization. Int J Appl Earth Obs Geoinf 12(2):S194–S202

Lu YM, Chen XW (2007) On the false alarm of planar K-function when analyzing urban crime distributed along streets. Soc Sci Res 36(2):611–632

Monseny J, López R, Marsal E (2011) The mechanisms of agglomeration: evidence from the effect of inter-industry relations on the location of new firms. J Urban Econ 70(2–3):61–74

Okabe A, Yamada I (2001) The K-function method on a network and its computational implementation. Geogr Anal 33(3):271–290

Okabe A, Boots B, Sugihara K, Chiu SN (eds) (2000) Spatial tessellations: concepts and applications of voronoi diagrams. Wiley, New York

Okabe A, Okunuki K, Shiode S (2006) SANET: A toolbox for spatial analysis on a network. Geogr Anal 38(1):57–66
Ruiz M, Lopez F, Paez A (2010) Testing for spatial association of qualitative data using symbolic dynamics. J Geogr Syst 12(3):281–309
Shekhar S, Huang Y (2001) Discovering spatial co-location patterns: a summary of results. In: Jensen CS et al (eds) SSTD 2001, vol 2121, 2001. LNCS, pp 236–256
Shiode S (2008) Analysis of a distribution of point events using the network-based quadrat method. Geogr Anal 40(4):380–400
Sierra R, Stephens R (2012) Exploratory analysis of the interrelation between co-located boolean spatial features using network graphs. Int J Geogr Inf Sci 26(3):441–468
Wang GH (2010) Research on knowledge spillover and the learning of enterprise in industrial clusters. Science Press, Beijing
Wu XH (2010) Research on regional cluster of manufacturing in China. Economic Science Press, Beijing
Xie ZX, Yan J (2008) Kernel density estimation of traffic accidents in a network space. Comput Environ Urban Syst 32(5):396–406
Yamada I, Thill J (2004) Comparison of planar and network K-functions in traffic accident analysis. J Transp Geogr 12(2):149–158
Zhang X, Mamoulis N, Cheung D, Chou Y (2004) Fast mining of spatial collocations. In: Proceedings of the ACM SIGKDD international conference on knowledge discovery and data mining, Seattle, USA, pp 384–393
Zhou WL (2010) The concentration, diffusion and policy decisions of manufacturing, evidence from guangdong province. Economic Science Press, Beijing

How to Model Roads in OpenStreetMap? A Method for Evaluating the Fitness-for-Use of the Network for Navigation

Xiang Zhang and Tinghua Ai

Abstract OpenStreetMap (OSM) is mostly a good map for viewers to look at but it lacks of sufficient quality in certain applications like navigation. Quality issues are usually related to how roads are 'drawn' (modeled) by OSM contributors. First of all, this paper identifies several issues in tagging and modeling OSM road network by case studies, and also gives suggestions for contributors and routing service providers. As a key contribution, this paper proposes a methodology to evaluate OSM roads that does not rely on reference data or ground truth. The evaluation aims not only to identify errors in OSM data, but also to give more intelligent suggestions based on the information available in the spatial context of the problematic data. More specifically, named roads are recognized based on the concept of "stroke". Missing or incorrect names can be found by outlier detection within the scope of the named roads. Such an idea can be widely applied to detect inconsistent tags and provide intelligent suggestions for data correction.

Keywords OpenStreetMap · Inconsistency detection · Data enrichment · Natural road recognition · Stroke

1 Introduction

Crowd sourced geographic content has enabled free access to geographic data world-wide, and especially in regions where official data are difficult/expensive to obtain. OpenStreetMap (OSM) is among others an excellence example which provides free access to the world's base map data. Often the data are 'drawn' (digitized) on screen by contributors of various backgrounds. Hence the data consistency can be problematic. While the quality, especially the completeness of OSM data is a major concern in recent literature, the fitness-for-use of the data seems to be less studied for certain applications (e.g. navigation).

X. Zhang (✉) · T. Ai
School of Resource and Environmental Sciences, Wuhan University, Wuhan, China
e-mail: xiang.zhang@whu.edu.cn

F. Harvey and Y. Leung (eds.), *Advances in Spatial Data Handling and Analysis*, Advances in Geographic Information Science,
DOI 10.1007/978-3-319-19950-4_9

Navigation and spatial analysis rely heavily on the geometric and semantic modeling of spatial data. However, heterogeneity in geometric and semantic modeling of OSM data is still a general issue for those who want to use OSM data. Moreover, during general editing processes existing OSM data often get destroyed by inexperienced users.

Specifically, the OSM-based routing services cannot always provide satisfactory results. For example, the generated paths may be unnecessarily long (Mondzech and Sester 2011), or the suggested routes can be totally wrong (e.g. Fig. 2), due to inappropriate modeling of the network, missing segments, or incorrectly tagged traffic direction or turn restrictions. On the other hand, the generated turn-by-turn navigation guide can be problematic if descriptive information (e.g. street names) is missing or wrong. At the same time, these are all affected by how people draw and tag roads.

Currently, data validation and proofing is mainly carried out by OSM contributors. However, as most OSM contributors are not necessarily the users of OSM data (Haklay 2010), relying on the contributors for quality control has several limitations as the OSM project evolves. First, it is not clear yet whether people are willing to revise the work of others (Neis et al. 2012a). Even if they will, the assessment relies heavily on visual inspection. This is a fairly ineffective and tedious. Also, according to the OSM tile disk usage,[1] only a tiny portion of OSM map was used/viewed by its users. This implies that using visual inspection may never finish the assessment for the whole dataset. Second, some quality issues can only be identified when the data is used in specific applications. Topological errors of road network, for instance, can be hard to find using bare eyes. Third, identifying the data errors is only the first step, while correcting them will need local knowledge or ground truth information, which can hardly be accomplished by remote contributors.

Therefore we argue that, in the OSM practice, modeling street and road network suitable for navigation should not rely entirely on OSM contributors. Rather, automated solutions that encompass expert knowledge is advocated to help discover domain-specific problems in data (missing items, misleading and inconsistent contents, errors, etc.) and to give more intelligent suggestions for correcting the problems.

The objective of this paper is to develop techniques that can spot situations that have a high possibility of committing data errors. We mainly deal with two types of errors in the OSM network and describe evaluation techniques that are independent from third party data. First, we detect situations where a dual carriageway changes its representation from double lines to a single line. Changing from one representation to another may not itself be an issue; mixing different representations up would cause serious problems in routing (see Sect. 2.2 for details). Second, we evaluate whether the attributes (tags) are consistently distributed along named roads. For example the name should be consistent for a set of consecutively

[1]http://wiki.openstreetmap.org/wiki/Tile_Disk_Usage.

connected road segments (i.e. named road): a segment can be highly suspicious if its name is inconsistent with other segments connecting to it. The core problem is how to organize named roads from discrete segments. In this paper, we recognize such entities by exploiting the concept of "stroke" and adapt it to OSM data.

1.1 Previous Work on OSM Data Quality

There are two basic approaches to quality assessment for OSM data, namely external and internal evaluation. External evaluation is a process where OSM data are cross-checked against external sources (third-party data, field survey, etc.). Internal evaluation, on the contrary, does not rely on any ground truth information; it assesses the data quality regarding the specifications and in relation to other objects in spatial context.

Currently more studies have concentrated on the general issues in spatial data quality, mostly on completeness and coverage of the OSM data. Many studies compared OSM data with professional or official data, and showed that the completeness of OSM data is at least equal to that of the official data in urban areas (Haklay 2010; Girres and Touya 2010; Zielstra and Zipf 2010). However, completeness also has limitations. For instance, even if an area is fully covered by road network, if the network is topologically not well structured, it is not useful in routing and navigation. For detailed comparisons to third-party reference data, data matching of individual objects becomes necessary (Girres and Touya 2010). If the discrepancy between two datasets is large, the matching becomes difficult and more advanced techniques may be required (Zhang et al. 2014).

In addition, Mooney et al. (2010) proposed measures that can be used to evaluate OSM data quality without referencing to ground truth. These include spatial sampling ratio and density of attribute tags.

On the other hand, few have been done from an applied perspective, i.e., how useful is the data in certain applications. An exception is a study by Mondzech and Sester (2011). They calculated shortest distance between arbitrary points on both OSM and ATKIS datasets and compared the resulting paths, and assume that the one with shorter total distance has better quality in routing application.

Goodchild and Li (2012) proposed that the quality of VGI can be approached from a geographic perspective. For example, one can make use of the first law of geography, fractal nature of geographic phenomena, etc.

1.2 Existing Tools for Inspecting OSM Road Quality

Schmitz et al. (2008) showed that the connectivity of OSM network has improved since the release of OSM-based online routing services. Hence the availability of inspecting tools would also help to improve the OSM network quality.

Although many tools are available for checking and spotting the quality of OSM data, only a few of them focused on the quality of road network. Among others, the Relation Analyzer analyzes OSM relations for identifying gaps in the relations. Besides, the NoName website checks if roads have no name tags, if buildings and residential areas are not tagged with addresses, etc. It can be used to detect missing names, but it cannot tell whether it is an unnamed street or the name is missing. Furthermore, it cannot give more intelligent suggestions for data correction.

In this paper, we describe an evaluation methodology that does not rely on external sources. The evaluation aims not only to identify errors in OSM data, but also to give more intelligent suggestions based on the information available in the spatial context of the problematic data.

2 Problems of OSM Data in Navigation

Modeling OSM streets that meets navigation requirements has several difficulties. First, OSM roads may not form fully topologically structured linear network (Neis et al. 2012b) and junctions may be modeled in very different ways. This is partly due to the lack of strict specifications and the VGI nature (Brando and Bucher 2010; Girres and Touya 2010). The raw OSM data have to be preprocessed before they are processed by routing algorithms. Second, few contributors are aware of how a routing algorithm works. Even if one knows the general principle of routing algorithms, there are algorithm-specific problems.

In the following, we demonstrate some of the typical issues of OSM road data in vehicle navigation. Although other traffic modes (e.g. cycling, pedestrian) are also relevant, many traffic constraints (one-way traffic, turn restrictions, etc.) are relaxed in those cases. Nonetheless, some of the issues described in the following also apply to the other traffic modes. To test the accessibility of OSM road network in this paper, we used several online routing services that are built on top of OSM data.

2.1 Issues in Tagging OSM Roads

Feature attributes are recorded and attached to features using a mechanism called "tagging". In OpenStreetMap a tag is a <key, value> pair. For example, "highway" = "primary" can be used to describe that a linear feature has an attribute field "highway" and the value is "primary".

Tags are commonly used by the Web users to organize and classify Web contents using their own way. Such a classification mechanism is also called "folksonomy". OpenStreetMap adopts a similar architecture in describing the semantic aspect of geographic features. Different from the free-form folksonomy used in other Web 2.0 sites (e.g. Flickr), OSM project suggested a dynamically growing set of tags, which are proposed, discussed and agreed upon by the OSM community. In the above example, "highway" is a suggested key (field) for linear features, for which a classification of highway levels is also suggested (e.g. "motorway", "trunk", "primary", "secondary", etc.).

However, OSM does not force the contributors to strictly follow the suggested tags, and the community is open to accepting new tags (both new keys and values are allowed). This semi-controlled vocabulary can be problematic when OSM data is used in navigation. Then, tags contributed by volunteers do not undergo a rigorous quality checking process. Data that contain errors can be submitted to the OSM database, long before the errors can be found and corrected by other volunteers: sometimes it needed as long as e.g. 29 days to identify and correct the errors (Neis et al. 2012a). In this paper, we identify the following types of tagging issues.

- Missing tags. Examples include missing names, highway classes, one-way tags, etc. Usually, missing information is easier to detect but harder to correct without cross-checking against some reference data. For example, it is difficult to tell if a road name is really missing during the acquisition process, or it is actually an unnamed road.
- Incorrect tags. Sometimes misleading (incorrect) tags can be more serious than missing tags. This can cause confusion when using machine-generated turn-by-turn guide and the names on the route are wrong. In addition, highway classes may not be tagged properly, as there is no unique mapping between various local schemes and the OSM scheme. This will change the calculated routes dramatically because the calculation of different travelling options (fastest route, shortest route, etc.) relies on such information.
- Inconsistent tags. There are at least two senses of inconsistency. First, tags are usually used in combination. For example, with "highway" = "[…]_link" tagged (meaning it is a ramp), "oneway" = "yes" must be enforced. There are many such constraints, but currently both a comprehensive list and validation tools seems to be still lacking. Second, tags are not compatible with surrounding tags. For example, a road name is obviously wrong according to its adjacent roads. Another example is that a one-way road is tagged in an opposite direction with other roads connecting to it.
- Missing turn restrictions. Turn restriction is a special type of tags in OSM data model. It specifies the relation that one road is allowed (or not allowed) to travel to or turn to another road at a junction. Turn restrictions are essential in navigation and are crucial when driving on the road. Missing turn restrictions are common in OSM (Neis et al. 2012b) and can only be corrected with ground truth information.

The missing of detailed information is regarded as one of the major quality issue in OSM data (Neis and Zielstra 2014). In this paper, we describe a method that automatically detects missing, incorrect and inconsistent items in a more intelligent manner. In addition to identifying problematic data, we also suggest possible ways

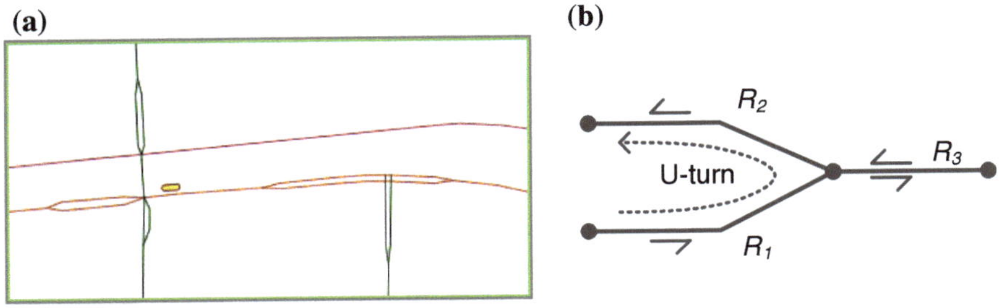

Fig. 1 Mixing up single-line and double-line representations for the same road in OSM data (**a**) could lead to problems at junctions, e.g., a false U-turn is allowed by routing algorithms (**b**)

for correction. By exploiting spatial relations among geographic features, data items can be checked to see if they are consistent with others in context.

2.2 Issues in Modeling Dual Carriageways

Dual carriageways, or divided highways, are normally modeled as double-line representations in commercial spatial data for vehicle navigation. However, as there is no strict specification as to how OSM roads should be modeled, the choice is left to the contributors. A related issue is that, the same named road (a geographic entity consisting of several connected segments of the same name) may be created or edited by several people. In this case the same named road may be alternately modeled by single and double lines (Fig. 1a).

Such a mixing-up modeling has a functional side-effect on navigation. That is, at the junction where the double lines join the single line (e.g. Fig. 1b), a U-turn from R_1 to R_2 is automatically allowed by routing algorithms (see discussion in Sect. 2.4), But such a U-turn is probably incorrect, because segments R_1, R_2 and R_3 belong to the same divided road. U-turns are usually not allowed on divided road.

Note that similar problems also occur when highway links, ramps join the divided highways modeled by single-lines. Figure 2 demonstrates some incorrect routing plans. The route plans are the results of several online routing planners, such as MapQuest Open[2] and OSRM.[3]

This is essential because roads such as motorway, divided highway, dual carriageway were modeled by single lines rather than double lines and these roads connect to other one-way roads. In Fig. 2a the route makes a sharp U-turn from the ramp to the divided road (modeled by a single line) and drives on the opposite direction (assuming a driving on the right system), which is forbidden in reality.

[2]http://open.mapquest.com/.

[3]http://map.project-osrm.org/.

Similar cases are shown in Fig. 2b, c. Figure 2d depicts a real-world case as demonstrated in Fig. 1b.

There are two ways to correct the problem:

1. by change the single-line to double-line representation can avoid the problem;
2. by adding turn restrictions at problematic junctions.

Both solutions require the problematic areas being detected in the first place. It should be noted that it is difficult to know which road is divided road in isolation, but it becomes easier by considering a group of road segments at the mixing-up junction (e.g. Fig. 2d). This is further discussed in Sect. 3.1.

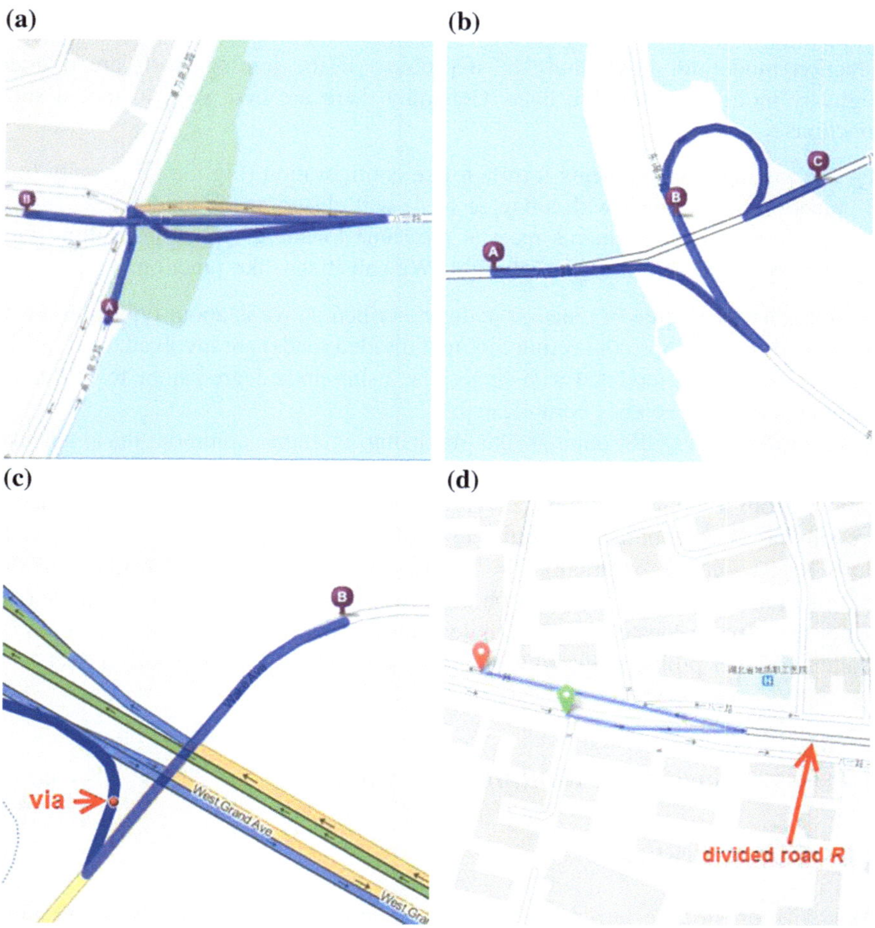

Fig. 2 Wrong routes found by some online route services: ramps connecting to a divided road modeled by a single-line representation could be problematic: areas in Wuhan, China (**a**) (**b**) (**d**); a similar problem in Oakland, USA (**c**)

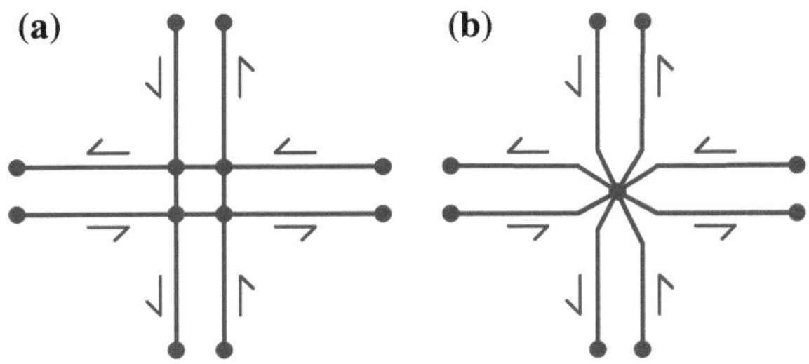

Fig. 3 Two approaches to modeling junctions: **a** #-like modeling and **b** star-like modeling

2.3 Issues in Modeling Junctions

Junction modeling should only be a problem if the joining roads are divided highway modeled by double lines. Generally there are two ways to model such junctions.

1. The joining road segments form a #-like configuration (Fig. 3a). Typically four junction nodes (each with a degree of 4) will cluster at the junction.
2. The joining road segments meet at the same location, typically resulting in a junction node with a degree of eight. We call it star-like junction.

Note that the degree of junction nodes described above is about typical cases. If more roads (e.g. ramps connecting different divided roads) are involved, or if one of the divided roads is modeled with single line, using node degree alone to determine the junction type becomes complicated.

In professional GIS data, #-like modeling is more common than star-like modeling; it is rare to see star-like junctions in commercial spatial data for navigation. From a topological point of view, however, the two modeling approaches are similar. Both approaches require to explicitly specify prohibitive turn restrictions (e.g. adding a "restriction" = "no_u_turn" tag). The major concern in the star-like modeling is that the junction node may become a heavily tagged node (i.e. many tags are co-located here), thus leading to data editing overhead and visual clutter. Another problem related to the star-like modeling is discussed in Sect. 2.4.

2.4 Issues Specific to Routing Algorithms

2.4.1 Turn Restrictions

As shown previously, turn restrictions are of great importance in real-world navigation. However, not all OSM-based routing services currently available support this type of information. We exemplify some of the issues in Fig. 4.

Figure 4 shows that some routing services did not seem to consider turn restrictions. There are two possible reasons. First, the routing algorithm does not incorporate turn restrictions. For example, the routing service GraphHopper[4] currently does not support turn restrictions (Fig. 4d). MapQuest Open and OSRM integrate turn restrictions in their algorithms and thus give reasonable results (Fig. 4b, c). OpenRouteService.org,[5] which bases their service on OSRM routing algorithm, fails to give correct result (the result is exactly the same to Fig. 4d). This may be explained by the fact that OpenRouteService.org is out of synchronized with current OSM data and the turn restrictions shown in Fig. 4a is not available in their local data copy.

2.4.2 Geometric Calculations

Another algorithm-specific issue is how routing algorithms determine whether to make U-turn at junctions. For example, some algorithms calculate U-turns based on the geometry of road junctions, and therefore, the calculated routes can be dramatically affected by the way the junctions are modeled (e.g. star-like junctions described in Sect. 2.3).

To calculate U-turn at star-like junctions should in principle be possible. However, some routing algorithm does not seem to allow a U-turn at a star-like junction. For example, MapQuest seems to take the assumption that if the two road segments intersect at a sharp angle, U-turn is not allowed even if there is no such restriction (c.f. Fig. 5b, c). This is normally not a problem at a #-like junction (Fig. 5a). We found many such examples. Some other routing service adopts an "open-world assumption". That is, if there are no restriction tags, all roads joining at star-like junctions are equally navigable (Fig. 5d, e).

We claim that making turning decision based on geometric criteria is not very reliable in the OSM world. A more reliable way to this problem is to adopt an "open world assumption" and leave the rest of the problem to OSM data. Domain-specific (turn restrictions in navigation domain) and detailed information should enter OSM database before it could be used in any application with satisfaction.

3 Enriching Data to Support Quality Evaluation

Before detecting the inconsistencies in data, this section introduce some data enrichment techniques used to support the evaluation methodology. These include junction analysis (Sect. 3.1) and natural road detection (Sect. 3.2).

[4]http://graphhopper.com/.

[5]http://openrouteservice.org/.

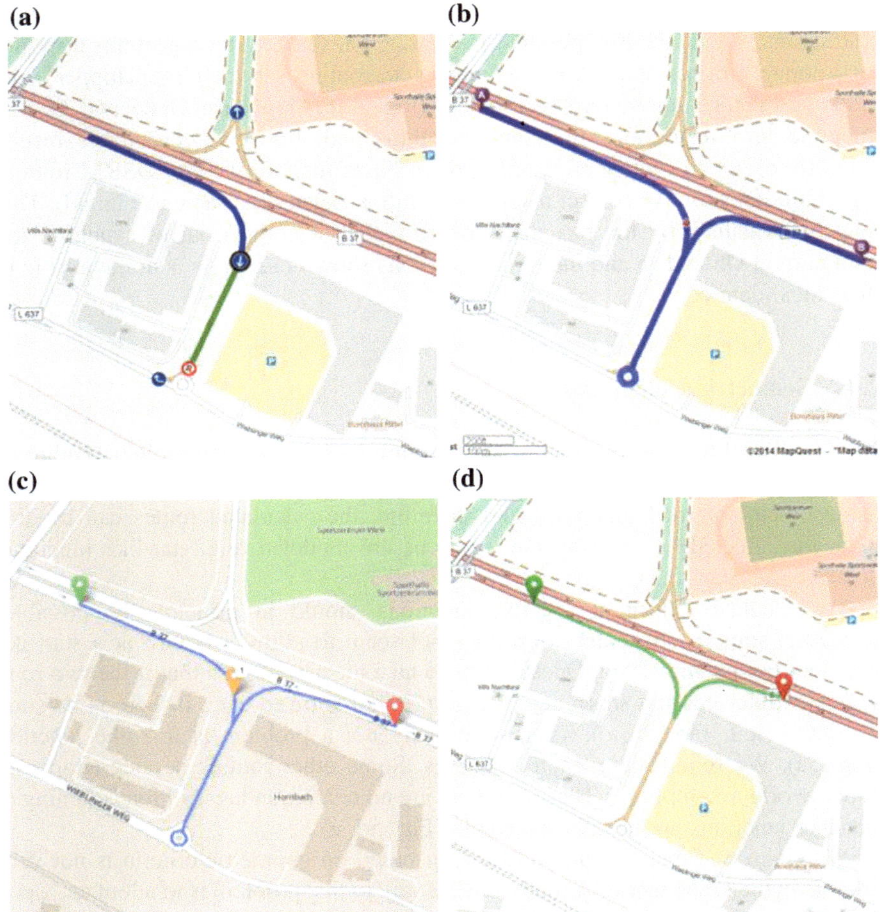

Fig. 4 Turn restrictions tagged in OSM data (**a**) and calculated route plans by different online router (**b**, **c**)

3.1 Junction Analysis

As discussed in Sects. 2.2 and 2.3, junction degree plays an important role in identifying the mixing-up junction where single-line represented road join double-line represented roads as well as star-like junctions. For mixing-up junctions, we simply assume they have node degree of three. This is an idealized assumption which has to be extended for real data.

Degree analysis from a pure topological point of view is not adequate for detecting mixing-up junctions. Other information such as traffic directions, road classes and spatial configuration of junctions should also be considered. For example, to distinguish between J_1 and J_2 in Fig. 6b (both have a degree of 3),

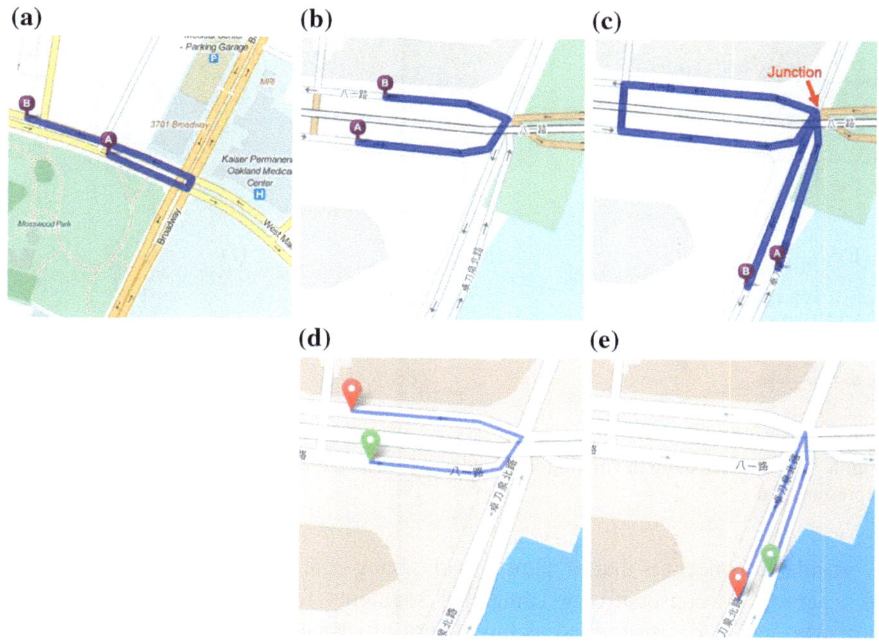

Fig. 5 U-turn at junctions is algorithm dependent: routing by MapQuest Open (**a**–**c**) and by OSRM (**d**–**e**) (no turn restriction tagged at the junctions in these situations)

directional configuration should be considered. Only when two one-way roads join a two-way road can be regarded as a potential candidate. In addition, topological analysis cannot tell Fig. 6a–c. To distinguish between the two, angle between two one-way roads should be calculated to see if it is less than a predefined threshold (e.g. 90°). A special case is when a two-way road joins a roundabout. To filter such cases, special tag of roundabout should also be used.

More complicated cases could emerge from the combination of several types of junctions, such as those depicted in Fig. 6e, f. Currently we are not handling such types of mixing-up junctions. We detect star-like junctions if node degree is greater than six.

In the stage of quality checking, the detected junctions can be used to inform OSM contributors for bug fixing (e.g. adding turn restrictions or changing corresponding single lines to double lines).

3.2 *Forming Natural Roads by Stroke Detection*

The basic idea of the detection here is that segments in a named road should be consistent with each other (e.g. road names, directions should be homogenous). The problem is that name road (a geographic concept) is not explicitly modeled in OSM data.

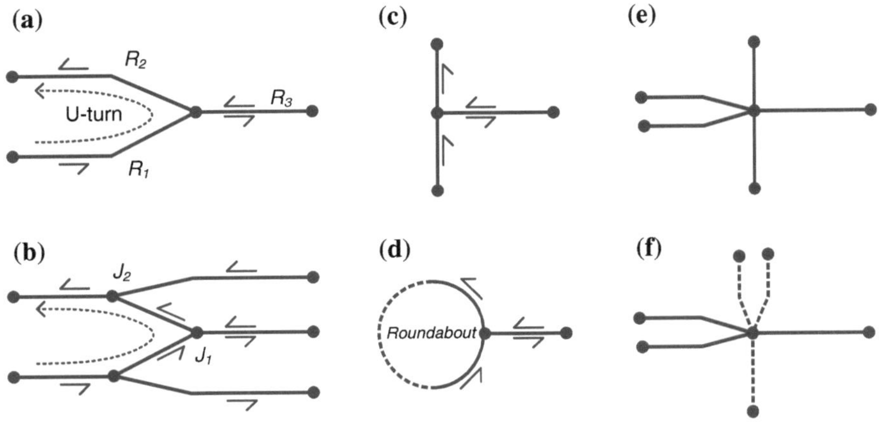

Fig. 6 Cases in detecting the mixing-up junction where single-line representation join double-line representation

A related concept is that of natural roads. Jiang et al. (2008) define a natural road as a group of consecutively connected segments following good continuation property, which is also termed as the pattern of strokes (Thomson and Richardson 1999). Forming natural roads (or strokes) is meaningful because roads are also constructed as curvature-poor as possible (Heinzle et al. 2007), obeying similar principles to stroke forming. Moreover, named roads are usually contained in natural roads (Jiang et al. 2008). Consequently, by detecting natural roads one obtains name roads.

Several strategies for stroke forming are described in Jiang et al. (2008). Every-best-fit and self-best-fit principles are relevant in our context. In the former, optimal connections in terms of maximum angle of deflection at each junction are determined (Fig. 7b). Strokes formed with this principle are unique and optimal. While in the latter, optimal connection is determined according the starting road; every starting road selects the one that has the maximum angle of deflection to itself (Fig. 7b). Hence the strokes detected are not necessarily optimal and depends on the sequence of starting roads. We employed every-best-fit principle as it can generate the longest possible strokes.

3.2.1 Handling Imperfect Data

The main problem of current stroke building methods is that they are designed for roads modeled by centerlines. As stated previously, OSM network is more complicated. Strokes may interrupt or connect to incorrect roads at junctions such as mixing-up and star-like junctions. To overcome this problem, we incorporate road directions in the stroke forming algorithm. That is, road segments in a stroke are either two-way roads or in the same direction. The experiments gave promising

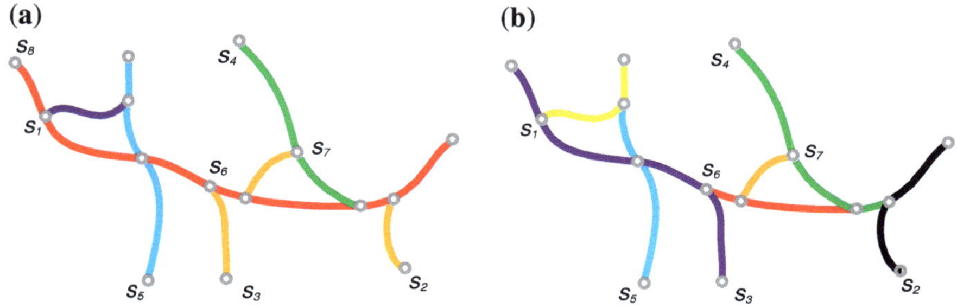

Fig. 7 Detect natural roads from discrete segments (curve between two nodes) using every-best-fit **a** versus self-best-fit principles **b** in the stroke forming algorithm (after Jiang et al. 2008)

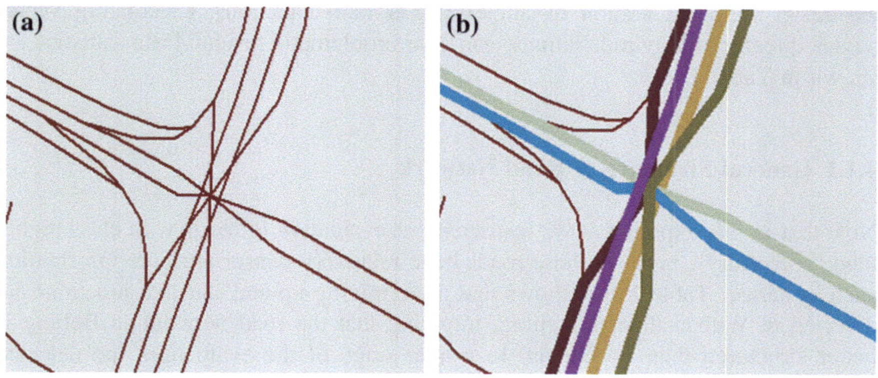

Fig. 8 Stroke detection at complex junctions **a**; results by considering road directions (maximum angle of deflection alone will fail to detect expected strokes at such junctions)

results (e.g. Fig. 8b). Without this treatment the roads in diagonal positions would have been recognized as one stroke.

4 Identifying Inconsistencies in OSM Network: Sample Results

This section describes the experiment and reports sample results of identified inconsistencies.

4.1 Experiment Setting and Processing of Data

For this study we used two datasets. One dataset covers the area of Wuhan. The other dataset covers the area of Beijing. Both are shape files downloaded from

Table 1 Datasets description and some detection results

	Wuhan dataset	Beijing dataset
No. of road segments	8319	16,591
No. of roads higher than secondary class	1985	5668
No. of two-way roads[a]	432	416
No. of mixing-up junctions detected	78	35
No. of star-like junctions detected	89	0
No. of strokes detected	387	507

[a]"oneway" = "no" selected from roads with class higher than "secondary"

BBBike.org. Before further processing, we transformed the original data (in WGS 84 datum) to a projected coordinate system with 3-degree Gauss-Kruger. This ensures a valid calculation of angles (for stroke detection). Calculating vector angles directly in longitude/latitude can give problematic results. Data statistics are shown in Table 1.

4.1.1 General Structure of Road Network

Note that in this experiment we restricted our evaluation to roads with class higher than "secondary", because these roads have relatively denser attribute information such as names. Table 1 also shows that more mixing-up and star-like junctions are detected in Wuhan than in Beijing, implying that the road network in Beijing is better structured than in Wuhan. In a later stage of the evaluation, the detected mixing-up junctions have a high probability of lacking extra tagging. As a result, they can be reported to OSM contributors, and the suggested correction could be: "no_u_turn" restrictions needed.

4.2 Processing of Natural Roads (Strokes)

We applied the presented stroke algorithm to both datasets (roads with class higher than "secondary"). For Wuhan dataset we obtained 387 strokes and for Beijing 507. The strokes are visualized in Fig. 9 (with random color).

4.2.1 Characteristics of Detected Strokes

First of all the number of strokes grouped by the number of segments contained exhibits a power law distribution for both datasets (Fig. 10). Strokes with fewer segments constitute the majority of the population. It seems that strokes detected in Beijing dataset contain much more segments (the longest stroke contains 251 segments in Beijing versus 35 segments in Wuhan). This can be explained by the

Fig. 9 Strokes recognized from OSM roads; only those higher than secondary class are used in the calculation: Wuhan (*left*) and Beijing (*right*)

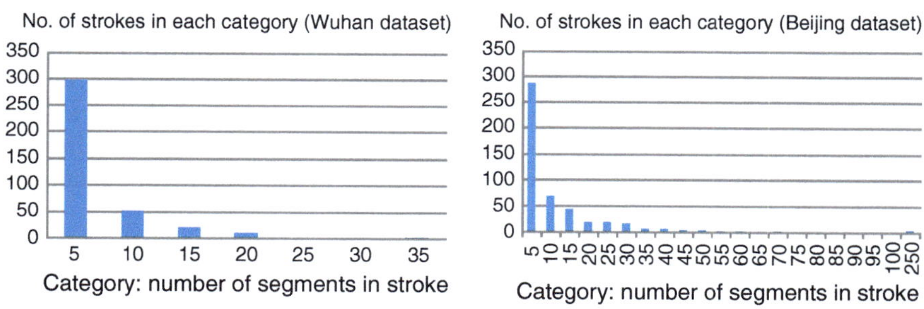

Fig. 10 Distribution of strokes in terms of how many segments the strokes contain

fact that the traffic system in Beijing is more developed than in Wuhan, and the Beijing dataset covers a larger area. More motorways and trunks are built in Beijing, so longer and smoother strokes are formed.

Besides, the plot of stroke length (in kilometers) against the number of segments contained shows that in general the length of stroke increases as it contains more segments. Although lengths may vary within each column, the shortest stroke in each column (the baseline of the graph) is increasing (Fig. 11).

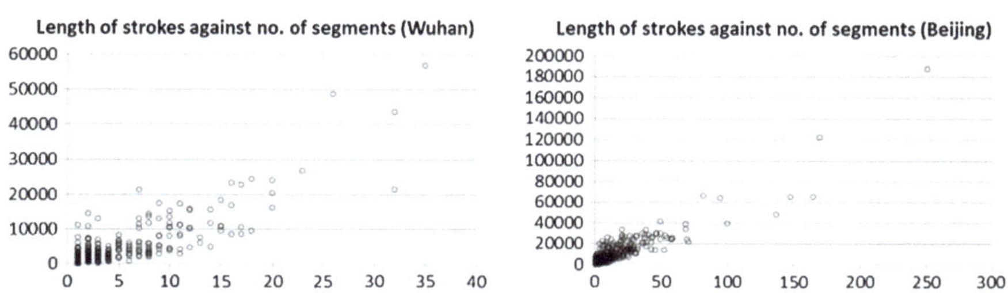

Fig. 11 Plot of stroke length against the number of segments contained

4.3 Inconsistency Detection

In this section, we evaluate whether the attributes (tags) are consistently distributed along named roads. We focus specifically on street names. If a street name is interrupted (e.g. by blank item or another name) and then occur again, we regard it as a missing or incorrect name.

First of all, we have to transform attributes (tags) in forms of strings to numbers for subsequent calculations.

4.3.1 Mapping Words to Numbers

Many attributes are tagged by words, such as road names, highway classes, one-way tags, etc. For each stroke, we map occurrence of names randomly to [0, *n*], with 0 indicate the name field is blank and *n* indicate the number of occurrence. Highway classes have an inherent order, so we map them to numbers according to Table 2. For one-way tag and bridge tag, "yes" is mapped to one and "no" to zero.

Note that, the numbers in the mappings are only meant to indicate the difference between items; at the same time it facilitates a quick visual plotting of the nominal variables. Other mapping schemes are also possible.

4.3.2 Examples of Inconsistent Names

Here we show two examples of detected inconsistencies. Figure 12 plots some attributes of two detected natural roads. We detected a named road by identifying the start and end of the same name. This process can be iterated so that several named roads can be detected if the stroke is long enough. The stroke in Fig. 12a contains 13 segments and a long named road is identified (highlighted in red dotted strip); the stroke in Fig. 12b contains 30 segments and one longer and two shorter named roads are detected.

Inconsistent names are outliers or interruptions in a named road. A value of zero indicates a missing name (blue rectangle); other non-zero values indicate that the names encoded are likely wrong (blue circles). In Fig. 12(b) we detected a large number of missing names for the longer named road. With this approach, we are also able to tell if a road is missing its name or it is an unnamed road. These detected inconsistencies have a high probability of committing errors. Hence detailed information as to how the data is possibly wrong should be made available to contributors. For both missing and incorrect names, the suggested correction could be to make the problematic name consistent with the detected named road that contains the problematic name. Similar approaches can be devised for other properties such as road direction, maxspeed, pavement type, etc.

Table 2 Mapping OSM highway classes to numerical values for visualization purposes

Motorway	Motorway link	Trunk	Trunk link	Primary	Primary link	Secondary	Secondary link	Tertiary	Tertiary link	Unclassified	Road	Residential	Service	Track
10	9.5	9	8.5	8	7.5	7	6.5	6	5.5	5	4	3	2	1

Only roads that are traversable by vehicles are considered in this mapping

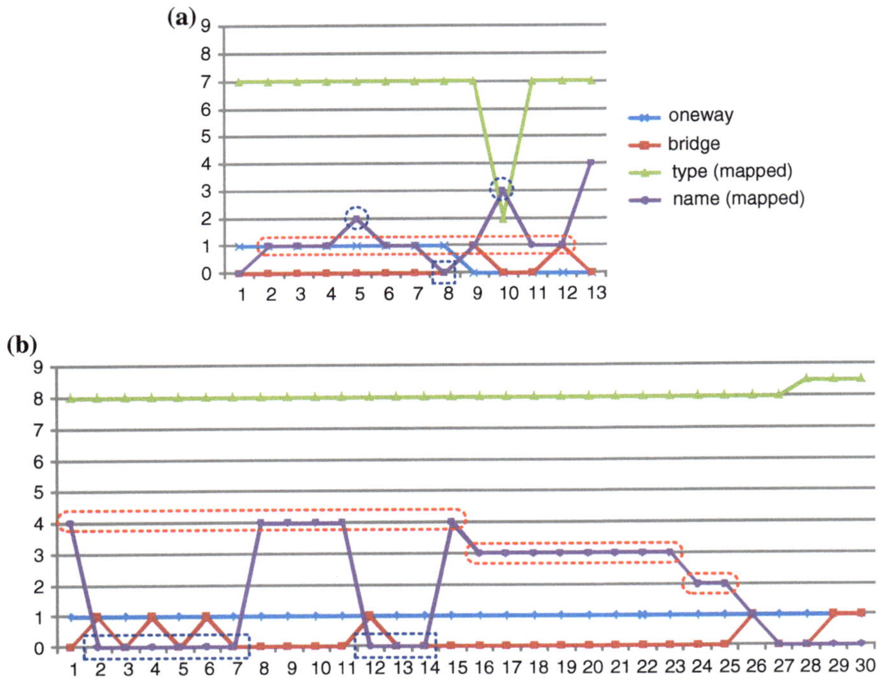

Fig. 12 Identifying inconsistent names in strokes by outlier detection (*red strip* detected named roads; *blue* inconsistent names)

5 Conclusions and Future Work

This paper summarizes several quality issues in OSM data for navigation application. The nature of OSM data collection leads to inconsistent modeling of e.g. roads (including their geometries and tags). An ideal solution would be to enforce uniform specifications as volunteers are contributing the content, but this can be difficult in OSM practice. Automated checking procedures that are able to spot problems and send revision requests to more advanced users or experts is therefore preferable. For navigation vendors, we suggest that the data should at least be cleaned before it can be used in applications like navigation. Additionally vendors should minimize their assumption about the data, e.g., calculations assuming geometric criteria as used in commercial data may fail in OSM data.

This paper proposes an approach to detect some of the issues without the use of ground truth information. The evaluation is not only able to identify errors in OSM data but also able to give more intelligent suggestions based on the information available in the spatial context of the problematic data. The notion of "stroke" is exploited to identify inconsistent attributes/tag distributed along the length of named roads. Finally, case studies are carried out to demonstrate our approach and to showcase the mentioned quality issues. There are several aspects that need further investigation:

- Currently we are working on shape files, which contain far less information than the original OSM file. Only names, highway class, one-way tag, bridge tag, maxspeed is available. To further analyze turn restrictions, we have to work on the original OSM file, or the PostgreSQL import.
- Forming strokes usually involves two levels of processing: (1) lower-level processing (perceptual only) and (2) higher-level processing (perceptual and semantic integrated). Currently only lower-level processing is considered. Further improvement should be made to the algorithm (including the calculation of angle of deflection)
- Currently we did not detect divided highways modeled by double lines. This should give more insight into the quality of OSM network.
- Another issue would be how to scale the described techniques up for larger datasets, e.g., for the whole OSM data or for calculations at a regional or national level. Out-of-core algorithms or parallel computing may be useful directions.

Acknowledgments We thank the anonymous reviewers for their valuable comments which substantially improved this paper. This work was financially supported by National Natural Science Foundation of China (Grant No. 41301410) and China Postdoctoral Science Foundation funded project (Grant No. 2013M531742).

References

Brando C, Bucher B (2010) Quality in user generated spatial content: a matter of specifications. In: Proceedings of the 13th AGILE international conference on geographic information science, pp 11–14

Girres JF, Touya G (2010) Quality assessment of the French OpenStreetMap dataset. Trans GIS 14 (4):435–459

Goodchild MF, Li L (2012) Assuring the quality of volunteered geographic information. Spat Stat 1:110–120

Heinzle F, Anders K-H, Sester M (2007) Automatic detection of pattern in road networks—methods and evaluation. In: Proceedings of joint workshop visualization and exploration of geospatial data (IAPRS), XXXVI—4/W45

Haklay M (2010) How good is volunteered geographical information? a comparative study of OpenStreetMap and ordnance survey datasets. Environ Plan 4(37):682–703

Jiang B, Zhao S, Yin J (2008) Self-organized natural roads for predicting traffic flow: a sensitivity study. J Stat Mech 7:P07008

Mondzech J, Sester M (2011) Quality analysis of OpenStreetMap data based on application needs. Cartographica 46(2):115–125

Mooney P, Corcoran P, Winstanley AC (2010) Towards quality metrics for OpenStreetMap. In: Proceedings of the 18th SIGSPATIAL international conference on advances in geographic information systems, ACM, New York, NY, USA, pp 514–517

Neis P, Goetz M, Zipf A (2012a) Towards automatic vandalism detection in OpenStreetMap ISPRS. Int J Geo-Inf 1:315–332

Neis P, Zielstra D, Zipf A (2012b) The street network evolution of crowdsourced maps: OpenStreetMap in Germany 2007–2011. Future Internet 4:1–21

Neis P, Zielstra D (2014) Recent developments and future trends in volunteered geographic information research: the case of OpenStreetMap. Future Internet 6:76–106

Schmitz S, Neis P, Zipf A (2008) New applications based on collaborative geodata—the case of routing. In Proceedings of XXVIII INCA international congress on collaborative mapping and space technology, Gandhinagar, Gujarat, India, pp 4–6

Thomson R, Richardson D (1999) The 'good continuation' principle of perceptual organization applied to the generalization of road networks. In: Proceedings of the ICA, Ottawa, Canada, Session 47B

Zhang X, Ai T, Stoter J, Zhao X (2014) Data matching of building polygons at multiple map scales improved by contextual information and relaxation. ISPRS J Photogram Remote Sens 92:147–163

Zielstra D, Zipf A (2010) A comparative study of proprietary geodata and volunteered geographic information for Germany. In: Proceedings of the 13th AGILE international conference on geographic information science

Development of a Nationwide Road Change Database for the U.S. in the Post-Recession Years (2008–2012)

Christopher L.B. Brown and Xiaobai Angela Yao

Abstract Previous transportation network growth research in the United States has focused on long-term historical changes across selected metropolitan areas. However, short-term comparative explorations, especially those considering all levels of the nationwide road hierarchy, have been neglected. This study aims to create a comprehensive road change database of the United States through compilation of U.S. Census Bureau TIGER/Line GIS datasets from 2008 to 2012. Annual and four-year road changes were derived using Python geoprocessing scripts, and aggregate road change statistics were developed for each state, metropolitan statistical area, and county nationwide. Preliminary analyses found that the changes exhibited moderate spatial autocorrelation and revealed regional differences. Counties surrounding Atlanta, Georgia produced highly statistically significant outliers—Forsyth County experienced road growth of over thirty-eight percent—suggesting that anomalous expansion processes have uniquely shaped this area.

Keywords Transportation road network · Spatial database · Data accuracy · Change detection · GIS · Spatial analysis

1 Introduction

1.1 Background and Aims

The growth of metropolitan road networks in the United States, although often indirectly referenced in academic literature, has been largely neglected as a primary object of scholarship. According to a recent review of transportation growth

C.L.B. Brown (✉)
Macon, GA, USA
e-mail: christopher.lbb@gmail.com

X.A. Yao
Department of Geography, University of Georgia, Athens, GA, USA

F. Harvey and Y. Leung (eds.), *Advances in Spatial Data Handling and Analysis*, Advances in Geographic Information Science,
DOI 10.1007/978-3-319-19950-4_10

modeling (Xie and Levinson 2009), only a handful of empirical studies have investigated temporal change in transportation network supply. The researchers attributed the gap to a dearth of computer processing power and historical data. However, while these considerations may apply to complex modeling prescriptions and temporal ranges exceeding the maturity of geographic information systems (GIS) technology, they need not preclude future research. As microprocessor densities have increased exponentially according to Moore's law for the past four decades (Moore 1998) and recent historical road network data have been continually archived by both the U.S. Census and some state or local government agencies, the impediments to scholarship on comparative metropolitan road network growth are steadily being deconstructed. Enabled particularly by the public-sector advancement of longitudinal GIS data provisioning, this study aims to develop a system for automated road change detection in the U.S. road network and to identify road changes across different aggregation levels, including states, metropolitan statistical areas (MSAs), and counties, for the time period between 2008 and 2012. Road change statistics produced from the database have a wide range of applications, including habitat fragmentation, urban morphology, regional science, and suburban real estate markets after the "Great Recession."

1.2 Related Work

This study of transportation network growth differs from previous analyses due to its nationwide scope, local scale, and temporal resolution, as well as its innovation in automated data extraction and analysis. It requires data assimilation across the entire United States and over a time period of four years, involving the processing of over a terabyte of data. Although the theoretical construction of a comprehensive spatiotemporal road database may be largely self-evident, the nontrivial effort required for data acquisition and change detection, combined with limited information on accurate, complete, and regularly updated GIS data sources, has appeared to construct a strong barrier of entry to the task. Furthermore, after amassing the data, one must determine how to extract new linear road features given positional or temporal uncertainty. A thorough survey of the literature failed to identify any previous transportation network growth research of comparable scope, scale, and resolution.

However, several types of transportation literature closely approximate different components of the national transportation network growth process, as shown in Table 1; as previously mentioned, these analyses are limited by either scale, scope, or temporal situation. The ideal database, advanced in this study, addresses these structural shortcomings by exhibiting all three characteristics: complete road hierarchy, nationwide coverage area, and temporal network archival.

Many economic transport studies focus on comparative nationwide transportation, and often utilize national GIS data sources such as the National Transportation Atlas Database, which excludes data for local roads. These studies may incorporate

Table 1 Example taxonomy of road databases in transportation literature

	Complete hierarchy	Nationwide coverage	Temporal continuity
Economic transport		X	
Historic transport			X
Regional planning	X		X
National planning		X	X
Road ecology	X	X	

economic statistics to support hypotheses on interstate commerce (Ham et al. 2005) and may have a temporal component as well, but usually limit the analysis to a regional scope due to the large volume of data required (Healey and Stamp 2000).

Prior research on historic transport typically describes the evolution of historical transportation networks, but tends to either focus on one metropolitan area or present a general survey of the nation. As much of the data acquired on historic road networks is derived from temporally inconsistent sources such as aerial/satellite imagery or digitized historical maps, historical transportation research has more difficulty in drawing analytical conclusions based on time series, instead preferring a more theoretical or descriptive approach. Furthermore, a considerable portion of historical transport studies are not explicitly concerned with changes over time. Instead, historical road data are often used to observe a particular time period, rather than engage in temporal comparison (Gregory and Healey 2007).

Although many urban growth models, which are verified according to their ability to reproduce historical conditions at a given time, only consider major roads in a single metropolitan area (Clarke and Gaydos 1998; David and Karamalaputi 2003; Hu and Lo 2007), several cities now provide complete GIS road archives. Therefore, regional planning literature now has the ability to incorporate temporal changes, yet the different database schemata, record attributes, and data assimilation methods complicate comparative metropolitan analyses. Current and future advancements in metropolitan data provisioning should augment planning research on the effects of urban sprawl (Kenworthy and Laube 1999; Ciscel 2001), as well as inform concerns of systemic racism (Jaret et al. 2006), inequity of transportation systems and services (Levinson 2002; Bullard 2003; Sanchez et al. 2003) and job accessibility (Henderson 2004; Weber and Sultana 2007). Likewise, national transportation planning has benefited, and will continue to reap insight, from the inclusion of temporal data in freight analyses, although unlike air and sea routes, the locations of domestic ground transportation routes remain relatively constant (Feyrer 2009).

Road ecology, likely the most specialized field of study on the list, assesses the ecological effects of the road network on various species, and may also use landscape ecology metrics to analyze habitat patches outlined by roads and their buffer zones (Forman and Alexander 1998). Although road ecology research has not yet embraced temporal road data, it has presented a static study that includes all roads in the United States (Riitters and Wickham 2003). This research compiled all of the county-based road networks in the commercial Tele Atlas/GDT Dynamap/2000 GIS dataset, making this analysis one of the first to utilize and process such a large linear road representation.

In the spirit of analyzing all available road data, our research aims to retain the original linear nature of the road centerlines, while including five years of annual revisions for the nationwide network. Unlike the Riitters and Wickham (2003) study, the raw data in our research are subjected to spatiotemporal accuracy assessments. Furthermore, the procedures for creating a functional GIS database out of these different temporal layers, and the Python programming routines utilized to extract information about road network change over time, may inform future temporal studies with nationwide scope and/or complete road networks.

2 Methods

2.1 Source Data and Spatiotemporal Accuracy Assessment

Although public-sector GIS data acquisition has matured over the past three decades, differences in accuracy, completeness, boundary definitions, attribute categories, and implementation have discouraged nationwide, comparative studies of metropolitan transportation network expansion. Therefore, in selecting an appropriate national data source, one must consider data quality issues such as logical consistency, positional accuracy, temporal accuracy, semantic accuracy, and completeness.

Four freely available national data projects were evaluated, including efforts from the United States Geological Survey (USGS), the United States Department of Transportation (USDOT), the United States Bureau of the Census, and the crowd-sourced OpenStreetMap project. The USGS Digital Line Graph was the least suitable for the proposed database; revision years were infrequent and not standardized across the country, obfuscating any possible comparative conclusions. Furthermore, the boundaries of the datasets do not coincide with any political boundaries, and the vast majority of the included road features are digitized at a scale of 1:100,000, which would prove to be quite inaccurate for analyzing individual urban or suburban road changes. Likewise, the USDOT National Transportation Atlas Database (NTAD) was also digitized at an inappropriate scale. Additionally, the NTAD dataset purposefully excludes many local roads, choosing instead to emphasize national, state, and arterial roads. However, the NTAD dataset does feature annual road updates, making it highly suitable for national-scale temporal analyses if hierarchical completeness is not required.

The United States Census Bureau road network dataset retains the same annual temporal resolution as NTAD, yet is complete and logically consistent. The U.S. Census MAF/TIGER (Master Address File/Topologically Integrated Geographic Encoding and Referencing) data were originally developed to support internal geocoding operations for the decennial census; therefore, all roads are included with standardized attribute information. Another preferred data provider could be OpenStreetMap, a free and open source worldwide road database originating from the United Kingdom. However, while excellent for countries where GIS data is restricted or commercialized, OpenStreetMap data in the United States, like the NTAD datasets, are largely derived from the United States Census Bureau MAF/TIGER roads. Moreover, the datasets do not follow the de facto shapefile standard for vector GIS data, and temporal extraction cannot be readily performed on the raw data, particularly due to the difficulty of creating and then processing a historical "snapshot" of the database features. Most importantly, the crowd-sourced nature of the database allows anyone to add or edit road features, which potentially allows spatiotemporal accuracies to vary widely across the nation. Therefore, due to these accuracy issues, the Census Bureau MAF/TIGER datasets were selected for the source data.

2.1.1 Positional Accuracy

In order to gain a better understanding of the positional accuracy of the MAF/TIGER dataset, specifically the TIGER/Line streets dataset, independent positional accuracy assessments were performed, including both absolute and relative accuracies. The datasets were also subjected to testing for completeness and temporal accuracy, using both local GIS data and reference aerial imagery. All the assessments were performed for a small case study area: the Atlanta-Sandy Springs-Marietta Metropolitan Statistical Area (Atlanta MSA). The majority of the MSA exhibits a highly curvilinear road network, making GIS road datasets particularly susceptible to line generalization errors, as shown in Fig. 1. Therefore, the Atlanta MSA typifies a "worst-case" scenario for minimizing positional accuracies along its linear features. Furthermore, the area has experienced explosive growth throughout the past two decades, with rapid construction of roads to support residential developments; these road network modification underscore the importance of this MSA for temporal accuracy assessment.

The absolute accuracy assessments were performed using a set of 150 sample points randomly distributed along the length of the roads, as displayed in Fig. 2. Because differential GPS was not available, 1 m resolution 2010 U.S. Department of Agriculture (USDA) National Agriculture Imagery Program (NAIP) aerial images were used as reference data. Errors were quantified by measuring the distance from each sample point on the network to the corresponding reference location on the imagery. The accuracy of local data from the Georgia Department of Transportation (GDOT) was also determined for use as a comparative benchmark, enabling a relative accuracy assessment for TIGER positional and generalization errors. The frequency distribution of errors for each dataset may be viewed in Fig. 3. The displacement value for each sample point was measured as the straight-line distance perpendicular to the tangent line of the NAIP road centerline. While the distribution of TIGER/Line errors is approximately normal, the GDOT

Fig. 1 Sample road network from the Atlanta MSA (*source* 2010 USDA NAIP)

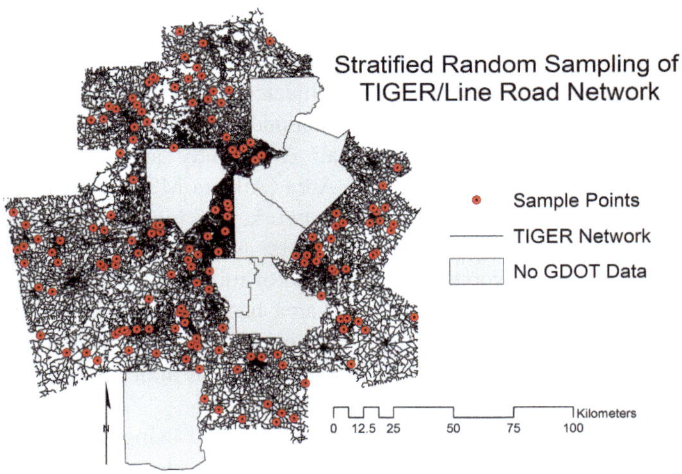

Fig. 2 Sample locations for positional accuracy assessment

distribution has a pronounced peak at a 1 m displacement, indicating that more of the locally digitized streets fell within a one meter tolerance of the NAIP road centerline—the same as the resolution of the imagery. Conversely, between four and ten meters, TIGER/Line data has a slightly higher error frequency, especially using metrics such as the National Standard for Spatial Data Accuracy (NSSDA) 95th percentile method. Thus the positional accuracy of the local GDOT data is slightly superior to that of the nationally collected TIGER data. However, despite the long tail of the TIGER/Line error distribution, the overall NSSDA 95th

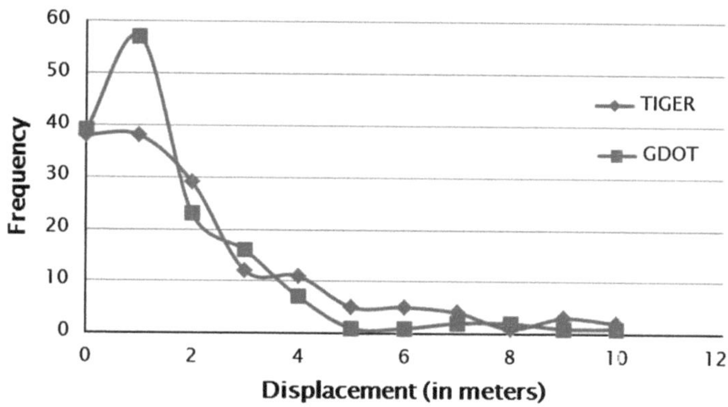

Fig. 3 Comparative displacement error distributions, accuracies truncated at 10 m

Table 2 NSSDA accuracy percentiles for displacement error distributions (in meters)

Percentile	GDOT	TIGER
99.9th	9.85	120.95
99th	8.51	22.75
95th	4.55	7.55

percentile statistic was found to just meet the TIGER standard of 7.6 m, as shown in Table 2. Therefore, the positional accuracy of the TIGER/Line dataset may be deemed of sufficient quality for road change detection.

2.1.2 Completeness Accuracy

The relative completeness accuracy assessment was performed with GDOT data as the benchmark, based on the assumption that local knowledge should better inform accurate road locations and be more complete. The data were evaluated for both errors of omission (type II error) and commission (type I). To determine errors of omission, a relative completeness accuracy assessment was performed using the concept of epsilon bands (Perkal 1966), in which the TIGER/Line dataset was buffered with its inherent error. The buffer size was set to 25 m, a value slightly above the 99th percentile of the TIGER/Line error distribution. The overall completeness was calculated as the GDOT length intersecting the TIGER/Line buffer divided by the total GDOT length. Table 3 reports the errors of omission for each year of available data.

To determine errors of commission, the buffer intersection process was reversed. The GDOT dataset was buffered by 10 m, representing the 100th percentile of the sample error distribution. Then the total TIGER/Line length was divided by the TIGER/Line length intersecting the GDOT buffer, and 1 was subtracted from the quotient. This is reported in the column of Initial Error in Table 4. However, a

Table 3 Errors of omission by year

Year	Omission error (%)	TIGER completeness (%)
2008	4.24	95.76
2009	9.29	90.71
2010	4.71	95.29
2011	3.51	96.49
2012	4.10	95.90
Average	5.29	94.71

Table 4 Errors of commission, corrected for positional error and temporal lag

Year	Initial error (%)	Meta-error (%)	True commission error (%)
2008	34.42	60.00	13.77
2009	19.65	70.00	5.90
2010	22.49	46.67	12.00
2011	46.61	58.89	19.16
2012	14.46	58.89	5.95
Total	25.46	58.89	10.47

closer investigation found that a considerable percentage of the commission errors was incorrectly selected, which we call false commission error or meta-error. The false commission errors encompassed attribute, positional, and temporal errors, all of which may easily masquerade as commission error in a buffer intersection-based GIS analysis. For instance, attribute errors occurred in instances where driveways or service roads were incorrectly characterized as local roads in the original TIGER/Line files. The latter two types of positional and temporal error unequivocally distorted the commission error, though, and were removed from the final commission error estimates by sampling the error distributions from 2008 to 2010, given that temporal accuracies could not be verified for 2011 and 2012 using the 2010 imagery. A stratified random sampling of these three years yielded 90 sample points of the commission errors; each of these samples was characterized by either true or false error, and the percentage of false error was removed from the original commission error estimates. The average false error was applied to the years 2011 and 2012 to facilitate interpretation of the dataset as a whole. The final true commission error, simply defined as roads present in the dataset where none should exist, decreased to around 10 %.

2.1.3 Temporal Accuracy

Due to the significant influence of temporal lag on the errors of commission, the TIGER/Line and GDOT datasets were both tested for temporal accuracy. The inaccuracy of the initial commission estimates was sufficient cause to question the use of the GDOT dataset as an accurate temporal reference for assessing the quality of the TIGER/Line database. For both road representations, there is a distinct possibility that certain errors of omission may not be caught until the next release of the dataset, resulting in a dataset exhibiting temporal lag; it may be extremely difficult for data providers to ensure that their data is temporally accurate at the moment of distribution, especially for rapidly growing areas such as the Atlanta MSA (Epstein et al. 2002). A similar phenomenon of spatiotemporal mismatch in national census data has already been documented in the neighboring Canadian context (Schuurman et al. 2006). To address these concerns and determine whether temporal errors are more prevalent in GDOT data, the previous 90-point stratified error sample was re-examined, eliminating positional error from consideration and specifically examining temporal errors. Once the percentage of commission error due to temporal lag was verified, this mismatch value was multiplied by the total commissioned length for the TIGER/Line dataset, which is reported in Table 5. Likewise, the procedure was reversed for the GDOT dataset, testing it against the TIGER/Line dataset. Table 6 shows the result for GDOT data. With a cross-examination of Tables 5 and 6, it was evident that the rates of temporal errors are similar for the two datasets. This implies that there were many more instances of temporal mismatch in the GDOT dataset, as the effects were multiplied by the sheer volume of GDOT data that had not been updated to the improved TIGER/Line temporal accuracy. This finding was unexpected since the GDOT data exhibited significantly better measures of positional accuracy. After the qualitative evaluation and quantitative assessment of various accuracy measures, especially with respect to temporal accuracy, the TIGER/Line data were deemed most applicable for the nationwide spatiotemporal database.

Table 5 Temporal error in TIGER commissioned length (km)

Year	Mismatch (%)	Commissioned length	Temporal error
2008	33.33	2456	819
2009	50.00	1396	698
2010	23.33	991	231
Total		4843	1748

Table 6 Temporal error in GDOT commissioned length (km)

Year	Mismatch (%)	Commissioned length	Temporal error
2008	23.33	321	75
2009	53.33	740	395
2010	26.67	226	60
Total		1287	530

2.2 Database Creation and Change Detection

TIGER/Line data for the entire United States from 2008 to 2012 were downloaded from the U.S. Census Bureau using batch downloading software. After saving the county shapefiles in separate temporal folders, a prototypical "snapshot" spatio-temporal database was produced, in which each geographic area has different data files corresponding to discrete time periods (Langran 1992). However, this form of spatiotemporal representation has two key disadvantages: it does not explicitly demonstrate dynamic changes over time and duplicates elements that have not changed over time, exacerbating data storage problems (Pelekis et al. 2004). The conversion of this extensive static database into a dynamic database requires the accurate identification of changes between each of the annual time "snapshots," ultimately creating a set of four change files (2008–2009, 2009–2010, 2010–2011, and 2011–2012) that capture new additions to the road network (Armenakis 1992). While this process appears to be quite straightforward and simple to implement, the inherent positional error of the TIGER/Line database has introduced significant difficulties for the static to dynamic conversion process.

In order to perform the conversion for all counties in the United States, an analytical geoprocessing framework was created with Python scripts to operate in the ESRI ArcMap environment. By default, relevant data are accessed using a predefined file storage hierarchy. The heart of the programming framework is the toolbox that contains sixteen Python scripts to enable automated geoprocessing (using the arcpy module to interface with ArcGIS) and workspace management (using the os module to interface with the file system), as displayed in Fig. 4. The scripts were separated into toolsets at the levels of individual road segment, county, and tract. The first group handles data preprocessing and change detection, while the following two groupings assist with the aggregation of road network change to the county and tract level, respectively. Graphical user interfaces allow the user to pick the starting and ending years of analysis (for example, Find New Roads between 2008 and 2012) and the geographic area of concern (either the entire U.S. or by an individual state). Importantly, for each county within the area of interest, the first year (2008) is buffered by a specified error tolerance, creating epsilon bands. Here a value of 25 m was used, a conservative estimate of the 99th percentile radial error of the TIGER dataset. Errors introduced by the reference imagery (±3.67 m with a 95 % confidence interval) were quantified by assuming a two-dimensional Gaussian probability distribution; due to error propagation, the reported accuracy of the 25 m buffers decreased to 95.83 %. These error buffers around the 2008 roads are then erased from the most recent (2012) roads, yielding only those parts of the road network unique to the 2012 roads dataset. Given that the vintage 2012 dataset is simply an updated version of the 2008 roads, we may

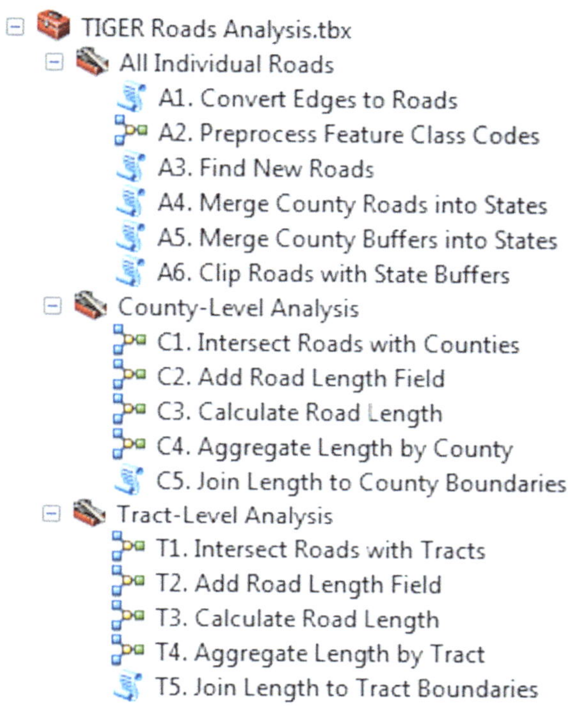

Fig. 4 Toolbox for change detection and summarization

assume that these datasets are already co-referenced, and that 95.83 % of new road length derived from the buffer erasure represents real road change.

The Find New Roads analysis would seem to be fairly straightforward, but the datasets are so large that they exceed the memory limitations presented by ArcMap and exacerbate pre-existing memory leaks in many of the geoprocessing tools. Therefore, the Find New Roads tool was split into three different hierarchical scripts (national, state, and county), allowing the lower-level scripts to terminate once their jobs are finished; this structure clears up more system memory and enables the script to run longer without hitting a runtime error. The national-level script translates the selected state name into a two-digit TIGER/Line code to dynamically search for the correct source files in the right locations; the state-level script manages the output folders for all the buffer, erase, and merge operations; and the county-level script performs the buffer and erase geoprocessing operations for each individual county road file, then terminates. After completing the Find New Roads script, the change files are placed in an output geodatabase folder, each uniquely identified by the state, from and to dates, and the buffer tolerance used. Once all scripts in the toolbox have been run in order, the entire national road network change geodatabase is populated with length statistics and joined to the appropriate unit of analysis to create density statistics.

3 Results and Findings

By summarizing the road change database, it was discovered that 289,021 km of new roadway have been constructed in the United States between 2008 and 2012, with each state on average experiencing an increase of 2.7 %. Four states, including North Dakota, Iowa, Kansas, and Nebraska, and the District of Columbia (0.64 %) experienced a growth rate of below 1 %, while four other states in the contiguous United States, including Connecticut, Maine, Delaware, and Georgia, experienced a growth rate of above 5 %. Figure 5 depicts the general spatial distribution of the changes in road network length by state.

At the MSA level, Table 7 lists the top 25 MSAs for road network growth between 2008 and 2012. The greatest impacts are exhibited in large metropolitan areas where population increases are small compared to the normalized road network change. The geographic distributions of new roads at the MSA level can be observed in Fig. 6, where both percent increase in length and the density of the new road additions are displayed. While the percent increase method of normalization tends to inflate the importance of MSAs with few total roads, the new road segment density measure is more sensitive for MSAs with a small total land area.

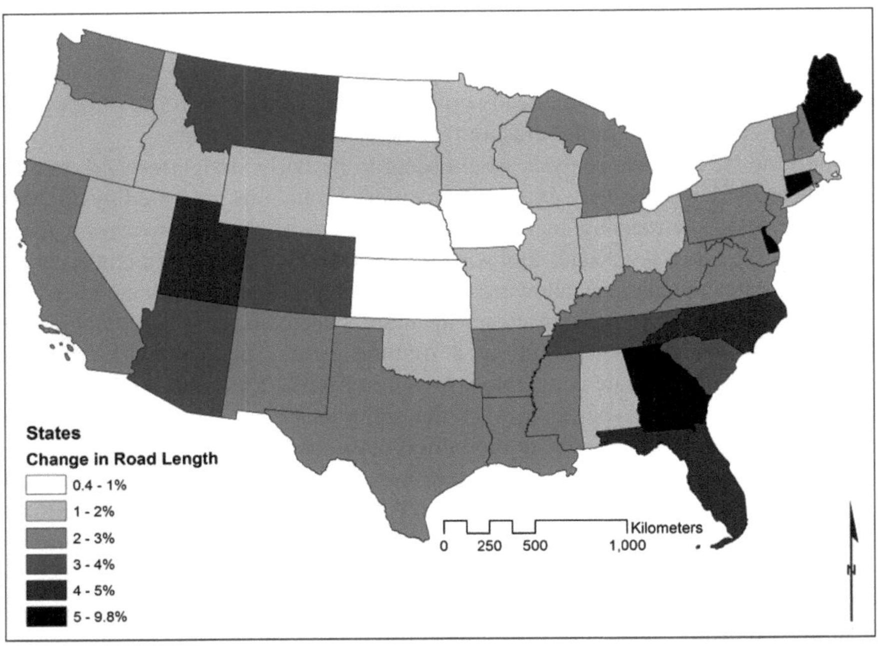

Fig. 5 Percent increase in road network length by state

Table 7 Top twenty-five MSAs for road network growth in the United States (2008–2012)

Metropolitan statistical area	%Δ Roads	%Δ Population	Population (2012)
Myrtle Beach-N. Myrtle Beach-Conway, SC	20.57	8.22	282,024
Lake Havasu City-Kingman, AZ	16.17	1.53	203,142
Gainesville, GA	14.34	4.39	185,055
St. George, UT	13.86	6.72	144,656
Atlanta-Sandy Springs-Marietta, GA	11.67	1.01	5,440,004
Dover, DE	11.19	6.09	167,540
Orlando-Kissimmee-Sanford, FL	10.65	7.89	2,223,674
McAllen-Edinburg-Mission, TX	10.57	9.40	805,975
Savannah, GA	10.08	8.18	361,941
Bridgeport-Stamford-Norwalk, CT	9.33	3.31	933,733
Jacksonville, NC	9.29	8.84	184,001
Pensacola-Ferry Pass-Brent, FL	8.75	1.79	461,227
Bangor, ME	8.74	0.09	153,509
Olympia, WA	8.26	5.86	258,653
Brunswick, GA	8.24	10.12	113,448
Athens-Clarke County, GA	8.12	3.42	196,425
Austin-Round Rock-San Marcos, TX	8.01	10.89	1,834,303
Dalton, GA	7.65	6.30	142,751
Warner Robins, GA	7.59	8.26	146,067
Houston-Sugar Land-Baytown, TX	7.59	8.32	6,177,035
Naples-Marco Island, FL	7.26	5.02	332,528
San Jose-Sunnyvale-Santa Clara, CA	7.15	4.63	1,894,388
Burlington, NC	7.13	4.04	153,672
Las Vegas-Paradise, NV	7.01	4.46	1,997,659
Hartford-West Hartford-East Hartford, CT	6.88	1.95	1,214,400

At the county level, Fig. 7 displays the spatial distribution of the new road additions; several clustered areas of spatial autocorrelation can be noted. Using the percent change normalization method, a Moran's I statistic of 0.43 was found to be statistically significant at a significance level of 0.0×10^{-6}. This moderately positive spatial autocorrelation value was further explored using local Moran's I statistics—several significant clusters were found at a significance level of 0.01. Figure 8 breaks down each county or equivalent by z-score to indicate the relative strength of spatial clustering. The strongest cluster of road network growth was

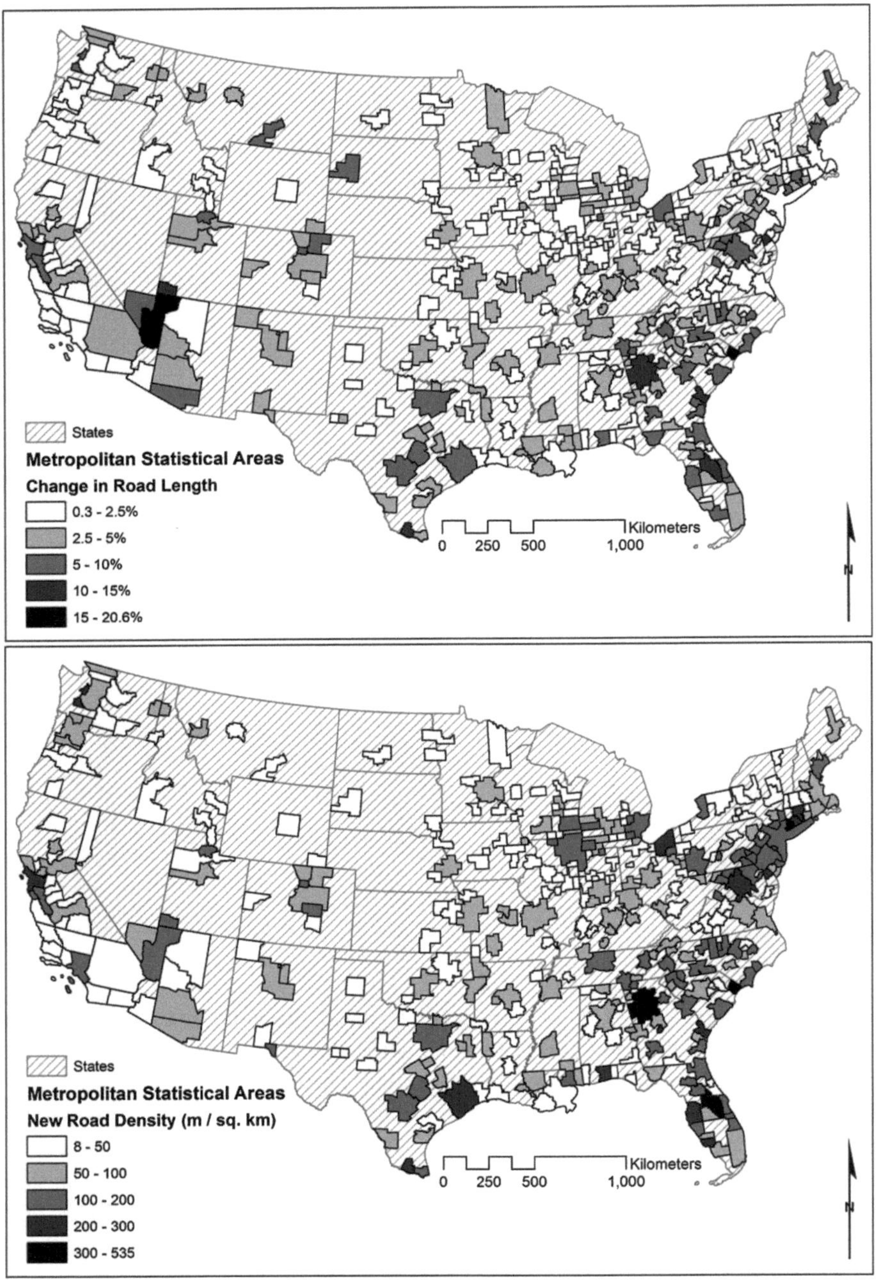

Fig. 6 Percent increase in road length and new road density by MSA

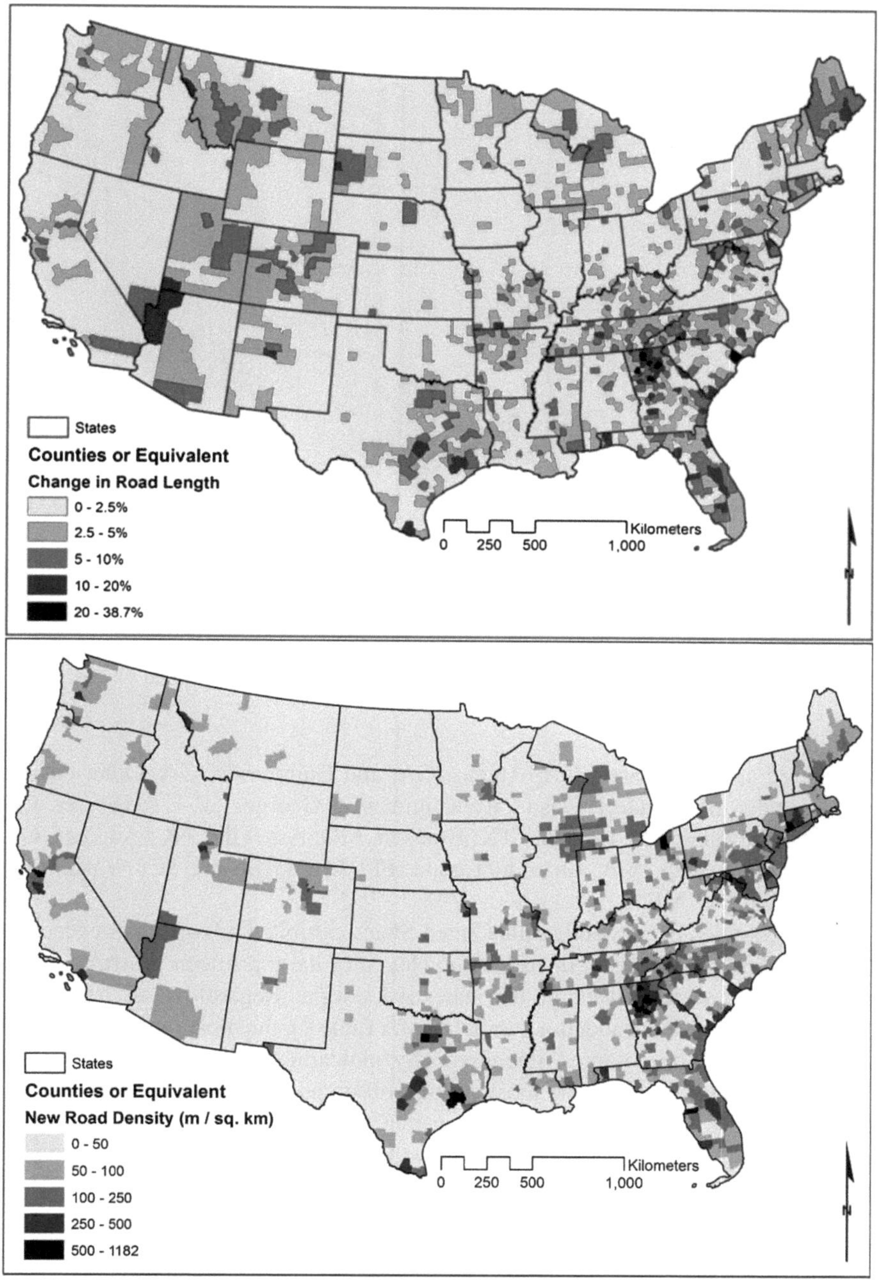

Fig. 7 Percent increase in road length and new road density by county or equivalent

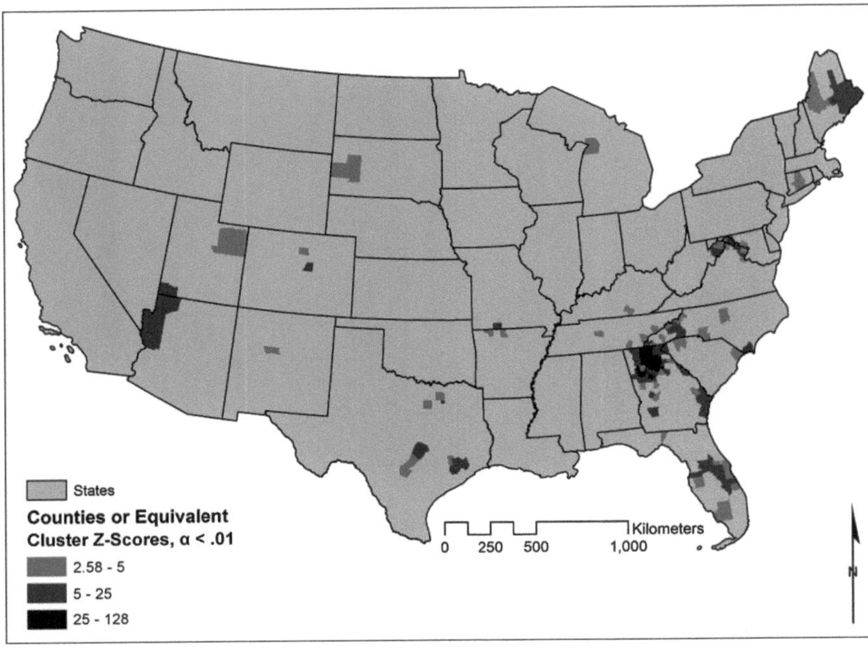

Fig. 8 Statistically significant clusters of road network growth and local Moran's I Z-scores

discovered around Atlanta, GA, Athens, GA, and Gainesville, GA. Other clusters were revealed near Lake Havasu City/Kingman/St. George, AZ-UT; Denver, CO; Austin, TX; Dallas, TX; Houston, TX; Branson, MO; Asheville, NC; Augusta, GA; Savannah/Brunswick, GA; Orlando/Tampa, FL; Myrtle Beach, SC; Washington, DC; and Bangor, ME.

Three counties in the contiguous United States exhibited growth rates exceeding 25 %, as shown in Table 8. Interestingly, eleven of the top fifteen growth counties by percentage were located in the state of Georgia. Regardless of the method chosen to normalize the added network length (Table 9), the most rapid road growth occurred in Georgia counties, unequivocally indicating an agglomeration around the city of Atlanta. Perhaps the most striking observation was that Forsyth County, GA placed first using both normalization methods; future research should further investigate this locus of road network growth.

Table 8 Top twenty-five counties for new road length, normalized by percent change (2008–2012)

County or Equivalent	Percent change in roads	%Δ Population	Population (2012)
Forsyth County, GA	38.65	12.98	187,927
Barrow County, GA	29.72	4.16	70,165
Gilmer County, GA	28.68	0.00	28,212
Henry County, GA	24.08	7.17	208,622
Paulding County, GA	23.55	5.73	144,920
Jackson County, GA	22.89	1.81	60,497
Polk County, NC	21.36	−0.60	41,160
Jefferson County, WV	21.15	4.55	54,558
Horry County, SC	20.57	8.22	282,024
Pickens County, GA	20.40	−0.56	29,365
Union County, GA	19.98	0.70	21,434
Douglas County, GA	19.35	3.44	133,957
Gwinnett County, GA	19.21	7.67	840,575
Forest County, PA	18.29	2.53	7659
Effingham County, GA	18.01	5.17	53,359
Rockwall County, TX	17.56	12.31	83,028
Cherokee County, GA	17.54	6.10	220,973
Columbia County, GA	17.40	11.96	131,563
Walton County, GA	17.07	2.42	84,598
Pasco County, FL	17.06	2.02	470,631
Oconee County, GA	16.74	5.77	33,525
Osceola County, FL	16.72	10.00	287,924
Lee County, GA	16.67	2.17	28,720
Mineral County, MT[a]	16.37	−0.34	4163
Fort Bend County, TX	16.30	15.27	625,853

[a]A majority of new roads were within national forest boundaries

4 Conclusions

This study constructed a comprehensive spatiotemporal road change database and developed a methodological framework for automated data extraction and processing. The source data were carefully chosen from the public domain for semantic accuracy, and positional and temporal accuracy assessments were performed. The quality of TIGER/Line data remains a major concern; however, we have confidence that the positional, temporal, and attribute accuracy will steadily improve.

The construction of the database allowed aggregate descriptive statistics to be derived for each state, MSA, county, and census tract nationwide. Rankings and visualizations have revealed significant road network expansion from 2008 to 2012,

Table 9 Top twenty-five counties for new road length, normalized by density of New Network additions (2008–2012)

County or Equivalent	New road density (m/km^2)	%Δ Population	Population (2012)
Forsyth County, GA	1181.99	12.98	187,927
Gwinnett County, GA	1012.61	7.67	840,575
Broomfield County, CO	944.13	9.65	58,311
Barrow County, GA	757.30	4.16	70,165
Henry County, GA	719.02	7.17	208,622
Manassas Park City, VA	709.86	24.54	15,871
Clayton County, GA	662.29	1.42	265,810
Tarrant County, TX	652.74	7.78	1,881,445
Douglas County, GA	601.95	3.44	133,957
St. Louis City, MO[a]	600.03	0.36	319,112
Harris County, TX	598.72	8.01	4,253,963
Rockwall County, TX	574.56	12.31	83,028
Cuyahoga County, OH[a]	544.59	-1.97	1,266,049
Fulton County, GA	541.34	9.95	977,129
Fairfield County, CT	535.14	3.31	933,733
Paulding County, GA	524.42	5.73	144,920
Contra Costa County, CA	523.96	5.37	1,078,257
Fort Bend County, TX	521.25	15.27	625,853
Gilmer County, GA	504.83	0.00	28,212
Pasco County, FL	502.59	2.02	470,631
Alexandria City, VA	501.74	10.45	146,839
Jefferson County, WV	500.04	4.55	54,558
Clarke County, GA	482.42	3.92	120,310
Prince William County, VA	476.06	14.76	430,100
Fairfax County, VA	475.21	7.14	1,118,683

[a]A majority of new roads were due to gross positional inaccuracies in the interstate highway system from 2008

with the largest county road growth rate exceeding 38 %. Many new roads were clustered in the southeastern U.S., particularly near Atlanta, Georgia.

Continued research should be conducted on the causes and effects of road network expansion, as it is a phenomenon that is still debated in transportation and related literature. For instance, a better understanding of the fine-scale interactions between race and road development, especially for low-density, "sprawling" residential areas, would be beneficial for urban social geographers. Spatial and temporal lag models, along with added information on human behavior in metropolitan real estate markets, may be useful in further explanatory analyses.

References

Armenakis C (1992) Estimation and organization of spatio-temporal data. In: Proc Can Conf GIS 1992, Ottawa, Canada
Bullard RD (2003) Addressing urban transportation equity in the United States. Fordham Urban LJ 31(5):1183
Ciscel DH (2001) The economics of urban sprawl: inefficiency as a core feature of metropolitan growth. J Econ Issues 35(2):405–413
Clarke KC, Gaydos LJ (1998) Loose-coupling a cellular automaton model and GIS: long-term urban growth prediction for San Francisco and Washington/Baltimore. Int J Geogr Inf Sci 12 (7):699–714
Epstein J, Payne K, Kramer E (2002) Techniques for mapping suburban sprawl. Photogramm Eng Remote Sens 63(9):913–918
Feyrer J (2009) Trade and Income: exploiting time series in geography. No. w14910. National Bureau of Economic Research
Forman RTT, Alexander LE (1998) Roads and their major ecological effects. Annu Rev Ecol Syst 29:207–231
Gregory IN, Healey RG (2007) Historical GIS: Structuring, mapping and analyzing geographies of the past. Prog Hum Geogr 31(5):638–653
Ham H, Kim TJ, Boyce D (2005) Assessment of economic impacts from unexpected events with an interregional commodity flow and multimodal transportation network model. Transp Res Part A: Policy Pract 39(10):849–860
Healey RG, Stamp TR (2000) Historical GIS as a foundation for the analysis of regional economic growth: theoretical, methodological, and practical issues. Soc Sci Hist 24(3):575–612
Henderson J (2004) The politics of mobility and business elites in Atlanta Georgia. Urb Geogr 25 (3):193–216
Hu Z, Lo CP (2007) Modeling urban growth in Atlanta using logistic regression. Comput Environ Urb Syst 31(6):667–688
Jaret C, Adelman RM, Reid LW (2006) Suburban sprawl, racial segregation, and spatial mismatch in metropolitan America. Sociation Today 4(2)
Kenworthy JR, Laube FB (1999) Patterns of automobile dependence in cities: an international overview of key physical and economic dimensions with some implications for urban policy. Transp Res Part A: Policy Pract 33(7):691–723
Langran G (1992) Time in geographic information systems. Taylor & Francis, Bristol PA
Levinson D (2002) Identifying winners and losers in transportation. Transp Res Rec: J Transp Res Board 1812:179–185
Levinson D, Karamalaputi R (2003) Induced supply: a model of highway network expansion at the microscopic level. J Transp Econ Policy 37(3):297–318
Moore GE (1998) Cramming more components onto integrated circuits. Proc IEEE 86(1):82–85
Pelekis N, Theodoulidis B, Kopanakis I et al (2004) Literature review of spatio-temporal database models. Knowl Eng Rev 19(03):235–274
Perkal J (1966) On the length of empirical curves. Mich Interuniv Community Math Geogr 10:17–20
Riitters KH, Wickham JD (2003) How far to the nearest road? Front Ecol Environ 1(3):125–129
Sanchez TW, Stolz R, Ma JS (2003) Moving to equity: addressing inequitable effects of transportation policies on minorities. Civil Rights Project at Harvard University
Schuurman N, Grund D, Hayes M et al (2006) Spatial/temporal mismatch: a conflation protocol for Canada Census spatial files. Can Geogr-Geogr Can 50(1):74–84
Weber J, Sultana S (2007) Journey-to-work patterns in the age of sprawl: evidence from two midsize southern metropolitan areas. Prof Geogr 59(2):193–208
Xie F, Levinson D (2009) Modeling the growth of transportation networks: a comprehensive review. Netw Spat Econ 9(3):291–307

Revising Self-Best-Fit Strategy for Stroke Generating

Jing Tian, Fuquan Xiong, Yingzhe Lei and Yifei Zhan

Abstract The strokes in road networks refer to a set of connected and non-branching road segments that follow the principle of good continuity. Generating strokes plays an important role in road network generalization, topological analysis, pattern recognition, and schematic map generation. In this study, the self-best-fit strategy for generating strokes was improved by prescribing road segment processing order based on the importance of the road segments. The importance of the road segments was determined by four parameters: length, degree, closeness and betweenness. The road networks of Detroit and Birmingham were used for experiments. Different stroke generating strategies were compared in terms of network functionality and visual recognition. In terms of network functionality, the improved self-best-fit strategy is superior to the every-best-fit strategy, and in terms of averages, it is superior to the self-best-fit strategy as well as the self-fit strategy. From a visual recognition perspective, the improved self-best-fit strategy tends to generate longer strokes with global property compared to the every-best-fit strategy.

Keywords Road network · Stroke · Self-best-fit · Global efficiency

1 Introduction

The strokes in road networks, also known as self-organized natural roads, refer to connected and non-branching road segments that follow the principle of good continuity (Thomson and Richardson 1999; Jiang et al. 2008; Zhou and Li 2012).

J. Tian
Key Laboratory of Geographic Information System, Ministry of Education,
129 Luoyu Road, 430079 Wuhan, China
e-mail: yutaka-2010@163.com

J. Tian · F. Xiong · Y. Lei (✉) · Y. Zhan
School of Resource and Environment Science, Wuhan University,
129 Luoyu Road, 430079 Wuhan, China
e-mail: 2012301130022@whu.edu.cn

F. Harvey and Y. Leung (eds.), *Advances in Spatial Data Handling and Analysis*, Advances in Geographic Information Science,
DOI 10.1007/978-3-319-19950-4_11

Since its establishment, the idea of strokes has had significant influence on road network studies. Generating strokes has become the primary topic of research such as road network generalization (Edwardes and Mackaness 2000; Chaudhry and Mackaness 2001; Zhang 2004; Tomko et al. 2008; Chen et al. 2009; Liu et al. 2009; Touya 2010; Yang et al. 2011), topological analysis (Jiang and Claramunt 2004; Porta et al. 2006; Jiang 2007), schematic map generation (Li and Dong 2010), and pattern recognition (Heinzle and Ander 2007). To generate a stroke, the key step is to establish concatenation rules and generation strategies. Concatenation rules, including geometric rules, thematic rules, and hybrid rules, determine whether two road segments can be connected (Zhou and Li 2012). Geometric rules derive from the Gestalt principle of good continuity, and they usually call for the deflection angle of the two road segments to be smaller than the threshold. Thematic rules typically apply to two roads which have the same name or class. Hybrid rules consider both geometric and thematic rules. Generation strategies include every-best-fit (EBF), self-best-fit (SBF), and self-fit (SF) strategies (Jiang et al. 2008). The EBF strategy compares pairs of road segments and selects the pairs in which the two road segments best fit each other for concatenation. For example, if the deflection angle of segment *a* is the smallest with segment *b*, while segment *b* also has the smallest deflection angle with segment *a*, then segment *a* and *b* will be concatenated. The SBF strategy begins with a starting road segment and selects a neighboring road segment that best fits the first one for concatenation, without considering if the starting road segment is the best fit for the selected segment. For instance, at the same intersection, segment *a* has the smallest deflection angle with *b*, and segment *b* has the smallest deflection angle with segment *c*, while segment *c* also has the smallest deflection angle with segment *b*. In such situations, if the concatenation starts with segment *a*, then segments *a* and *b* will be concatenated rather than segments *b* and *c*. The SF strategy also begins with a starting road segment, but randomly selects a neighboring road segment as long as the concatenation criterion is met. Theoretically, the stroke sets generated through the EBF strategy are definite. Based on the correlation between network metrics and traffic flow, Jiang et al. (2008) ascertained that the SBF strategy is the optimal strategy, as it better presents self-organized natural roads. However, it can be difficult to apply this strategy in practice. For instance, the generalization and analysis of road networks are based on definite stroke sets, while the SBF strategy generates indefinite stroke sets. With a single stroke generating process, the generated stroke set cannot be judged. In addition, if none of the generated results is ideal, multiple runs will be needed until an ideal one is found. In the work of Jiang et al. the average of the results from 20 runs was used. This impedes the application of the SBF strategy.

In this study, attempts were made to modify the SBF strategy by setting the starting road segment and the new starting segment after a stroke is built according to the importance of the road segments.

This paper is organized as follows: Sect. 2 introduces the SBF strategy and the specific approach used to modify it. Section 3 compares the stroke generating results of the improved SBF strategy with other strategies in terms of network functionality and visual cognition. Section 4 presents the conclusion.

2 Modification Approach

2.1 The Self-Best-Fit Strategy

The SBF strategy starts with a random road segment *R* and searches for its best-fit road segment *R_best* (the road segment with which the concatenation rules are satisfied to the greatest extent, e.g., having the smallest deflection angle), and then search for a best-fit segment for *R_best*. After generating a stroke, a road segment is randomly chosen from the unprocessed road segments and the above process is repeated, until all segments have been processed.

The strategy is detailed as follows with geometric rules applied (i.e. only the deflection angles between segments are considered). For a road network as shown in Fig. 1a, assuming road segment *R*1 as the starting segment, then *R*1 and *R*5 are concatenated, forming a stroke *S*1. Then, road segment *R*2 is chosen as the starting segment, and *R*2 and *R*3 are concatenated into stroke *S*2. Figure 1b illustrates the two strokes. If *R*5 is assumed to be the starting road segment, *R*5 and *R*1 are concatenated into *S*1, and the new starting segment *R*4, along with *R*2, are concatenated into *S*2, as shown in Fig. 1c.

Obviously, the diversity of stroke sets result from the different choices of the starting road segment, as well as the choices for the new starting road segment after a stroke is generated.

2.2 Modification

To improve the SBF strategy, this study tried to determine a starting segment and the new starting segment after a stroke is generated. The starting segment is determined based on the importance of road segments. There are four methods to identify the importance of road segments, including: (1) Importance by Rule, which

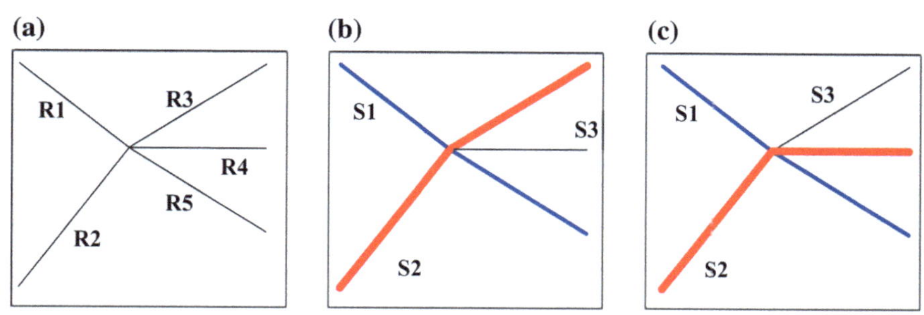

Fig. 1 Illustration of SBF strategy

means to define the priorities of attribute parameters to determine the importance of road segments (Chen et al. 2009); (2) Importance by Weight, which means to weight the attribute parameters, and the weighted sum of a road segment's parameters is its importance (Yang et al. 2011); (3) Importance by Parameter Transformation, which means to transform or integrate one or more parameters to generate new parameters for describing the importance of road segments (Tomko et al. 2008); (4) Importance by Impact, which means to remove a road segment from the road network and determine its importance based on its impact on the network (Touya 2010). In this study, the weighting method by Yang et al. (2011) was chosen to determine the importance of road segments. With this method, the parameters of a road segment, including length, degree, closeness, and betweenness, are integrated into a single parameter, *Level*. The parameters are dynamically weighted using the CRITIC method (Diakoulakli et al. 1995) in different road networks. The parameters are based on the dual graph, which turns the road segments into nodes and the intersections into edges.

Length is the length of the road segment. *Degree* measures the number of segments that connect to the given segment. *Closeness* measures the shortest distance from the given segment to all other segments, which is defined by Eq. 1.

$$C_j^C = \frac{1}{\sum_{j=1, j \neq i}^{n} d_{ij}} \tag{1}$$

where n is the number of road segments in road networks and d_{ij} the shortest path length from i to j.

Betweenness measures the mediator effect of the given road segment in a road network, which is defined by Eq. 2.

$$C_i^B = \sum_{j \neq k \neq i}^{n} \frac{n_{jk}(i)}{n_{jk}} \tag{2}$$

where n_{jk} is the total number of shortest paths from j to k, and $n_{jk}(i)$ is the number of those paths that pass through i.

Before the integration, weights of the four parameters are calculated according to specific functions of the CRITIC method (Diakoulakli et al. 1995), which relate to the variance and correlation coefficient of each parameter.

After the importance of the road segments are determined, the improved strategy chooses the most important road segment as the starting segment, search for its best-fit segment to generate a stroke, and then chooses the most important road segment in the unprocessed segments as the new starting segment to generate a new stroke. This is repeated until all road segments have been processed.

3 Experiments

3.1 Data and Preprocessing

The road networks of Detroit and Birmingham were downloaded from OpenStreetMap as the input of the experiments. As this study aimed to improve the SBF strategy, it focused on single lines, thus patterns such as dual carriageways, subsidiary facilities of roads, and complex road junctions were simplified. Stroke generation for dual carriageways can be found in the study of Yang et al. (2011). In terms of concatenation rules of road segments, only geometric rules were adopted in this paper. The first reason for this is that the study of Zhou and Li (2012) shows that with a combination of geometric concatenation rules and the EBF strategy, the generated strokes will lead to better map generalization results. Although adding thematic rules to concatenation rules may improve the accuracy of the results, the improvement is not statistically significant. The other reason for using geometric rules in this study is that only geometric rules were considered in the research of Jiang et al. (2008). Geometric rules refer to the reflection angles between road segments with a threshold of 60°, an empirical value concluded in the study of Jiang and Liu (2009). Admittedly, using other concatenation rules is also simple.

3.2 Experiment Design

Zhou and Li (2012) designed an evaluation index to evaluate stroke generating approaches by counting two types of errors at road intersections. Its core idea is to evaluate stroke generating results of different strategies according to their consistence with map generalization results. However, for indefinite strategies like SBF and SF strategies, different results generated from multiple processes were not reported. Jiang et al. (2008) evaluated the strategies by assessing whether the generated strokes can well reflect the correlation of network metrics and traffic flow. For SBF and SF strategies, the stroke generating process was repeated 20 times each to determine the average for the evaluation. Meanwhile, these two evaluation methods are specific to the applications, and need special data (multi-scale data and traffic flow data). As a result, these methods are not used in this study because of the lack of data associated with them.

In this paper, the improved-SBF strategy was compared to other strategies in terms of network functionality and visual cognition.

Evaluating stroke generating results in terms of network functionality, as yet, has not been studied. This paper attempted to employ global efficiency, which is commonly used in complex network theory, to evaluate stroke generating results. Global efficiency of network, proposed by Latora and Marchiori (2001), describes the way the nodes in a network interconnect, thus reflecting the degree of smoothness of the information dissemination in the network. For a network G, the efficiency ε_{ij} between nodes *i* and *j* is inversely proportional to the shortest distance from node *i* to node *j*, i.e. $\varepsilon_{ij} = 1/d_{ij}$. The average efficiency of network *G* is defined by Eq. 3.

$$E = \frac{1}{N(N-1)} \sum_{i,j \in G, i \neq j} \varepsilon_{ij} = \frac{1}{N(N-1)} \sum_{I,J \in G, i \neq j} \frac{1}{d_{ij}} \tag{3}$$

If *Gid* is the complete graph of network *G*, the global efficiency of network *G* can be defined by Eq. 4.

$$E_{glob} = \frac{E(G)}{E(Gid)} \tag{4}$$

It was assumed that the road network is represented as a dual graph, in which road segments are represented by nodes of the network and the intersection relations of road segments are represented by network edges. In this network, the length of the shortest path between nodes *i* and *j* is the minimum number of edges connecting them. The average efficiency of the complete graph of such a network is 1, thus it is also the global efficiency of the network. The average efficiency is calculated by Eq. 2, where *N* is the total number of road segments and d_{ij} is the minimum number of steps connecting nodes *i* and *j*, i.e. the path length. The value of global efficiency ranges from 0 to 1.

Because the SBF and SF strategies generate uncertain results, repeated experiments were carried out 100 times and the average global efficiency values were calculated for comparison purposes. Experiments of improved SBF and EBF strategies were also carried out 100 times to verify whether the stroke generating results were unique.

For visual cognition, the stroke generating results of the improved SBF and EBF strategies were compared through visual inspection because both strategies usually generate unique results. Strokes that were longer and more likely to fit the human visual perception were thought to be the better ones.

3.3 Results

The results of the repeated stroke generating processes demonstrated that the improved SBF strategy usually produces unique results, for stroke generating results of 100 repeated experiments having the same global efficiency value.

Through global efficiency, the stroke generating results of four strategies for two cities were compared. In Fig. 2, the first column shows the global efficiency values of the stroke generating results of the improved strategy of this paper, while the 2nd, 3rd, and 4th columns show those of the EBF, SBF, and SF strategies. For Detroit's road network, the improved SBF strategy had a global efficiency value (0.2148) higher than the EBF strategy (0.2111). In terms of the averages, the global

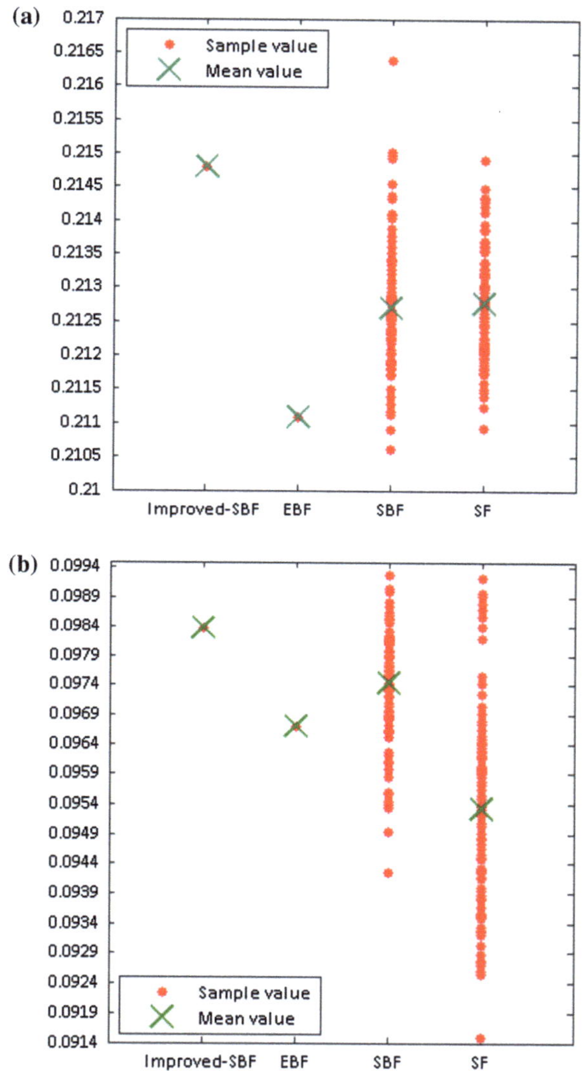

Fig. 2 Comparative analysis of four strategies based on global efficiency: **a** Detroit; **b** Birmingham

efficiency value of the improved SBF strategy (0.2148) was higher than those of SF strategy (0.2127) and SF strategy (0.2128). For single stroke generating processes, the global efficiency value of the improved SBF strategy (0.2148) was lower than that of the best results generated by SBF strategy (0.2164) and SF strategy (0.2149) in this study. For Birmingham's road network, the improved SBF strategy had a global efficiency value (0.0984) higher than the EBF strategy (0.0967). In terms of the averages, the global efficiency value of the improved SBF strategy (0.0984) was higher than those of the SBF strategy (0.0974) and SF strategy (0.0953). In terms of single stroke generating processes, the global efficiency value of the improved SBF

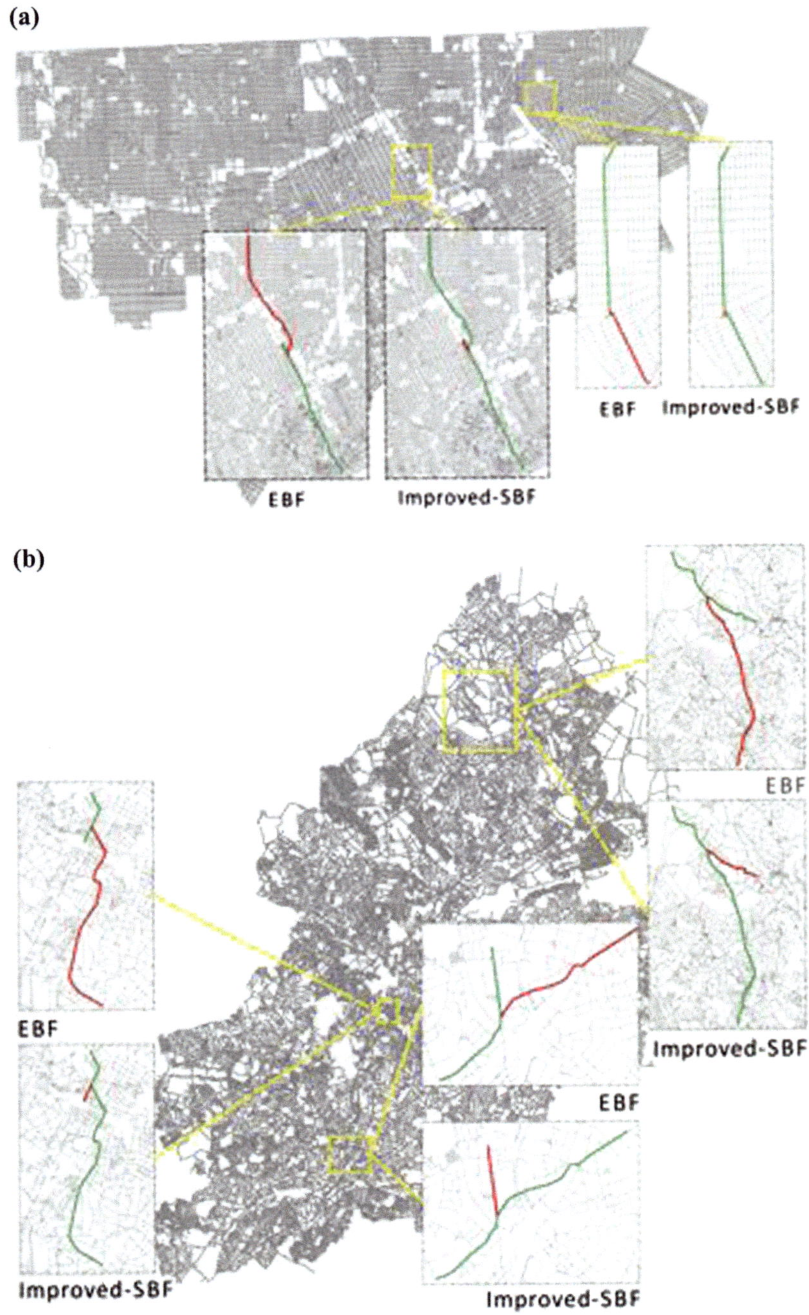

Fig. 3 Stroke generating results by the improved SBF strategy and EBF strategy: **a** Detroit; **b** Birmingham

strategy (0.0984) was lower than that of the best results generated by SBF strategy (0.0998) and the SF strategy (0.0992) in this experiment.

For visual inspection, major differences of strokes generated by the improved SBF strategy and EBF strategy are presented in Fig. 3. It can be seen that the improved SBF strategy of this paper generated longer strokes, which better comply with the cognitive rule.

Analyzing the above results, it was found that:

1. The stroke sets generated by the improved SBF and EBF strategies are usually unique, whereas those generated by the SBF and the SF strategies are not unique. This is as expected.
2. The improved SBF strategy is better than the EBF strategy, and better than the SBF strategy and the SF strategy in terms of averages from the perspective of network functionality. However, for single stroke generating processes, the improved SBF strategy is inferior to the optimum results of the SBF and SF strategies.
3. The improved-SBF strategy generates strokes that better fit the human visual perception rule, which is longer with global property compared with the EBF strategy.

4 Conclusion

In this study, the SBF strategy for generating strokes of road networks was improved. The basic idea was to prescribe the processing order of road segments based on their importance. This method can be further generalized; segments with the largest value of a certain attribute parameter can be chosen for segment concatenation, and this approach reflects the idea of ‘best fit’ as well.

The contribution of the study is two-fold: (1) A modification of the SBF strategy was proposed that generates better stroke results in terms of network functionality and visual cognition; (2) A new method to evaluate stroke generating results was also proposed.

Further research may focus on the following two aspects: (1) Due to the fact that generating strokes is usually the first step in analyzing and generalizing road networks, empirical studies should be carried out in road network generalization and analysis for the improvements made in this study. (2) Generating strokes is a typical combination optimization problem. The improvement made in this study is actually a heuristic method which made use of optimization theories, and approaches to solving this issue should be the focus of future studies.

Acknowledgments The authors express their special thanks to three anonymous reviewers for their valuable comments. The study is supported by ‘National Science Foundation for Fostering Talents in Basic Research of the National Natural Science Foundation of China (Grant No. J1103409)’ and ‘Innovation and Entrepreneurship Training Project for College Students of Wuhan University (Grant No. S2014446)’.

References

Chaudhry O, Mackaness W (2001) Rural and urban road network generalization deriving 1:250000 from OS mastermap. Available via DIALOG. http://www.era.lib.ed.ac.uk/bitstream/1842/1137/1/ochaudry001.pdf

Chen J, Hu YG, Li ZL, Zhao RL, Meng LQ (2009) Selective omission of road features based on mesh density for automatic map generalization. Int J Geogr Inf Sci 23(8):1013–1032

Diakoulakli D, Mavrotas G, Papayannakis L (1995) Determining objective weights in multiple criteria problems: The critic method. Comput Oper Res 22(7):763–770

Edwardes AJ, Mackaness WA (2000) Intelligent generalization of urban road networks. In: Proceedings of GIS research UK conference, York

Heinzle F, Ander KH (2007) Characterising space via pattern recognition techniques: identifying patterns in road networks. In: Mackaness WA, Ruas A, Sarjakoski LT (eds) Generalisation of geographic information: cartographic modelling and applications. Elsevier, Amsterdam, pp 233–253

Jiang B (2007) A topological pattern of urban street networks: universality and peculiarity. Phys A 384:647–655

Jiang B, Claramunt C (2004) Topological analysis of urban street networks. Environ Plann B 31:151–162

Jiang B, Liu C (2009) Street-based topological representations and analyses for predicting traffic flow in GIS. Int J Geogr Inf Sci 23(9):1119–1137

Jiang B, Zhao SJ, Yin JJ (2008) Self-organized natural roads for predicting traffic flow: a sensitivity study. J Stat Mech-Theory Exp P07008

Latora V, Marchiori M (2001) Efficient behavior of small-world networks. Phys Rev Lett 87 (19):198701-1

Liu XJ, Zhan BF, Ai TH (2009) Road selection based on Voronoi diagrams and "strokes" in map generalization. Int J Appl Earth Obs 12:S194–S202

Li ZL, Dong WH (2010) A stroke-based method for automated generation of schematic network maps. Int J Geogr Inf Sci 24(11):1631–1647

Porta S, Crucitti P, Latora V (2006) The network analysis of urban streets: a dual approach. Phys A 369:853–866

Thomson RC, Richardson DE (1999) The "good continuation" principle of perceptual organization applied to the generalization of road networks. In: Proceedings of 19th international cartographic conference, Ottawa

Tomko M, Winter S, Claramunt C (2008) Experiential hierarchies of streets. Comput Environ Urban 32(1):41–52

Touya G (2010) A road network selection process based on data enrichment and structure detection. Trans GIS 14(5):595–614

Yang BS, Luan XC, Li QQ (2011) Generating hierarchical strokes from urban street networks based on spatial pattern recognition. Int J Geogr Inf Sci 25(12):2025–2050

Zhang QN (2004) Road network generalization based on connection analysis. In: Proceedings of 11th international symposium on spatial data handling. Leicester, UK

Zhou Q, Li ZL (2012) A comparative study of various strategies to concatenate road segment into strokes for map generalization. Int J Geogr Inf Sci 26(4):691–715

A Spatio-Temporal Decision Support Framework for Large Scale Logistics Distribution in the Metropolitan Area

Wei Tu, Qingquan Li, Xiaomeng Chang, Yang Yue and Jiasong Zhu

Abstract Rapid growing urbanization and explosive e-business expect effective logistics distribution service in the metropolitan area. Because of traffic control, commuting peak and unpredictable traffic accidents, traffic states in the metropolitan area fluctuate sharply, leading to the unacceptable logistics service delay in our daily life. To overcome this problem, a spatio-temporal decision support (STDS) framework is developed to facilitate large scale logistics distribution in the metropolitan area. It consists of a traffic information database, a spatio-temporal heuristic algorithm module, many intelligent mobile apps and a cloud geographical information science (GIS) based logistics server. The spatio-temporal heuristics algorithm is to optimize logistics vehicle routing with the historical traffic information. The mobile apps guide the deliverymen in the real-time logistics. The cloud GIS based logistics server integrates traffic information, client demands, vehicle information, the optimization of vehicle routing and the monitoring of logistics processes. The STDS framework has been implemented in a GIS environment. Its performance is evaluated with large scale logistics cases in Guangzhou, China. Results demonstrates the effectiveness and the efficiency of the developed STDS framework. The STDS framework could be widely used in the logistics distribution in metropolitan area, such as the express delivery, e-business, and so on.

W. Tu · Q. Li · X. Chang · Y. Yue · J. Zhu
Shenzhen Key Laboratory of Spatial Smart Sensing and Services, College of Civil Engineering, Shenzhen University, 518060 Shenzhen, China

W. Tu · Q. Li · X. Chang · Y. Yue · J. Zhu
Key Laboratory for Geo-Environment Monitoring of Coastal Zone of the National Administration of Surveying, Mapping and GeoInformation, Shenzhen University, 518060 Shenzhen, China

W. Tu (✉)
College of Information Engineering, Shenzhen University, 518060 Shenzhen, China
e-mail: tuwei@szu.edu.cn

Q. Li · X. Chang
State Key Laboratory for Information Engineering in Surveying, Mapping and Remote Sensing, Wuhan University, 129 Luoyu Road, 430079 Wuhan, China

F. Harvey and Y. Leung (eds.), *Advances in Spatial Data Handling and Analysis*, Advances in Geographic Information Science,
DOI 10.1007/978-3-319-19950-4_12

Keywords Spatio decision support · Heuristic algorithm · Vehicle routing · Traffic information

1 Introduction

Rapid growing urbanization and e-business contribute to the outbreak of logistics service in China (Hu and Sheng 2014). According to the anniversary day (11/11/2013)'s report of Tmall/Taobao, the biggest B2C and C2C company in China, about 140 million parcels were ordered in 24 h and waited delivery on that day. In metropolitan areas in China, e.g. Beijing, Shanghai, Guangzhou, Shenzhen et al., tens of thousands of parcels are delivered by large logistics companies (Fang et al. 2013; Tu et al. 2013, 2014a, b). Such large amounts of customer demands facilitate the development of the logistics industry. However, many problems related to the logistics distribution are also coming, like the overflow of the warehouse, the high transportation cost, the unacceptable delivery delay on customers, the tough labor strength for deliverymen, etc. Therefore, efficient logistics operation is needed to not only reduce transportation cost, but also provide high quality logistics service to associated people.

Nowadays, logistics distribution in the metropolitan area confronts great challenges. Firstly, the daily logistics service suffers from severe delay. Sometimes, the delivery delay is almost unacceptable for up to several hours. Usually, traditional logistics scheduling approaches ignore the fluctuation of urban traffic and design logistics vehicle routing plan (Laporte 2009; Zeimpekis et al. 2007a, b). This naive assumption results in a big gap in the routing between the logistics plan and the reality (Kong et al. 2013). That is mainly due to the complex metropolitan transportation. Factors like the commuting, traffic control, and traffic accidents disturb normal traffic flow, which heavily affects the travel time on the road (Fleischmann et al. 2004; Tu et al. 2010; Novaes et al. 2011). Such drawback significantly lowers logistics service. Secondly, effective decision tool for urban logistics is scarce. Logistics involves many data, like warehouse inventory, customer orders, expected service time windows, the topology of transport network, and dynamic traffic information. The collection and integration of a variety of data are not easy. Moreover, right decision making with these data as fast as possible is very difficult. This leads to the absence of effective spatio-temporal support decision tools in the dynamic transportation for logistics operators, deliverymen and customers.

In recent years, rich dynamic traffic information in the metropolitan area has been produced by floating car system (Tu et al. 2010; Ehmke et al. 2012; Li et al. 2011), video-based traffic monitoring system (González et al. 2011), or loop-based traffic monitoring system (Wu and Coifman 2014). Such useful traffic information is the cornerstone of routing in the dynamic transportation environment in the metropolitan area. Hence, the classical vehicle routing problem (VRP), which design vehicle routes to serve geographical scattered customers in a static transportation network, gradually shifts to the dynamic vehicle routing problem (DVRP) to design distribution plan in a dynamic transportation network (Klaus 2009; Novaes et al. 2011; Pillac et al. 2013). The coming question is how to optimize to the high quality logistics solutions in the dynamic transportation network as fast as possible. Metaheuristics that makes use of spatio-temporal information to speed up the optimization is a promising approach (Pillac et al. 2013).

Meanwhile, the decision support system (DSS) has gained much attention from logistic science, operation research and geographical information science (Miller and Shaw 2001). Due to the spatial nature of logistics, geographical information system (GIS) has been the foundation of the spatial decision support system (SDSS) for logistics (Weigel and Cao 1999; Taniguchi and Thompson 2002). It not only collects and processes logistics data, but also designs logistics solutions with operational research models (Keenan 1998). Graphical interfaces and digital maps are also used to provide a friendly user experience, for example, ArcLogistics by ESRI, TransCAD by Caliper. Usually, the heuristics algorithm for the VRP is embedded in the SDSS to deal with logistics vehicle routing. This tight coupling manner must maintain a large mount of data and cost a lot of money to update. Recently, it transfers to a loose coupling web-based SDSS (Santos et al. 2011). Moreover, the SDSS for logistics in a static transportation network should be updated with spatio-temporal traffic information for a more accurate logistics service.

Attempting to deal with these challenges, this contribution developes a spatio-temporal heuristics algorithm to solve large scale logistics vehicle routing in dynamic transportation network. A spatio-temporal decision support framework integrating spatio-temporal traffic information, heuristics algorithms and intelligent logistics server and apps is designed to facilitate the metropolitan logistics distribution. The experiment on a real logistics case in Guangzhou, China is conducted to validate the performance of the developed heuristic algorithm and framework. Results indicates its effectiveness and efficiency.

The remainder of this paper is organized as follows. Section 2 reviews related literatures. Section 3 introduces the spatio-temporal heuristics algorithm for solving the dynamic vehicle routing problem. Section 4 demonstrates the STDS framework for logistics distribution in the metropolitan area. Section 5 reports the experiment and the results of real logistics cases. Section 6 draws valuable conclusions.

2 Literature Review

The section reviews heuristics algorithm for DVRP and the SDSS for logistics.

2.1 Heuristic Algorithm for Dynamic Vehicle Routing Problem

The DVRP models and designs the vehicle routes in the real transportation network by considering side constraints (Larsen et al. 2008). Due to the NP hard property, heuristics algorithms reporting near-optimal solutions are the promising approach to solve the DVRP in a reasonable time.

According to the principle, heuristic algorithms for the DVRP are divided into two categories, namely, local search and population search. Local search explores the solution space following a single trajectory (Laporte 2009). Usually, the exploration is done by relocating a node in a route or swapping nodes between routes. Many intelligent strategies have been developed to guide the local search process to keep the balance between the diversification and the intensification, e.g. simulated annealing (SA) (Tu et al. 2012; 2014a, b), tabu search (TS) (Nguyen et al. 2013), large neighborhood search (LNS) (Hong 2012), and etc. Population search simultaneously explores multi parts of the solution space, including genetic algorithm (GA) (Haghani and Jung 2005; Mendoza et al. 2009), ant colony optimization (ACO) (Rizzoli et al. 2007), particle swarm optimization (PSO) (Kennedy and Eberhart 1995). They evolve to achieve high quality solutions by simulating nature behaviors.

These advanced heuristics algorithms have been able to solve the DVRP with several hundreds of customers (Hong 2012). To deal with larger DVRP instances, more efficient heuristics is needed. This paper proposes a spatio-temporal heuristics to solve large scale DVRP with thousands of customers in the metropolitan area.

2.2 Spatial Decision Support System for Logistics

Spatial decision support system makes use of spatial information and spatial intelligence to assist decision making in complex spatial problems (Densham 1991). It gives us a framework to integrate database management with graphical interface, operation research model, and the expert knowledge of decision makers. Combined with the enterprise resource planning (ERP) system, the SDSS for logistics facilitates the order collection, routing design, logistics scheduling and the monitoring with the use of GIS spatial data handling techniques (Keenan 1998; Santos et al. 2008).

The VRP's heuristics algorithm is the kernel of the logistics operating. Recently, SDSS for urban logistics has achieved a big advance with modern heuristics algorithms. Mendoza et al. (2009) integrated the ERP with an optimization engine to support logistics operation for waste management in the urban area. They used a clustering based evolutionary-based heuristics algorithm to design shorter vehicle routes. Santos et al. (2011) developed a web based SDSS using Google map for logistics operations, which loosely couples with spatial data from third parties. System users just need an internet browser to access at anywhere at anytime. However, these SDSS frameworks are still rooted in the static transportation. Traffic fluctuation is ignored (Ehmke et al. 2012). To approach better logistics service, the vehicle routing should consider the dynamic transportation. Fleischmann et al. (2004) collected many traffic data from the stationary measurement facilities and specially equipped vehicles for logistics distribution in the metropolitan area of

Berlin, Germany. Ehmke et al. (2012) tested the usability of floating car data for logistics distribution planning. Their results validated the effectiveness of spatio-temporal traffic information. To effectively facilitate the logistics distribution in the metropolitan area, this paper develops a spatio-temporal decision support framework for logistics by integrating large volume of historical traffic information and a useful heuristics algorithm.

3 A Spatio-Temporal Heuristics Algorithm for Dynamic Vehicle Routing Problem

The DVRP aims at designing routes to serve geographically scattered customers, considering variable travel time, time window, service duration, vehicle capacity, and so on. Multiple objectives, such as the total number of used vehicles, the total routes length and the total travel time, are minimized. Due to the NP-hard property, it is very difficult to solve (Laporte 2009).

This section introduces the spatio-temporal heuristic algorithm for large scale DVRP. It uses the spatio-temporal information to speed up the optimization process. The workflow of the proposed algorithm is depicted as Fig. 1, including a construction phase and an improvement phase. The construction phase is to build a good initial solution with the spatio-temporal clustering. The improvement phase iterates to improve the solution quality. Details about the algorithm are described as below.

Input: the dynamic vehicle routing problem, transportation network *TNet*, and traffic data *TD*,
the number of useable vehicles *K*,
the maximum number of iteration *Imax*,
the maximum number of two non-improvement iterations *Zmax*.
Output: the found best solution *Sbest*.

The construction phase:
Step1: Read problem data and cluster customers into *K* group.
Step2: Design the vehicle routing for the *k* (*k*=1, 2, …, *K*) group.
Step3: Report the initial solution *S0*.
The improvement phase:
Step4: Set *s*= *S0*, *s*= *Sbest*.
Step5: Generate a random value and find the ejected nodes with it.
Step6: Eject involved nodes from current solution *s* and insert them back to create a new solution .
Step7: If < (1+)*s*, *s*=. If < *Sbest*, *Sbest*=.
Step8:If the total iteration number is less than *Imax* and the iteration number between two non-improvement iterations is less than *Zmax*, go to *step5*. Else, report *Sbest* and exit.

Fig. 1 The workflow of the spatial-temporal heuristic algorithm for the DVRP

3.1 The Construction Phase

The construction phase groups customers to many clusters and designs a route for each cluster. The clustering is based on a spatial-temporal distance between customers with time window, which weights the traveling time $\bar{t}_{ij}$, the start time window violation $(e_j - e_i - \bar{t}_{ij})^+$ and the end time window violation $(l_i + \bar{t}_{ij} - l_j)^+$, where i, j denotes customers, $\alpha_1, \alpha_2, \alpha_3$ are parameters and set to be equal, $\alpha_1 + \alpha_2 + \alpha_3 = 1$. $\bar{t}_{ij}$ denotes the mean travel time from i to j during the time window $< e_i, l_i >$ of historical traffic information. e_i, e_j are the start time window of i, j. l_i l_j are the end time window of i, j. The formal definition is as Eq. 1. Using this definition, the bottom-to-up clustering algorithm is developed to separate customers into many groups. Firstly, each customer is initialized as a single cluster and its k nearest neighbors is listed. Then, all customer-neighbor pairs are stored in a priority queue L according to the spatio-temporal distance. Then, the front pair in L is popped. The involved clusters in the popped pair will be merged if the two customers in the pair belong to different clusters and the sum demand of their clusters is less than vehicle capacity Q. Otherwise, the merging will not be done. After merging, involved customers' cluster label is updated. This process will continue until L is empty.

$$std_{ij} = \alpha_1 \bar{t}_{ij} + \alpha_2 (e_j - e_i - \bar{t}_{ij})^+ + \alpha_3 (l_i + \bar{t}_{ij} - l_j)^+ \quad (1)$$

After the clustering, each customer is assigned to a cluster smaller than the vehicle capacity Q. The routing in a cluster is done to serve customers with a sequential insertion operation. Firstly, customers will be sorted in a list according to the start time window. Then, an empty route r is initialized. The front node in the list will be inserted the best place in the route r. The insertion will be repeated until customers in the cluster are routed. After all customers are routed, the initial solution S_0 is reported.

3.2 The Improvement Phase

This phase uses the local search to improve the solution quality. It is guided by some micro routes structures in the current solution. Three types of unpromising spatial-temporal features are used to guide, including time window violation nodes, long distance nodes and small routes' nodes. To increase the diversity, random nodes are also used.

1. Time window violation nodes

In current solution, logistics service may arrive early or late at a customer due to the traffic congestion. These nodes are named time window violation nodes N_t. The formal definition is as Eq. 2. The total objective will increase for this local route structure. This paper will remove such nodes from their current position.

$$N_t = \{i | a_i < e_i || a_i > l_i, i \in V_c\} \tag{2}$$

2. Long distance nodes

The temporal solution in the local search may contain long distance segments. A few of them may be hard to be broken. We eject end points of long distance segments and name it as long distance nodes N_d. The formal definition is as Eq. 3, where d_{ij} is the distance from i to j, and $\bar{e}$ is the mean length of segments in the solution.

$$N_d = \{i, j | d_{ij} > 3\bar{e}, i \in V_c, j \in V_c\} \tag{3}$$

3. Small route's nodes

A small route has few of nodes and is hard to be modified. This article ejects all nodes in the small routes, which is defined as Eq. 4, where $|R|$ is the number of customers in a route R, and RN_{min} is a threshold set to 3.

$$N_s = \{i || R | \leq RN_{\min}, i \in R, i \in V_c\} \tag{4}$$

4. Random nodes

To increase the diversification of local search, a few of random nodes N_r are also removed from current solution.

Local search explores the solution space by ejecting and inserting defined nodes. It consists of two steps, namely, a remove step and a rebuilding step. The remove step ejects certain nodes from the current solution. A roulette strategy is used to select which type of nodes to eject, as shown in Eqs. 5 and 6, where α_t, α_l, α_s, α_r are the probability of four types of nodes, f_k is the improved objective value that N_t, N_l, N_s, N_r has gained in the latest $4n$ iterations, n is the number of customers. At the beginning, their value is set to be equal. During the search, they are updated with their contributions to the objectives as shown in Eq. 6. The remove step randomly generates a value α between [0, 1], identifies the type of ejected nodes with it, and ejects the corresponding nodes.

$$\alpha_t + \alpha_l + \alpha_s + \alpha_r = 1 \tag{5}$$

$$\alpha_k = \frac{\sum f_k}{\sum f_t + \sum f_l + \sum f_s + \sum f_r} \quad (k = t, l, s, r) \tag{6}$$

The rebuilding step sequentially inserts all removed nodes. Firstly, all removed nodes are randomly pushed into a queue L. Then, nodes in front of L will be popped and inserted the place before or after a routed customer to minimize multiple

objectives, including the total used vehicles, the total routes length and the total service time. The insertion will be repeated until L is empty.
Local search will naturally drop into local minima. To overcome this problem, some worse solution is accepted to escape from local minima (as the step 7 in Fig. 1). The improving iteration will continue until a number of iterations I_{max} has been done or the objective is not improved in Z_{max} iterations. Finally, the best solution is reported.

4 A Spatio-Temporal Decision Support Framework for Logistics Distribution

A spatio-temporal decision support framework is developed to integrate traffic information, logistics data and the heuristics algorithm to facilitate logistics operators, deliverymen and customers. As shown in Fig. 2, it is comprised of four modules, namely, a remote accessing traffic information database, a spatio-temporal heuristic algorithm module, a cloud GIS based logistics server and many intelligent apps.

The traffic information database stores a time series of traffic snapshot to describe the traffic fluctuation in a metropolitan area. Each snapshot is comprised of traffic state descriptions for each road segment of the transportation network at the corresponding time. This description is measured by the mean travel speed on the road, provided by a floating car system covering the transportation network (Li

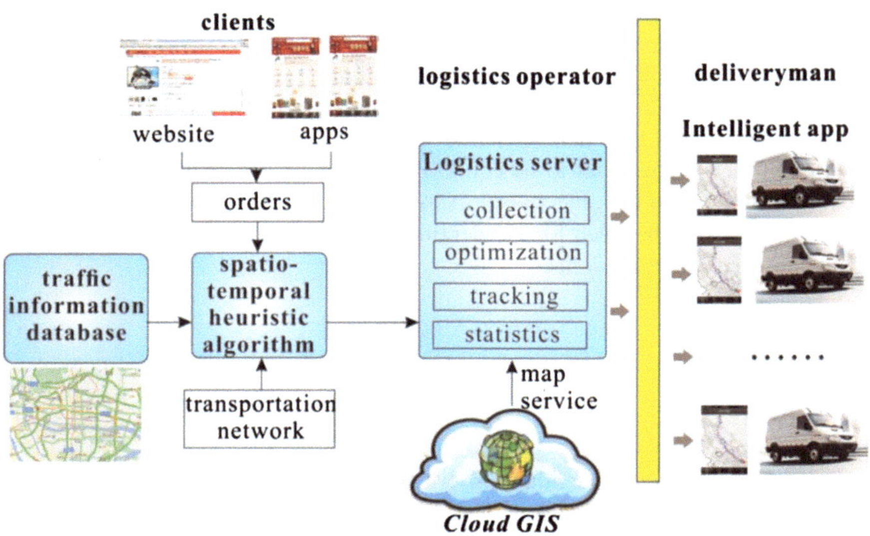

Fig. 2 The spatio-temporal decision support framework for logistics distribution

Fig. 3 User interfaces of logistics server and intelligent apps

et al. 2011). For example, from 8:55 to 9:00, the travel speed on the road segment 11,245 is 10.2 m/h. However, due to the start of commuting peak, the travel speed on the road segment 11,245 is down to 9.0 m/h between 9:00 and 9:05. By using travel speed, travel time from a location to another in this network is calculated.

The spatio-temporal heuristic algorithm module uses the proposed algorithm to design vehicles routing, considering dynamic traffic, client demand, vehicle capacity, et al. Details of the algorithm is referred to the Sect. 3.

The cloud GIS based logistics server integrates the spatio-temproal heuristic algorithm, the web map service (WMS) and the spatial database to collect logistics data, optimize vehicle routing, track vehicles and summarize logistics process. Client orders are collected from the seller's website or apps on mobile phones. The optimization uses the proposed heuristic algorithm to design high quality logistics plan. Tracking service monitors the real-time logistics progress. The summarization of the logistics process calculates the total routes length, the total service time, the total time delay, etc. Main user interfaces is as Fig. 3 (left).

An intelligent app is a mobile GIS embedded program giving the detailed logistics direction for a deliveryman [as Fig. 3 (right)]. Information about total number of assigned customers, goods id and route details are received from the server after the distribution planning. During the distribution, next customer and the route direction are also displayed.

5 Case Study in Guangzhou, China

After the implementation of the proposed algorithm and the STDS framework, its performance is tested by five logistics cases from a logistics company in Guangzhou, a metropolitan in south China, which distributes goods to thousands of customers each day. Traffic information is obtained from a remote database in the traffic command center at the Guangzhou transportation administration. What-if analysis is

Table 1 Computing results on real logistics cases in Guangzhou, China

Name	N	TD	Q	The construction phase					The improvement phase				
				TNV	TRL (/km)	TTT (/h)	TTD (/h)	CT (/min)	TNV	TRLh (/km)	TTT (/h)	TTD (/h)	CT (/min)
P1	4521	106,542	1000	154	12,967.8	862.3	175.9	1.6	109	5467.2	342.5	46.3	27.3
P2	4814	110,489	1000	187	14,080.5	888.4	189.3	1.8	115	5830.7	318.9	49.5	29.9
P3	4017	98,876	1000	139	13,027.9	793.4	162.8	1.5	100	5292.5	302.1	42.7	24.2
P4	4125	100,563	1000	189	12,532.4	787.6	172.7	1.6	103	5230.8	327.0	45.8	26.7
P5	4892	118,347	1000	202	14,218.2	912.9	182.2	1.9	125	5947.6	353.6	58.2	30.4
CNV				**871**					**552**				
CRL					**66,826.8**					**27,768.8**			
CTT						**4244.6**					**1644.1**		
CVD							**882.9**					**242.5**	
AT								**1.68**					**27.7**

Name name of problem. *N* number of customers. *TD* total demand. *Q* vehicle capacity. *TNV* total number of vehicles. *TRL* total routes length (/km). *TTT* total travel time (/h). *TTD* total time delay (/h). *CT* computing time(/min). *CNV* the cumulative number of vehicles; *CRL* the cumulative value of routes length (/km); *CTT* cumulative value of travel time (/h); *CVD* the cumulative value of delay (/h); *AT* average value of computing time (/min) conducted to test the logistics distribution case. To validate the quality, comparison is done with the hand-made distribution plan from by the logistics company.

Table 1 reports details of the results, such as problem name, customer number, total number of vehicles, total routes length, total travel time, total time delay and the computing time. Summary of the results is listed as the last five rows in the table. As Table 1 indicates, the construction algorithm reports the initial solution for logistics case about 4500 customers within 2.0 min. The final best results are given in 31.0 min. Compared to the initial solution, fewer vehicles and labors are cost. For the five test cases, the cumulative number of vehicles (CNV) is reduced from 871 to 552. The cumulative value of routes length (CRL) is shortened from 66,826.8 to 27,768.8 km. The cumulative value of travel time (CTT) is reduced from 3744.6 to 1644.1 h. Therefore, we can conclude that the improvement phase sharply improves the quality of the initial solution.

Table 2 reports the comparison between computed results and hand-made plans.. The total time delay of hand-made plans is missed for the unavailability of such information. As Table 2 indicates, the STDS framework uses 552 vehicles for the five cases, which is less than hand-made plans (631 vehicles). Considering the cumulative value of routes length, value of the STDS framework is 27,768.8 km, which is much shorter than hand-made plans (33,619.7 km). When it comes to the total travel time (TTT), the STDS framework takes 1644.1 h, which is also fewer than hand-made plans do (2172.7 h). Furthermore, the mean travel speed of the STDS framework's solution is (16.8 km/h = 27,768.8/1644.1). It is faster than the speed of the hand-made plan (15.7 km/h = 66,826.8/4244.6). In other words, the STDS framework's solutions keep away from some congested area in the metropolitan area. In conclusion, the STDS framework could not only promote the efficiency, but also reduce cost for logistics distribution in the metropolitan area.

Table 2 Comparison of results on logistics cases in Guangzhou, China to hand-made plans

Name	N	Result of hand-made plan			Results of the STDS framework		
		TNV	TRL (/km)	TTT (/h)	TNV	TRL (/km)	TTT (/h)
P1	4521	132	6550.7	445.3	109	5467.2	342.5
P2	4814	129	7056.9	410.6	115	5830.7	318.9
P3	4017	110	6402.8	400.8	100	5292.5	302.1
P4	4125	118	6280.4	445.2	103	5230.8	327.0
P5	4892	142	7328.9	470.8	125	5947.6	353.6
CNV		**631**			**552**		
CRL			**33,619.7**			**27,768.8**	
CTT				**2172.7**			**1644.1**

Name name of problem. *N* number of customers. *TNV* total number of vehicles. *TRL* total routes length (/km). *TTT* total travel time (/h). *TTD* total time delay (/h). *CNV* the cumulative number of vehicles; *CRL* the cumulative value of routes length (/km); *CTT* the cumulative value of travel time (/h)

6 Conclusion

The urbanization and the explosive e-business in China expect large scale, high dynamic logistics service. This contribution developed a spatio-temporal decision support framework for logistics distribution instances in the metropolitan area. This STDS framework integrates a traffic information database, a spatio-temporal heuristic algorithm, a cloud GIS based logistics server, and many intelligent mobile apps. The spatio-temporal heuristics algorithm is to optimize logistics vehicle routing with the historical traffic information. The cloud GIS based logistics server is to optimize vehicle routing and monitor the real-time logistics process. The mobile apps guide the deliverymen in the distribution. The STDS framework was implemented and its performance was evaluated with real logistics cases in Guangzhou, China. The results demonstrated the effectiveness and efficiency of the developed STDS framework.

The main contributions of this paper are two-folds as following:

- A spatio-temporal heuristics algorithm for large scale dynamic vehicle routing problem is proposed. Experiment in Guangzhou, China validated that the proposed algorithm could achieve a high quality solution for logistics instance with 4800 customers about 30 min.
- A spatio-temporal decision support framework is developed to facilitate logistics distribution in the metropolitan area. It loosely couples with spatio-temporal traffic information, the spatio-temporal heuristics algorithm, the cloud GIS based logistics server and many intelligent apps to provide friendly and efficient logistics service.

In the future, we will extend the proposed local search heuristic algorithm with real-time orders, which is more complex than the solved dynamic vehicle routing problem in this paper. Also, we plan to embed real-time traffic information into the spatio-temporal decision support framework to provide more reliable logistics service.

Acknowledgments This research was jointly supported by the National Science Foundation of China. (No. 41401444, 41371377), the Shenzhen Scientific Research and Development Funding Program (No. ZDSY20121019111146499, No. JSGG20121026111056204, No. JCYJ201208 17163755063), the Shenzhen Dedicated Funding of Strategic Emerging Industry Development Program (No. JCYJ20121019111128765), China Postdoctoral Science Foundation funded project (2014M560671) and the open research fund of State Key Laboratory of Information Engineering in Surveying, Mapping and Remote Sensing (No. 13S02). The authors also would like to thank the reviewers for their valuable comments and suggestions.

References

Densham PJ (1991) Spatial decision support systems. Geogr Inf Syst Principles Appl 1:403–412

Ehmke JF, Meisel S, Mattfeld DC (2012) Floating car based travel times for city logistics. Transp Res C Emerg Technol 21:338–352

Fang Z, Tu W, Li Q, Shaw SL, Chen S, Chen B (2013) A Voronoi neighborhood-based search heuristic for distance/capacity constrained very large vehicle routing problems. Int J Geogr Inf Sci 27:741–764

Fleischmann B, Gietz M, Gnutzmann S (2004) Time-varying travel times in vehicle routing. Transp Sci 38(2):160–173

González M, Dominguez D, Sánchez Á (2011) Learning sequences of sparse correlated patterns using small-world attractor neural networks: an application to traffic videos. Neurocomputing 74(14–15):2361–2367

Haghani A, Jung S (2005) A dynamic vehicle routing problem with time-dependent travel times. Comput Oper Res 32(11):2959–2986

Hong L (2012) An improved LNS algorithm for real-time vehicle routing problem with time windows. Comput Oper Res 39(2):151–163

Hu ZH, Sheng ZH (2014) A decision support system for public logistics information service management and optimization. Decision support systems. http://dx.doi.org/10.1016/j.dss.2013.12.001

Keenan PB (1998) Spatial decision support systems for vehicle routing. Decis Support Syst 22 (1):65–71

Kennedy J, Eberhart R (1995) Particle swarm optimization. In: Proceedings of IEEE international conference on neural networks, vol 1944. IEEE Press, New York, pp 1942–1948

Klaus P (2009) Logistics research: a 50 years' march of ideas. Logistics Res 1(1):53–65

Kong QJ, Zhao Q, Wei C, Liu Y (2013) Efficient traffic state estimation for large-scale urban road networks. IEEE Trans Intell Transp Syst 14:398–407

Laporte G (2009) Fifty years of vehicle routing. Transp Sci 43(4):408–416

Larsen A, Madsen OB, Solomon MM (2008) Recent developments in dynamic vehicle routing systems. In: Golden B, Raghava S, Wasil E (eds) The vehicle routing problem: latest advances and new challenges. Springer, Heidelberg, pp 199–218

Li Q, Zhang T, Yu Y (2011) Using cloud computing to process intensive floating car data for urban traffic surveillance. Int J Geogr Inf Sci 25:1303–1322

Mendoza JE, Medaglia AL, Velasco N (2009) An evolutionary-based decision support system for vehicle routing: the case of a public utility. Decis Support Syst 46(3):730–742

Miller H, Shaw SL (2001) Geographic information systems for transportation: principles and applications. Oxford University Press, USA

Nguyen PK, Crainic TG, Toulouse M (2013) A tabu search for time-dependent multi-zone multi-trip vehicle routing problem with time windows. Eur J Oper Res 231(1):43–56

Novaes AN, Frazzon E, Burin P (2011) Dynamic vehicle routing in over congested urban areas. In: Kreowski H-J, Scholz-Reiter B, Thoben K-D (eds) Dynamics in logistics. Springer, Berlin, pp 49–58

Pillac V, Gendreau M, Guéret C, Medaglia AL (2013) A review of dynamic vehicle routing problems. Eur J Oper Res 225(1):1–11

Rizzoli A, Montemanni R, Lucibello E, Gambardella L (2007) Ant colony optimization for real-world vehicle routing problems. Swarm Intell 1(2):135–151

Santos L, Coutinho-Rodrigues J, Antunes CH (2011) A web spatial decision support system for vehicle routing using google maps. Decis Support Syst 51:1–9

Santos L, Coutinho-Rodrigues J, Current JR (2008) Implementing a multi-vehicle multi-route spatial decision support system for efficient trash collection in Portugal. Transp Res A Policy Pract 42(6):922–934

Taniguchi E, Thompson RG (2002) Modeling city logistics. Transp Res Rec J Transp Res Board 1790(1):45–51

Tu W, Fang Z, Li Q (2010) Exploring time varying shortest path of urban OD pairs based on floating car data. In: IEEE GRSS the 18th international conference on geoinformatics, geoinformatics 2010. The Geographical Society of China, Beijing

Tu W, Fang Z, Li Q (2012) A fast algorithm for large scale vehicle routing routing optimization based on Voronoi diagram. J GeoInf Sci 14(6):781–787

Tu W, Fang Z, Li Q (2013) An empirical analysis of Voronoi neighborhood characteristics of heuristic solutions for capitated vehicle routing problems. Paper presented at the international symposium on recent advances in transport modeling, OPTIMUM2013. Golden coast, Australia

Tu W, Fang Z, Li Q, Shaw SL, Chen B (2014a) A bi-level Voronoi diagram-based metaheuristic for a large-scale multi-depot vehicle routing problem. Transp Res E Logistics Transp Rev 61:84–97

Tu W, Li Q, Fanng Z (2014b) Large scale multi-depot logistics routing optimization based on network Voronoi diagram. Acta Geodaet Cartogr Sin 43(10):1075–1082

Weigel D, Cao B (1999) Applying GIS and OR techniques to solve Sears technician-dispatching and home-delivery problems. Interfaces 29:112–130

Wu L, Coifman B (2014) Improved vehicle classification from dual-loop detectors in congested traffic. Transp Res C Emerg Technol 46:222–234

Zeimpekis V, Minis I, Mamassis K, Giaglis GM (2007a) Dynamic management of a delayed delivery vehicle in a city logistics environment. In: Zeimpekis V, Tarantilis C, Giaglis G, Minis I (eds) Dynamic fleet management. Operations research/computer science interfaces series, vol 38. Springer, US, pp 197–217

Zeimpekis V, Tarantilis CD, Giaglis GM. Minis I (2007b) Dynamic fleet management: concepts, systems, algorithms & case studies. Springer, New York

Part III
Spatial Modelling and Reasoning

Local Absolute Vertical Accuracy Computation of Wide-Coverage Digital Terrain Models

Gev Ben-Haim, Sagi Dalyot and Yerach Doytsher

Abstract The use of Digital Terrain Models (DTMs) as a computerized mapping and modeling infrastructure of our natural environment is rapidly and intensively growing. Many different sources of wide-coverage DTMs differ significantly in terms of geometric attributes, resolution and accuracy. These sources often describe the same coverage area by discrepant representation of the topography. Studies have shown that the inner-vertical accuracy of an individual wide-coverage DTM is not homogeneous, and cannot be expressed by a single global accuracy value. Still, in terms of quality analysis, most studies continue to investigate the accuracy of topographic models globally while ignoring inner-quality aspects. This paper suggests a method for spatial regional inner-accuracies analysis that is based on terrain characteristics and error-propagation computation via a simultaneous relative DTMs comparison mechanism. We present the evaluation of data-error characterization and local trend detection, which is an improvement in terms of DTM data quality analysis, since it addresses the vertical absolute accuracy measures on a regional scale—and not globally. The research presents the feasibility to identify regions within DTMs that show errors and data-reliability anomalies. The paper presents the outcome produced by applying this methodology, with geostatistical comparisons to the common global scheme, demonstrating the feasibility and effectiveness of the proposed algorithms.

Keywords Digital terrain models · Quality assessment · Spatial data · Accuracy · Geostatistics · Error modeling · Characterization

---

G. Ben-Haim (✉) · S. Dalyot · Y. Doytsher
Mapping and Geo-Information Engineering, Technion, Haifa, Israel
e-mail: gev@technion.ac.il

S. Dalyot
e-mail: dalyot@technion.ac.il

Y. Doytsher
e-mail: doytsher@technion.ac.il

F. Harvey and Y. Leung (eds.), *Advances in Spatial Data Handling and Analysis*, Advances in Geographic Information Science,
DOI 10.1007/978-3-319-19950-4_13

1 Introduction and Overview

1.1 Digital Topographic Representation

Terrain relief is a continuous spatial surface entity describing our natural environment, i.e., the topography. Morphological and dynamic analysis of the terrain provides geo-related useful and qualitative information. The fact that the reality of the terrain is continuous in nature, demands—for practical reasons—a somewhat reduced representation that is reliable and precise in the form of discrete data. The notion of terrain representation in the form of a digital (discrete) model is therefore used and is analyzed in effectiveness in computerized working environments.

Perhaps the most commonly used digital structure for storing and representing topographic spatial data is the Digital Terrain Model (DTM), or Digital Elevation Model (DEM). This model represents the continuous reality of our environment via a collection (set, matrix) of positional discrete samples, in the form of geometric primitives: points, lines, and sometimes surfaces (or a combination of these). Normally, these primitives are in the form of 2.5-Dimensions that store planar position and height. Accordingly, the continuous reality is translated and represented as a sparse and (to some extent) partial form of representation. Still, the aspiration is for this model to be as reliable and authentic to reality, while simultaneously taking into account its shortcomings (i.e., it is a model and not the reality). This digital infrastructure is an important requirement for establishing an efficient and computerized management of our environment (Weibel and Heller 1991). Many National Mapping Agencies (NMAs), as well as private companies and public bodies, are involved today in the establishment and maintenance of this type of infrastructure (Parry and Perkins 2000).

1.2 Accuracy Provisions of DTMs

In the digital mapping discipline, the term “accuracy” is mostly being used in a quantitative manner to conclude on how “close” the observed values (measurements) are to the “true” (reference) values—what error exists, in terms of quantitative and qualitative measures. It should be noted that although to some extent resolution contributes to the overall reliability of DTMs, the influence of this factor is not addressed in this paper. Consequently, great efforts and resources are channeled towards modeling processes that focus on the spatial modeling data quality. These processes ensure that the end product represents correctly the statistical and spatial distribution of data, while minimizing errors, together with having a qualitative and quantitative measure of these factors.

1.3 DTM Accuracy Measures

Most studies commonly divide the accuracy measures of DTMs in planar accuracy (X, Y) and vertical accuracy (Z). Since DTMs are mainly used for topographic analysis purposes, the accuracy of a DTM is usually referred to its height absolute accuracy. This is normally quantified by a single constant value, even if the DTM represents a large coverage area.

Various characteristics and attributes associated with the DTMs influence the magnitude and pattern of the computerized topography (Wilson et al. 2000), and as such have a direct effect on the quality and reliability. This exemplifies the need for quantitative assessment of the accuracy. Wide-coverage DTMs are produced via different means that evidently influence and contribute to different error characteristics in the data produced, represented and analyzed. Still, not much has been done on a local scale, i.e., aspiring to quantify these measures on a regional basis instead of a global one.

The majority of topographic accuracy analysis procedures rely on some data-reference that is considered more reliable. Some rely on knowledge regarding the acquiring systems that affect the observations and measurements, and as such, the quality of the data is analytically derived. When a reference is used, most likely a direct coordinate-based superimposition comparison is used to compare height-values based on the databases' mutual coordinate reference systems. This is not sufficient in the case of datum ambiguities, different data-characterization and inner-changing qualities, which results in an erroneous topographic analysis (Dalyot and Doytsher 2008). Still, most traditional approaches use a single global parameter to express vertical accuracies of large areas. Another reason also suggests that different areas present different topographic attributes, thus altering the qualities of the model.

Several approaches use overlapping DTM datasets that cover the same area for DTM manipulation, such as fusion, merging, and rubber-sheeting. The goal of such processes is to achieve continuous and complete terrain representation using all available data (Katzil and Doytsher 2005). For example, DTM fusion (or conflation) algorithms aspire to improve the quality of the merged DTM using several different databases (Dalyot 2010). As wide-coverage DTMs become more available and free to use (e.g., ASTER, SRTM), different types of data describing the same relief are increasingly common. Still, the topography represented in these wide-coverage DTMs show fundamental discrepant models in terms of scale, resolution, format, coverage area, etc. When comparing these DTMs, certain dissimilarities are evident. Quality analysis of DTMs rarely address such problems and deficiencies, let alone on a local level.

Data quality issues in DTMs, expressed as errors in data points, are the accumulated result of inaccuracies introduced in the sum of operations related to the terrain modeling processes as derived from the various sources. As presented by Li (1992), the factors are related to: (I) the characteristics of the terrain surface, (II) the methods used for constructing the DTM surface, (III) the three attributes (accuracy, distribution and density) of the raw data, and (IV) the characteristics of the resulting DTM surface. The characteristics (e.g., morphology) of the terrain define the degree of difficulty in trying to model it, such that these have a big effect on the accuracy of the resulting DTM representation.

This research study proposes a locally-based geostatistical algorithm implemented for analyzing the spatial regional vertical accuracies of DTMs. It requires no preliminary a priori knowledge regarding reference or accuracy. The proposed method is an improvement in terms of DTM data quality analysis; since it addresses the vertical absolute accuracy measures on a regional scale, and not only globally. As a result, the feasibility to identify regions within the different DTMs, that show errors and data-reliability anomalies in respect to their surroundings, is achieved. Moreover, the algorithm concept is straightforward and effective, and does not require heavy computational resources.

The structure of the paper is as follows: Sect. 2 gives a literature review on this topic. Section 3 presents the methodology of the proposed method, and the mathematical formulae of the geostatistical solution. Sections 4 and 5 present the data preparation and experimental results based on simulated and real data, presenting the correctness and reliability of the proposed approach. Finally, Sect. 6 outlines the conclusion and future outlook of this research.

2 State of the Art

2.1 Global Analysis of DTM Accuracy

There is a lot of research on the topic of global DTM quality, using well-established approaches that are based on morphometric and geometric modeling of existing data (e.g., Li 1993; McCullagh 1988; Karel et al. 2006). Several studies attempt to handle this issue deriving methods from the field of signal processing, such as the application of spectral analysis for DTM accuracy estimation (e.g., Tempfli 1980; Frederiksen 1980). In these papers, the global accuracy is derived from the measurement accuracy and the transfer function of the interpolation method. This function describes the ratio of the amplitudes of the input and output signal. Other approaches attempt to handle outliers and detaching accuracy measures under the assumption that there exist an underlying normal distribution (e.g., Atkinson et al. 2007; Aguilar et al. 2007b) and the coefficient of variation of the sample variance (Aguilar et al. 2007a). Many approaches attempt to tackle this problem, either on spatial extent or spectral one; however, all studies handle these measures as if they reflects on a global level—and not on a local one. Alternatively, Ben-Haim et al. (2013) suggested a method to compute absolute vertical accuracies of multiple DTMs based on height discrepancies existing among DTMs by using a simultaneous Least Squares Adjustment (LSA) process, proving that accuracies should be analyzed on a local scale.

2.2 Data Propagation

The classic theory of errors and the least squares theory was introduced by Gauss more than two centuries ago, introducing the idea that random error can be considered as normally distributed for an infinite number of observations (Bjerhammer 1973). Today it has come to be known as the Gaussian Distribution (GD) or Gaussian Error Propagation Law (GEPL), and is used in various scientific disciplines for error modeling (Chen and Murai 2008). Physical quantifications of geometrical quantities are being measured using these ideas, namely assuming that all measurements contain random error due to measurement conditions regarding instrument, human and environment aspects, i.e., stochastic errors.

In this domain, a theoretical-empirical model for modeling the accuracy of grid DTMs was presented (Aguilar et al. 2006). The theoretical component of the model integrates sample data accuracy by means of the error-propagation theory, while the empirical component models what is known as information loss. The empirical models' morphology is based on the product of two potential functions, one related to the terrain roughness and another related to the sampling density, (Aguilar 2007b) focused on DTM uncertainty assessment using approximate confidence intervals.

2.3 Wide-Coverage DTM Accuracy Assessment

The Shuttle Radar Topography Mission (SRTM) radar signal measurements depend on the terrain cover and is a complicated function of the electromagnetic and structural properties of the scattering medium (Bhang et al. 2007). Studies have shown that the absolute vertical accuracy estimation is approximately 5 m (Rosen et al. 2001; Sun et al. 2003). While this is a very optimistic value, recent studies (e.g., Kellndorfer et al. 2004; Brown et al. 2005) have demonstrated that this value is in the range of less than 15 m for the analyzed areas, with approximately 10 m relative height accuracy. Conversely, Advanced Spaceborne Thermal Emission and Reflection (ASTER) DTM uses stereo image data for production. The overall absolute vertical RMSE with 95 % confidence interval for close to 1000, 1 by 1°, tiles was less than 20 m (Aster GDEM Validation Team 2009). This was proved to mainly be due to anomalies and artifacts. In areas with high terrain roughness (slope greater than 10°), these values were even worse (Tighe and Chamberlain 2009). The variation existing in these values confirms that the accuracy analysis assessment should be carried out on a local scale—and not a global one.

3 Local Accuracy Assessment Methodology

3.1 Overview

In this paper we extend the concept suggested by (Ben-Haim et al. 2013) implementing a localized simultaneous comparison mechanism of multiple independent wide-coverage DTMs. The premise is that the relative height difference value of two corresponding grid points is derived from the absolute accuracy of both DTMs. Therefore, computing the absolute vertical accuracies of DTMs is geostatistically enabled using numerous height differences of several DTMs. This is a classical problem of an over-determined system of linear equations where a larger number of equations (DTM height differences) describes fewer unknowns (absolute accuracies of the DTMs). This system is described in Eq. 1, where A is the partial derivatives of the unknowns, L is the observations vector and x is the solution vector.

$$Ax = L \tag{1}$$

Solving the above system is an estimation process carried out by implementing an iterative mathematical approximation of weighted LSAs using relative height differences of each DTM pair extracted earlier from observations. Equation 2 depicts the calculation of a relative accuracy m_{RS} of a DTM pair—R and S—based on a series of height values h_i^R and h_i^S, respectively, where i depicts a specific position index in both DTMs, and N is the total number of mutual points in both datasets. This value is an element of observations vector L represented in Eq. 1.

$$m_{RS}^2 = \frac{1}{N}\sum_{i=1}^{N}(h_i^S - h_i^R)^2 \tag{2}$$

There are ½ n (n−1) equations representing the comparison combinations of n datasets existing in the process. The LSA model used for describing the observations is based on the theory of errors, where the relative accuracy of a model RS represents the accuracy combination of datasets R and S, i.e., m_R and m_S, respectively, as depicted in Eq. 3.

$$m_{RS}^2 = m_R^2 + m_S^2 \tag{3}$$

Matrix A is created using the above model, expressing the partial derivatives as a function of the unknowns. Finally, the solution vector x, which represents the absolute accuracies of all DTMs, is computed by solving the equation system, represented in Eq. 4.

$$x = (A^T PA)^{-1}(A^T PL) \tag{4}$$

The process is carried out without any preliminary knowledge regarding the a priori accuracies of the databases. The first iteration is executed using the identity matrix I (instead of P), where the solution is used for generating a weight matrix P on the consecutive iterations. The above calculation results with the extraction of the absolute accuracy of all DTMs participating in the process with respect to the area analyzed.

3.2 Local Accuracy Determination

As discussed earlier, vertical accuracies of wide-coverage DTMs are not uniform and homogeneous. Therefore, they are divided into regions to implement several region-scale LSA processes under the assumption that the localized height accuracies are relatively uniform, and can be considered as homogenous (schematically depicted in Fig. 1). In this paper, the separation of such regions is carried out by dividing each DTM into equal-sized regions having local vertical accuracy. A finer-resolution might suggest applying the above method using one point (pixel) only at a time. By doing so, the LSA solution might result in inaccurate biased accuracy values due to a small sample size. Hence, according to local topography exist and resolution, a mask of 13 by 13 grid points is taken around every examined point as a representative of its surrounding. The outcome of such a process is a smoothed point-scale accuracy analysis of all DTMs. Such analysis enables the extraction of polygonal sub-area accuracy based on the absolute accuracy values computed.

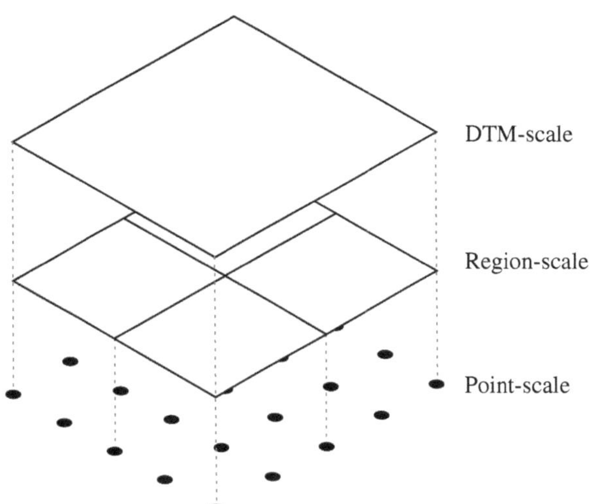

Fig. 1 Different levels of absolute accuracy analysis: DTM-scale (entire area), region-scale (local area) and point-scale

4 Data Preparation

4.1 Synthetic Data

To evaluate the proposed concept, synthetic data is used, thus having an a priori knowledge of the expected results. A source DTM was used as reference, having 25 m resolution produced by digitization of national maps of Israel. The chosen section has a coverage area of 1.85 by 1.85 km (3.4225 km^2), which translates to 75 by 75 (5625) grid points. From this DTM, four different DTMs were generated representing the same region, while adding noise to their height values. A normally distributed noise with a mean value of 0 m and variance values of (σ) ±2, ±4, ±12 and ±5 m was added to the four different datasets (indexed as 1, 2, 3 and 4, respectively), as depicted in Fig. 2.

These synthetic models are chosen to illustrate the performance of the suggested mechanism, where the unequal standard deviations represent accuracy of terrain models covering the same relief. In addition to the normally distributed added noise, two types of noise were later inserted to validate the feasibility of the proposed mechanism, as follow:

Gradual Noise: Gradual noise was inserted to dataset number 4 along the x-axis direction, i.e., as the value of x increased the magnitude of the noise also increased: from 0 m, and up to approximately +14 m. The equivalent weighted overall variance of dataset 4 is therefore ±9.56 m.

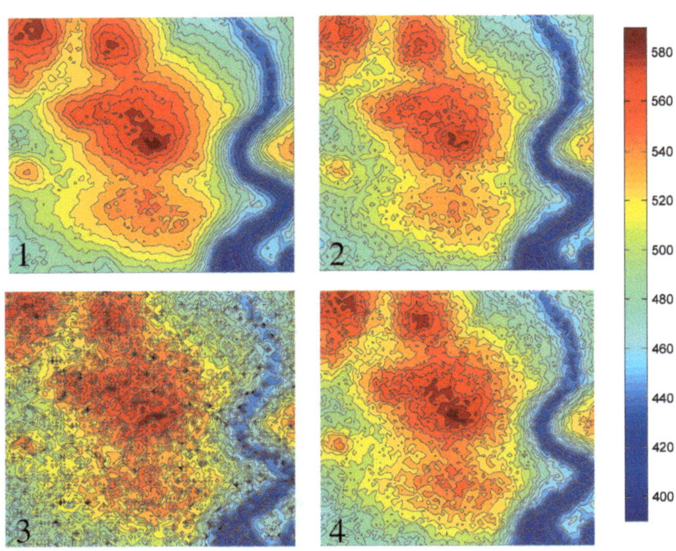

Fig. 2 Four generated synthetic DTMs represented by 10 m height interval contours (values in meters)

Regional Noise: Regional noise was locally inserted to DTMs 1, 2 and 3 to illustrate the effectiveness of the proposed method in extracting local distortions that might exist (i.e., region-scale). All four datasets were subdivided into 25 equally sized regions, representing local areas. A regional normally distributed noise with a mean value of 0 m and variance values (σ) of ±5, ±10 and ±2 was locally inserted into certain regions of datasets 1, 2 and 3, respectively, replacing the original noise (depicted in Table 3), such that certain areas within the examined region expressed fundamentally different accuracies.

4.2 Real Data

The suggested mechanism was also examined using real data. For this, four different DTMs covering the same area of 100 km^2 were used, as follow:

25 m DTM data: a 25 m resolution DTM of Israel generated from close to 100 1:50,000 topographic map sheets of Israel (Hall 2008). Contour layer of the maps was generated via scans, later interpolated and re-sampled to a 25-meters grid with Israel Transverse Mercator (ITM) coordinates. The vertical accuracy of this DTM is estimated at the range of 4–6 m.

50 m DTM data: a 50 m resolution DTM, referenced to ITM generated by the Survey of Israel (SOI). The geospatial data was collected by photogrammetric means via aerial photographs (at a scale of 1:40,000). The original horizontal resolution of this DTM is 5 m. SOI declare that the vertical accuracy of this DTM is 2 m (Raizman and Peled 2002).

Shuttle Radar Topographic Mission (SRTM) data: SRTM digital elevation data was acquired by a single-pass interferometry, processed from raw radar signals at intervals of 1 arc-second (approximately 30 m). The data for regions outside the United States was sampled at 3 arc-seconds (approximately 90 m) using a cubic convolution resampling technique. SRTM is projected to geographic coordinate system of the World Geodetic System 1984 (WGS84) datum (The Shuttle Radar Topography Mission Team 2007).

Advanced Spaceborne Thermal Emission and Reflection Radiometer (ASTER) data: ASTER Global Digital Elevation Model (GDEM) data was acquired by a multispectral imager collecting in-track stereo high-resolution images. It was formatted with 1 by 1° tiles with a resolution of 1 arc-second (approximately 30 m), referenced to WGS84 datum.

An area from all DTMs is depicted in Fig. 3, showing the different topography and morphology represented by the models.

In order to apply the suggested mechanism on the real data, all values are horizontally registered based on their given coordinates. The two geographic models were transformed and interpolated, to correspond to the ITM datum, such that all four models share the same reference system, resolution and coverage area (Blue asterisks depict morphological interest points extracted automatically). These

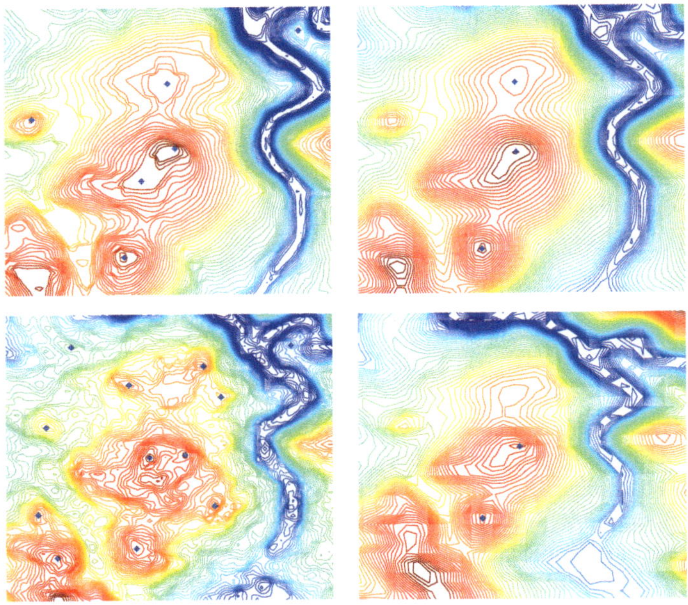

Fig. 3 Topography represented by all DTMs with 10 m height interval contours: 25 m ITM (*top left*), 50 m ITM (*top right*), ASTER (*bottom left*), SRTM (*bottom right*)

are planned to be used to validate the mutual referencing of all models, i.e., not relying on the given coordinates only, and will be carried out in future work.

5 Experimental Results

5.1 Synthetic Data Experiments and Analysis

5.1.1 DTM-Scale Analysis

Table 1 depicts the absolute vertical accuracy values of all four different DTMs computed with respect to the entire area (i.e., a single global process—DTM-scale), with no preliminary a priori accuracy knowledge. These results indicate the feasibility and fast convergence of the solution, where the last (3rd) iteration showed a value difference of several millimetres only. When compared to the errors and noise added (left column for each solution), it is clear that, although the proposed methodology works well and calculates the expected value, the local-regional distortions affect the global dataset accuracy value dramatically—a fact that strengthens the claim that accuracies must be treated regionally.

Table 1 Statistical results for four synthetic datasets with three noise levels

Dataset	Original noise (original accuracies)		Gradual noise		Regional noise	
	Known σ (m)	Computed σ (m)	Known σ (m)	Computed σ (m)	Known σ (m)	Computed σ (m)
1	2.00	1.98	2.00	2.05	3.40	3.44
2	4.00	4.01	4.00	3.92	4.29	4.21
3	12.00	12.07	12.00	11.99	9.87	9.91
4	5.00	5.09	9.56	9.55	5.00	5.05

5.1.2 Region-Scale Analysis of Gradual Noise

Table 2 depicts the vertical accuracy values calculated for 25 homologues regions (350 by 350 m each) for all DTMs with respect to gradual noise along the x-axis direction. Only after applying the LSA process locally, the increasing accuracy trends in dataset 4 are detected. The left hand side of the dataset has a vertical accuracy variance value of approximately ±5.00 (original noise), whereas the right hand side shows an increasing value of up to approximately ±14.00 m (maximum noise), when compared to the equivalent overall vertical accuracy value in Table 1 of ±9.55 m.

5.1.3 Region-Scale Analysis of Regional Noise

Table 3 depicts diverse regional noise of datasets 1, 2, 3 and 4 ('known' σ)—input, where Table 4 depicts the vertical accuracy values computed—output.

Table 2 Local vertical absolute accuracy of 25 equal size regions in datasets 1, 2, 3 and 4 with respect to regional noise (all values with ± sign and in meters)

1

2.36	2.10	2.30	2.18	1.47
2.32	2.55	0.79	2.33	1.62
1.36	1.19	1.92	0.86	2.61
1.38	2.38	2.86	2.85	2.94
1.88	2.19	1.41	2.39	2.27

2

3.85	4.10	3.73	3.87	3.97
3.85	3.63	4.01	3.58	4.14
4.30	4.06	3.57	4.43	3.77
4.00	3.93	3.70	3.48	3.72
4.25	3.66	3.98	3.62	4.26

3

10.97	12.75	11.56	11.30	12.39
11.65	11.38	12.63	12.31	12.49
12.56	12.27	11.14	11.92	12.12
11.84	12.16	10.88	12.21	12.08
12.84	11.28	12.09	11.34	13.20

4

5.38	7.09	8.61	10.75	14.55
5.58	6.35	9.06	11.09	13.45
5.42	6.31	8.40	11.13	13.72
5.34	6.37	8.32	10.78	13.80
5.10	6.43	8.80	10.71	13.97

Table 3 Original local vertical accuracies of 25 regions with respect to regional noise (all values with ± sign and in meters)

1

2.00	2.00	2.00	2.00	2.00
2.00	2.00	2.00	2.00	2.00
2.00	2.00	5.00	5.00	5.00
2.00	2.00	5.00	5.00	5.00
2.00	2.00	5.00	5.00	5.00

2

10.00	4.00	4.00	4.00	4.00
4.00	4.00	4.00	4.00	4.00
4.00	4.00	4.00	4.00	4.00
4.00	4.00	4.00	4.00	4.00
4.00	4.00	4.00	4.00	4.00

3

12.00	12.00	12.00	12.00	12.00
12.00	2.00	2.00	2.00	12.00
12.00	2.00	2.00	2.00	12.00
12.00	2.00	2.00	2.00	12.00
12.00	12.00	12.00	12.00	12.00

4

5.00	5.00	5.00	5.00	5.00
5.00	5.00	5.00	5.00	5.00
5.00	5.00	5.00	5.00	5.00
5.00	5.00	5.00	5.00	5.00
5.00	5.00	5.00	5.00	5.00

Table 4 Computed local vertical accuracies of 25 regions with respect to regional noise (all values with ± sign and in meters)

1

2.13	2.14	2.15	2.17	2.18
2.29	1.79	2.31	1.98	2.45
2.14	1.83	4.58	4.90	5.44
1.56	1.92	5.01	5.00	4.88
2.30	2.23	4.89	5.41	4.86

2

9.83	3.62	3.68	3.64	4.14
3.54	3.67	3.76	3.69	3.45
4.12	3.33	3.79	4.11	3.89
4.01	4.03	4.00	3.95	4.01
3.84	3.42	4.12	3.23	3.73

3

12.35	12.26	11.82	11.91	11.05
14.15	1.84	1.99	1.89	13.08
12.00	2.19	2.46	2.14	12.83
12.28	1.88	1.84	2.31	12.59
11.69	12.20	12.67	11.85	11.59

4

4.87	5.52	4.83	5.01	4.81
4.72	5.08	4.75	5.09	4.72
4.84	4.77	5.04	4.98	4.94
5.27	5.28	5.33	5.11	4.63
4.96	4.90	5.75	5.88	5.39

Accuracy trends are successfully detected in datasets 1, 2 and 3, while noise in 1, 2 and 3 did not affect the values computed for dataset 4. It is worth noting that each region is independent in the LSA process—it remains unaffected by inaccuracies existing in equivalent corresponding regions. For instance, the most northwest region of dataset 2 suffers from local distortion (in respect to the rest of the area) expressed by a vertical accuracy variance value of ±10.00 m, while its equivalent corresponding regions (datasets 1, 3 and 4) have only the original DTM accuracies (values of ±2.00, ±12.00 and ±5.00 m, respectively). Still, when the LSA process is carried out, the computed accuracy values of datasets 1, 3 and 4 are not affected, mainly due to iterative weighting criteria applied. It is evident that each region expresses different accuracy values, where a relatively high (or 'better') accuracy is depicted by a smaller σ value.

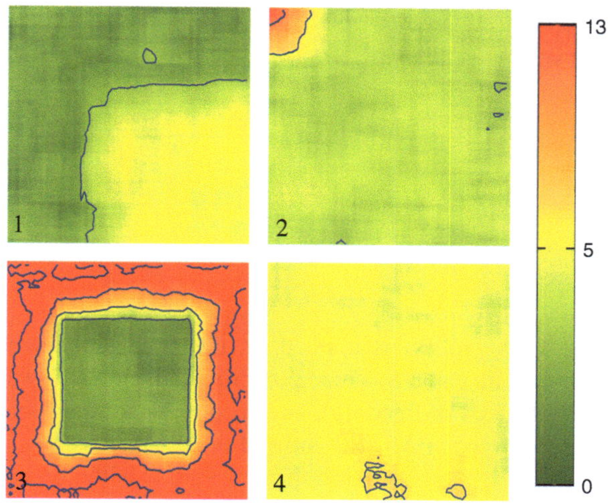

Fig. 4 Four generated synthetic DTMs with regional noise represented by stretched classification and 10 m height interval contours (values in meters)

5.1.4 Point-Scale Analysis of Regional Noise

Still, even regional accuracies are not always homogenous, thus a grid-point-scale LSA analysis is applied where the accuracy of each grid point is determined by the outcome of its 13 by 13 surrounding pixels, resulting in a smoothed representation of local accuracies, as depicted in Fig. 4. The region boundaries in this case are not taken into account. Contour representation shows the transition zones between low and high accuracy values, enabling the later automatic extraction of polygonal sub-areas expressing local accuracy ranges. It is clear that this visualization corresponds to the values computed and presented in Table 4.

The two synthetic analyses presented above express more precise local accuracy estimations for each area when compared to the overall global one. Hence, if one would treat the overall DTM as homogeneous, the accuracy measure may not always properly reflect the actual local accuracies exist. It is therefore necessary to address the existing local accuracies and error trends to derive more qualitative accuracy measures. This analytical conclusion is derived from the fact that the various databases have different data characteristics, namely data-production mechanisms, data acquisition sources, resolution, etc.

5.2 Real Data Experiments and Analysis

Two areas are examined, existing in the north mountainous and the south lowlands of Israel, representing two different terrain and morphologic characteristics: rough and relatively smooth, respectively. The distance between these two regions is roughly 160 km. The outcome of the transformation and interpolation to a unified reference system is roughly a coverage area of 10 by 10 km, with 25 m resolution.

Table 5 Statistical results for four real DTMs with respect to the entire area

DTM	Computed σ using 4 DTMs (m)		Computed σ using 3 DTMs (m)	
	North area	South area	North area	South area
25 m ITM	4.42	2.81	3.47	2.79
50 m ITM	5.07	2.05	5.07	2.35
ASTER 1 arc-second	10.39	10.74	–	–
SRTM 3 arc-second	3.15	2.65	4.09	2.4

The LSA process is applied on the entire coverage area, where the outcome is depicted in Table 5. This LSA process is applied twice: first, using all four DTMs (two degrees of freedom), and second, using only three best DTMs (no degrees of freedom). Since ASTER showed the poorest accuracies and error-trends in the first analysis, it was suspected that it might bias the outcome of the other three DTMs, thus these values were re-examined with the second implementation.

It is evident that the accuracies of global DTMs are not homogenous even within relatively small-scale areas, such as Israel. Therefore, the accuracy of an 'accurate' DTM might substantially vary, especially in areas where significant topographic changes occur. As discussed earlier, there exists a strong correlation between topographic behavior (or morphometric characteristics) and vertical accuracy. Statistically, models representing rough topographic areas are expected to have a greater variance value, and vice versa. When comparing the two areas expressed by the same DTMs, it is evident that the south area (smooth topography) showed relatively 'better' accuracies than the northern one (rough topography), proving the above statement. Eventually, the notion is to use as many DTMs as possible that cover the same area to generate a high-redundancy system. Even if a poor-accuracy DTM is inserted into the solution system (as in the case of ASTER), the weighting mechanism of the LSA process prevents the solution from being biased.

Grid-point-scale LSA analysis was implemented on the southern DTM of Israel to compute the absolute accuracy of each grid point in each DTM. The results were processed with a non-linear median filter to reduce noise. Figure 5 depicts the classification of polygonal accuracies into five classes.

Table 6 depicts the statistical mean and standard deviation values of all grid-point accuracies, expressing the overall vertical absolute accuracy value and the measure of dispersion from it.

Results show the feasibility of the method in extracting point-scale absolute vertical accuracies and their spatial positions, expressing local anomalies. Moreover, when compared to the equivalent overall vertical accuracy value in Table 5 (third column from the right) statistical mean values (Table 6) prove that even a single overall accuracy value can be reliably derived from such analysis. It is evident that the ASTER DTM demonstrate low (poor) accuracy with large values and deviations, in respect to the other DTMs participated. The resolution of a DTM affects the results, where relatively low-resolution DTMs (e.g., SRTM), show large

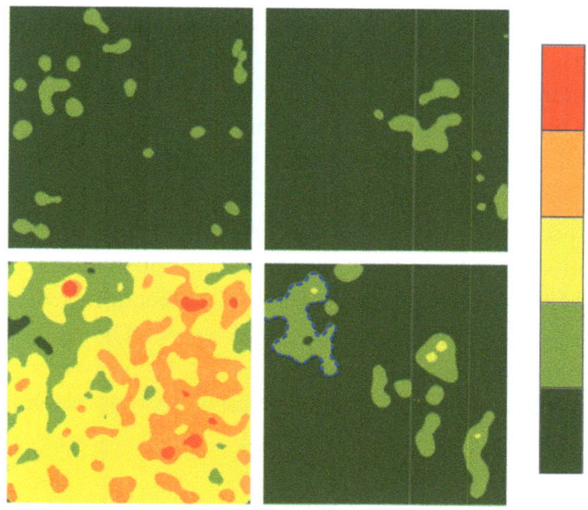

Fig. 5 Polygonal accuracy classification of four real southern DTMs based on point-scale analysis: 25 m ITM (*top left*), 50 m ITM (*top right*), ASTER (*bottom left*), SRTM (*bottom right*) (accuracy values in meters)

Table 6 Statistical results for point-scale accuracies of four real DTMs

DTM	Mean (m)	Standard deviation (m)
25 m ITM	2.57	0.90
50 m ITM	1.72	1.20
ASTER 1 arc-second	10.34	2.69
SRTM 3 arc-second	2.10	1.68

and smoothed accuracy polygons, while high resolution DTMs (e.g., 25 m ITM) show smaller ones. For example, the SRTM shows an area of approximately 7.5 km^2 (depicted in Fig. 5 as a blue dashed line), which represents a homogenous sub-area at the range of 4–8 m.

6 Discussion and Future Work

Because of diverse data sources and different production methods of wide-coverage DTMs, different data qualities, accuracies and representations exist; thus, the inner-accuracy of an individual DTM is not homogeneous. Using simultaneously multiple DTMs for applications requires addressing the issue of spatial analysis while investigating the characterization of the terrain.

This paper focused on the local analysis and calculation of absolute vertical accuracies of DTMs, by applying a multiple wide-coverage DTM comparison method using weighted LSA processes. The methodology proved to be reliable in computing global, regional and grid-point-scale absolute vertical accuracy values—even when local trends existed. This is carried out regardless of the need to use any preliminary a priori absolute accuracies (normally unknown). This method offered an original analytical methodology having the potential to replace the commonly used long and tedious pre-production processes that evaluated vertical accuracies of DTMs. The local continuous accuracy evaluation can be used to examine whether alternative post-production accuracy assessment procedures, usually relying on finite number of Ground Control Points (GCPs), meet the standards. As discussed above, relying solely on the straightforward coordinate-based superimposition comparison between all DTMs might not suffice in some cases for analyzing quality measures as proposed here, mainly when rapid topographic changes exist. Additional research is required to accomplish a proper horizontal registration, based on geomorphologic interest point identification, which is believed to enable a better and more reliable accuracy assessment. The feasibility to automatically extract accuracy-derived polygonal sub-areas was proven, when point-scale LSA process was applied. In future studies, such polygons, extracted from different models, will enable a comprehensive absolute vertical accuracy analysis of DTMs by regions and not only globally.

Acknowledgements The ASTER and SRTM data products were obtained through the online Data Pool at the NASA Land Processes Distributed Active Archive Center (LP DAAC), USGS/Earth Resources Observation and Science (EROS) Center, Sioux Falls, South Dakota. We express our gratitude to the Survey of Israel (SOI) for their generous help and assistance with providing the 25 and 50 m DTMs of Israel.

References

Aguilar FJ, Aguilar MA, Aguera F, Sanchez J (2006) The accuracy of grid digital elevation models linearly constructed from scattered sample data. Int J Geogr Inf Sci 20(2):169–192

Aguilar F, Agüera F, Aguilar A (2007a) A theoretical approach to modeling the accuracy assessment of digital elevation models. Photogramm Eng Remote Sens 73(12):1367–1379

Aguilar FJ, Aguilar MA, Aguera F (2007b) Accuracy assessment of digital elevation models using a non-parametric approach. Int J Geogr Inf Sci 21(6):667–686

ASTER GDEM Validation Team (2009) ASTER global DEM Validation Summary Report

Atkinson A, Ariza DJ, López FJ, García-Balboa JL (2007) Positional accuracy control using robust estimators.In: Proceedings of the 21st international cartographic conference, Spain

Ben-Haim G, Dalyot S, Doytsher Y (2013) A geostatistical approach for computing absolute vertical accuracy of digital terrain models. In: Fourth international conference on computing for geospatial research and application (COM.Geo), IEEE, California, USA, pp 32–39

Bhang KJ, Schwartz FW, Braun A (2007) Verification of the vertical error in C-band SRTM DEM using ICESat and Landsat-7, Otter Tail County, MN. IEEE Trans Geosci Remote Sens 46 (1):36–44

Bjerhammer A (1973) Theory of errors and generalized matrix inverses. Elsevier Scientific Publishing Company, New York

Brown CG, Sarabandi K, Pierce LE (2005) Validation of the shuttle radar topography mission height data. IEEE Trans Geosci Remote Sens 43(8):1707–1715

Chen X, Murai S (2008) Universal error propagation law. In: Proceedings of the international archives of the photogrammetry, remote sensing and spatial information sciences, Vol XXXVII, Part B2, pp 1125–1132

Dalyot S (2010) Hierarchical modeling and integration of topographical databases: algorithm development and geospatial capabilities. Ph.D. Thesis, Technion, Haifa

Dalyot S, Doytsher Y (2008) A hierarchical approach toward 3-D geospatial dataset merging. In: Representing, modelling and visualizing the natural environment: innovations in GIS 13, CRC Press, pp 195–220

Frederiksen P (1980) Terrain analysis and accuracy prediction by means of the Fourier transformation. Int Arch Photogramm Remote Sens 23(4): 284–293

Hall JK (2008) The 25-m DTM (digital terrain model) of Israel. Israel J Earth Sci 57(3/4):145–147

Karel W, Pfeifer N, Briese C (2006) DTM quality assessment. In: ISPRS technical commission II symposium, Vienna, pp 7–12

Katzil Y, Doytsher Y (2005) Spatial Rubber Sheeting of DTMs. In: Proceedings of the sixth geomatic week conference, Barcelona

Kellndorfer J, Walker W, Pierce L, Dobson C, Fites JA, Hunsaker C, Vona J, Clutter M (2004) Vegetation height estimation from shuttle radar topography mission and national elevation datasets. Remote Sens Environ 93(3):339–358

Li Z (1992) Variation of the accuracy of digital terrain models with sampling interval. Photogramm Rec 14(79):113–128

Li Z (1993) Theoretical models of the accuracy of digital terrain models: an evaluation and some observations. Photogramm Rec 14(82):651–660

McCullagh M (1988) Terrain and surface modelling systems: theory and practice. Photogramm Rec 12(72):747–779

Parry RB, Perkins CR (2000) World mapping today. University of Reading, Department of Geography, Cartographic Unit, 2nd edn, UK

Raizman Y, Peled A (2002) Technological aspects of the 3-dimensional photogrammetric updating of the Israeli national GIS. In: Symposium on geospatial theory, processing and applications, Ottawa

Rosen PA, Hensley S, Gurrola E, Rogez F, Chan S, Martin J, Rodriguez E (2001) SRTM C-band topographic data: quality assessments and calibration activities. IEEE Geosci Remote Sens Symp 2:739–741

Sun G, Ranson K, Kharuk V, Kovacs K (2003) Validation of surface height from shuttle radar topography mission using shuttle laser altimeter. Remote Sens Environ 88(4):401–411

Tempfli K (1980) Spectral analysis of terrain relief for the accuracy estimation of digital terrain models. I.T.C J 1980(3): 478–510

The Shuttle Radar Topography Mission Team (2007) The Shuttle Radar Topography Mission. Rev. Geophysics 45, RG2004. Available online at http://www2.jpl.nasa.gov/srtm/SRTM_paper.pdf

Tighe ML, Chamberlain D (2009) Accuracy comparison of the SRTM, ASTER, NED, NEXTMAP USA digital terrain model over several USA study sites. In: Proceedings of the ASPRS/MAPPS fall conference

Weibel R, Heller M (1991) Digital terrain modelling. In: Maguire DJ, Goodchild M, Rhind DW (eds) GIS: principles and applications, London, pp 269–297

Wilson JP, Repetto PL, Snyder RD (2000) Effect on data source, grid resolution and flow-routing method on computed topographic attributes. In: Wilson JP, Gallant JC (eds) Terrain analysis: principles and applications, Chapter 5, New York

Analyzing the Influence of Ethnic Composition and Immigrant Residents on the Spatial Distribution of Violent Crime

Matthew Quick and Jane Law

Abstract The spatial distribution of violent crime is influenced by small-area characteristics. The social disorganization theory proposes that neighbourhood-scale characteristics, including ethnic composition and immigrant residents, indirectly influence crime through social control. Recent spatial demographic changes in urban areas, including increased immigration and ethnic heterogeneity in city peripheries, have motivated reconsiderations of social disorganization. Using exploratory spatial data analysis and spatial regression methods, this research identifies violent crime hotspots and analyzes the influence of ethnic composition and immigrant resident concentration on violent crime in Toronto, Ontario, at the census tract scale. Results suggest that violent crime hotspots are located in downtown and north Toronto and that ethnic heterogeneity is positively associated with violent crime rate while immigrant resident concentration is negatively associated. This research provides novel insight into the spatial dimensions of crime and the effects of spatial demographic changes on violent crime and social disorganization in contemporary cities.

Keywords Spaital analysis · Violent crime · Spatial regression · Social disorganization · Demography

1 Introduction

The spatial distribution of crime in urban areas is associated with variations in demographic and socio-economic characteristics (Brantingham and Brantingham 1981). At the small-area or neighbourhood scale, the social disorganization theory is often employed to explain the spatial patterning of crime, hypothesizing that

M. Quick (✉) · J. Law
University of Waterloo, Waterloo, Ontario, Canada
e-mail: mquick@uwaterloo.ca

J. Law
e-mail: jane.law@uwaterloo.ca

F. Harvey and Y. Leung (eds.), *Advances in Spatial Data Handling and Analysis*, Advances in Geographic Information Science,
DOI 10.1007/978-3-319-19950-4_14

neighbourhood social dynamics are an important determinant of crime beyond individual motivations and behaviours (Sampson et al. 2002). Recently, many urban areas have experienced considerable spatial demographic change characterized by increasing immigration rates and a transition of high-immigrant neighbourhoods from city centres to the urban periphery. These changes may influence the location of crime hotspots as well as the neighbourhood characteristics associated with crime.

This research is motivated by two shortcomings. First, contemporary spatial demographic changes have complicated the operationalization of social disorganization variables, particularly ethnic heterogeneity and immigrant residents (Martinez and Lee 2000; Graif and Sampson 2009). Despite these characteristics being conceptually distinct, past research, for example, has assumed that the percent of immigrant residents in a neighbourhood is representative of its ethnic heterogeneity (Andresen 2006a, b). Second, investigations of social disorganization and crime do not often use appropriate spatial analysis methods, which could lead to misleading results and limited insight regarding the location of high crime areas and the spatial processes influencing crime (Graif and Sampson 2009; Kubrin and Weitzer 2003). This research addresses these shortcomings to identify violent crime hotspots and analyze the influence of ethnic composition and immigrant resident concentration on violent crime in Toronto, Ontario at the census tract scale.

Briefly, the social disorganization theory proposes that neighbourhood-scale structural characteristics including high population turnover, high ethnic heterogeneity, large immigrant populations, and economic deprivation lead to community-scale social disorganization (Warner and Rountree 1997). These structural characteristics are linked to crime through mediating neighbourhood-scale social processes including a limited ability to form and maintain social relationships, reduced social ties, and a collective inability to solve commonly experienced problems (Shaw and McKay 1942; Bursik 1988; Sampson and Groves 1989; Veysey and Messner 1999). These social processes result in a weakening of neighbourhood-scale informal social control and, consequently, increases in crime.

Because social disorganization attempts to explain the spatial variation of crime across small-areas rather than individual people, using spatial analysis to study neighbourhood-scale crime has both methodological and theoretical advantages (Lee and Martinez 2002; Kubrin and Weitzer 2003). Methodologically, spatial analytical methods recognize that crime offenses do not occur randomly in space and they allow for researchers to overcome the complications of using spatially autocorrelated data in regression models (Townsley 2009). Positive spatial autocorrelation of crime data, or when similar values are located in close proximity to each other, may result in biased coefficient or standard error estimates and incorrect inferences in ordinary least squares regression models due to error terms not being independent (Tita and Radil 2010b; Baller et al. 2001). Theoretically, the results of spatial analyses inform our understanding of the spatial processes related to social disorganization and crime (Ouimet 2000). For example, it is possible that crime hotspots uncovered through exploratory spatial data analysis methods are located close to commercial land uses (Kinney et al. 2008) or that high crime in one area appears to influence crime rates in adjacent areas even after accounting for neighbourhood socio-economic characteristics (Kubrin and Weitzer 2003).

Contemporary changes to the spatial demographic characteristics of urban areas suggest that past interpretations of social disorganization no longer reflect the current social processes related to crime (Martinez and Lee 2000). Toronto, Ontario, is a city that has been transformed by recent immigration and increases in social and cultural diversity (Siematycki and Isin 1997). Since the 1960s, Toronto has experienced historically high immigration rates with large proportions coming from non-European countries (Bourne and Ross 2001). Over the same time period, there has been an increase in the number of low-income neighbourhoods that are home to disproportionately high concentrations of recent immigrants (Hulchanski 2010). Spatially, low-income and high-immigrant neighbourhoods used to be located close to the centre of Toronto, however these populations have transitioned to the periphery of the city such that the largest concentrations of immigrant residents are now located in the inner suburbs (Hulchanski 2010; Siemiatycki and Isin 1997). Because of these profound spatial demographic changes, Toronto is a compelling case study for this research.

This research begins with a review of past studies that investigate the influence of immigration and ethnic composition on crime from a social disorganization perspective. Following this, we review the data and visualize the spatial distribution of violent crime, immigrant resident concentration, and ethnic composition. Next, exploratory spatial data analysis and spatial regression methods are reviewed and results are shown and discussed, focusing on how this research compares and contrasts with past interpretations of social disorganization. We also discuss the additional understanding gained from spatial analysis and the implications of results for policy makers.

2 Literature Review

2.1 Immigration and Crime

A number of theoretical approaches address the relationship between immigration and crime including economic and subcultural explanations that focus on labour shortages, illegitimate work opportunities, ethnic gangs, and social and cultural marginalization (Reid et al. 2005; Wortley 2009). From the social disorganization perspective, neighbourhoods with high immigrant populations are hypothesized to have an increased likelihood of crime (Shaw and McKay 1942; Stowell et al. 2009; Kubrin 2009). Large immigrant populations are considered a collective source of disruption because they tend to live in areas with high residential instability or population turnover (Lee and Martinez 2002; Ousey and Kubrin 2009). Limited socioeconomic resources, high pre-existing crime rates, and communication difficulties among neighbourhood residents in areas where there are large immigrant concentrations are also believed to contribute to limited informal social control and increases in social disorganization and crime (Stowell et al. 2009; Graif and Sampson 2009; Peterson and Krivo 2005).

Following individual-level research demonstrating that immigrants commit fewer crimes than non-immigrants (Ousey and Kubrin 2009) and critical observations that the disorganizing effect of immigrant populations was thought to be due to population turnover and not, in fact, immigrant residents themselves, recent research has shown that neighbourhoods with high immigrant concentrations often exhibit low crime (Martinez et al. 2010; Lee and Martinez 2009; Wortley 2009). For example, Sampson and Raudenbush (1999) have found that neighbourhood immigrant concentration was inversely associated with robbery and Graif and Sampson (2009) observed that immigrant populations were insignificantly associated with homicide in Chicago. Velez (2009) noted that an increasing immigrant presence in disadvantaged neighbourhoods was associated with decreases in homicide. Over time, cities with increasing immigration rates have been shown to exhibit decreasing violent crime and robbery trends (Ousey and Kubrin 2009; Stowell et al. 2009).

2.2 Ethnic Composition and Crime

Neighbourhoods with a highly mixed ethnic composition (i.e., high ethnic heterogeneity) are hypothesized to exhibit social disorganization, weakened informal social control, and an increased likelihood of crime (Warner and Rountree 1997). In highly heterogeneous neighbourhoods, it is believed that there is limited communication among residents because of language differences and also because socializing often occurs between residents of the same ethnicity (Kubrin 2008). Ethnic heterogeneity has also been identified as a characteristic that contributes to fractured neighbourhood social ties, complications regarding the establishment of common values, and a barrier to solving common community problems including crime (Bursik 1988; Lee and Martinez 2002; Kubrin 2009). Because social disorganization theorists assume that norms are shared among members of the same ethnicity, high ethnic heterogeneity introduces competing sets of norms that impede collective informal social control and increase the likelihood of crime (Warner and Rountree 1997).

Compared to the literature on immigration and crime, less research has focused directly on the influence of ethnic composition on crime. Sampson and Groves (1989) observed a positive relationship between ethnic heterogeneity and street robbery, total victimization, and burglary, and Warner and Rountree (1997) noted that ethnic heterogeneity was associated assault and burglary in Seattle at the census tract scale. Further, racial heterogeneity was found to be the strongest predictor of violent crimes in both cross-sectional and longitudinal analyses (Kubrin 2008).

One reason for the relative lack of literature focusing on ethnic composition or ethnic heterogeneity and crime may be due to the failure to conceptually distinguish between ethnic composition and immigrant concentration and employ variables that are representative of each concept (Kubrin 2009). Past studies have assumed that immigrant concentrations and ethnic composition can be inferred from the same variable measuring the concentration of immigrants (e.g., percent of immigrant residents) or the concentration of one ethnic group (e.g., percent of Black or Latino residents). For example, Andresen (2006a, b) and Ouimet (2000) measure the ethnic heterogeneity dimension of social disorganization through the percent of immigrant residents in a neighbourhood. Moreover, both Martinez et al. (2010) and Stowell et al. (2009) observe a negative association between the concentration of immigrant residents and crime, yet conclude that ethnic heterogeneity has a protective effect on crime even though heterogeneity was not analyzed. This distinction is important, particularly in the case of ethnic enclaves, which are neighbourhoods that have relatively low ethnic heterogeneity but have a high concentration of immigrant residents. In Canada, ethnic enclaves do not appear to be co-located with other characteristics of social disorganization such as poverty, which emphasizes the importance of the distinction between immigrant concentration and ethnic composition in this case study (Walks and Bourne 2006).

3 Data

In 2006, Toronto had a residential population of approximately two and a half million and was comprised of 524 census tracts. Three census tracts had incomplete Census data and were omitted from analysis (Statistics Canada 2006). For reference, census tracts are small-area units located in census metropolitan areas (populations greater than 50,000) and have permanent and recognizable boundaries with populations between 2500 and 8000 (Statistics Canada 2012). Regarding immigrant residents and ethnic composition, approximately half of Toronto residents are immigrants and the city is home to twenty percent of all Canadian immigrants and thirty percent of all recent immigrants to Canada (Statistics Canada 2006; City of Toronto 2013; Hulchanski 2010). Toronto is home to over two hundred ethnic groups and twenty of these ethnic groups are represented by at least two percent of the city's residents (City of Toronto 2013).

Crime offense data was extracted from the 2006 Uniform Crime Reporting Survey (UCR) at the census tract scale. UCR data is reported by municipal police services (i.e., Toronto Police Service) and aggregated and disseminated by the Canadian Centre for Justice Statistics. Violent crime types include robbery, criminal harassment, and uttering threats, among others (Statistics Canada 2011). Violent crime rate was calculated as the number of violent crime offenses divided by residential population. Residential population was chosen as the denominator because offenders often commit violent crimes in their own neighbourhoods (Morenoff et al. 2001; Steenbeck et al. 2012; Kubrin 2008) and as such, it best represents risk to potential targets (Wikstrom and Dolmen 1990). In total, there were 25,764 violent crime incidents in Toronto with an average of 10.35 violent crimes per 1000 people and about 49 crimes per census tract (Table 1). Violent crime was highest in downtown areas of Toronto (south, close to the centre of the city) and in the northwest (Fig. 1). Global spatial autocorrelation of violent crime was measured using Moran's I, which quantifies the total deviation from spatial randomness across the study region between negative one and positive one, where zero is spatial randomness, positive one is positive spatial autocorrelation, and negative one is negative spatial autocorrelation (Tita and Radil 2010b). Violent crime rate (per 1000 residents) was slightly spatially autocorrelated in Toronto ($I = 0.24$).

Four dimensions of social disorganization were analyzed: economic deprivation, family disruption, population turnover, and ethnic heterogeneity (Table 1) (Law and Quick 2013). Economic deprivation was operationalized through unemployment rate, percent of low-income families, and percent of families receiving government transfer payments. Family disruption was measured through the percent of lone-parent families. Population turnover was operationalized through one- and five-year residential mobility rates. The ethnic heterogeneity dimension of social disorganization was operationalized through index of ethnic heterogeneity and the concentration of immigrant residents. The index of ethnic heterogeneity measures the number of ethnic groups and relative size of groups in a census tract, and is shown in Eq. 1 where g is the proportion of each ethnic group, i. The index of ethnic heterogeneity ranges from zero to one, where one represents maximum heterogeneity (when there are equal proportions of each ethnic group represented in the census tract population) (Sampson and Groves 1989; Warner and Rountree 1997).

Table 1 Descriptive statistics for violent crime and social disorganization covariates at the census tract scale

	Mean	Std. Dev.	Min.	Max.
Census tract population	4766	1834	310	22,725
Violent crime (count)	49.36	38.81	7	469
Violent crime rate (per 1000 people)	11.37	14.46	1.05	225.69
Social disorganization				
1. Economic deprivation				
Unemployment rate (%)	7.61	2.7	1.6	18.8
Government transfer payment (%)	11.07	5.79	0.5	34.5
Low-income families (%)	14.9	8.48	0.9	56.6
2. Family disruption				
Lone parent families (%)	20.03	7.44	3.57	51.06
3. Residential mobility				
One-year movers (%)	15.06	5.93	3.73	53.85
Five-year movers (%)	43.91	11.93	12.9	87.33
4. Ethnic heterogeneity				
Immigrant residents (%)	48.27	15.37	13.52	79.12
Index of ethnic heterogeneity (0–1)	0.63	0.13	0.22	0.86

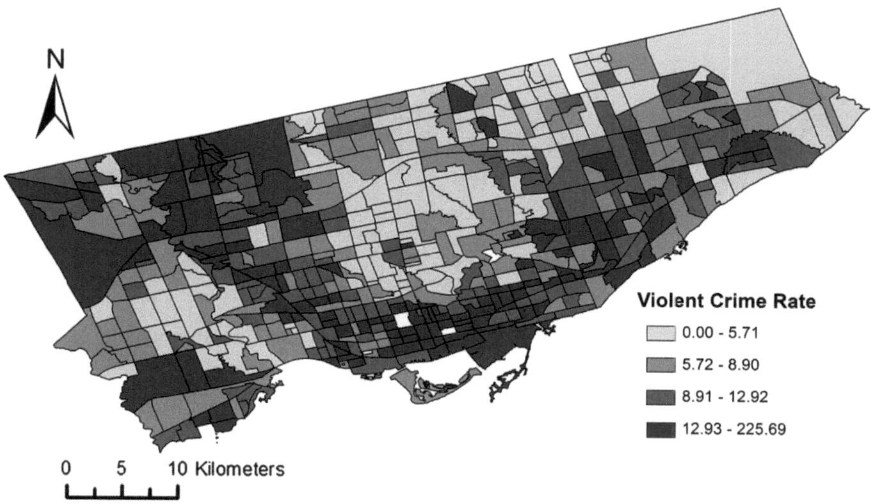

Fig. 14.1 Violent crime rate (per 1000 residents) at the census tract scale in Toronto

$$\text{Index of Ethnic Heterogeneity} = 1-\sum g_i^2 \quad (1)$$

Ethnic heterogeneity is highest in the northwest and east areas of Toronto (Fig. 2) and index of ethnic heterogeneity was moderately spatially clustered (I = 0.57, $p < 0.05$).

As defined by the 2006 Statistics Canada Census, immigrants are people who are, or have been, landed immigrants in Canada. Immigrant resident concentration was calculated as the [(number of immigrants in a census tract/total number of residents) × 100]. As noted, areas with high immigrant concentrations are the inner suburbs in the northeast, north, and northwest of Toronto (Fig. 3). Immigrant resident concentration was highly clustered as measured by Moran's I (I = 0.70, $p < 0.05$).

Socio-economic data at the census tract scale was obtained from the 2006 Statistics Canada Census. All Census data and GIS shapefiles were obtained from the University of Waterloo Geospatial Centre.

4 Modeling Approach

The first stage of our analysis was exploratory spatial data analysis to visualize violent crime hotspots (Ratcliffe and McCullagh 1999). We used local Moran's I, which decomposes global Moran's I into small-area observations and identifies hotspots as areas that reject the null hypothesis of spatial randomness by exhibiting high spatial autocorrelation (Anselin 1995). In the crime context, hotspots are

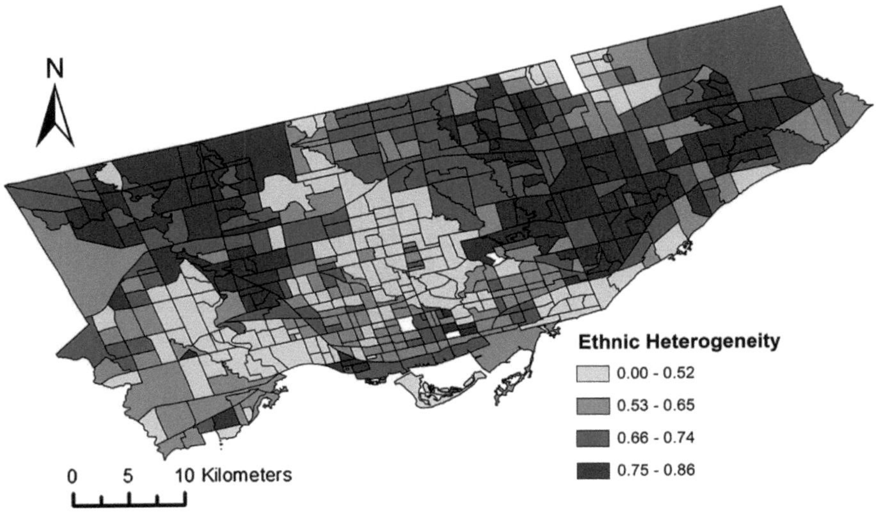

Fig. 14.2 Index of ethnic heterogeneity at the census tract scale in Toronto

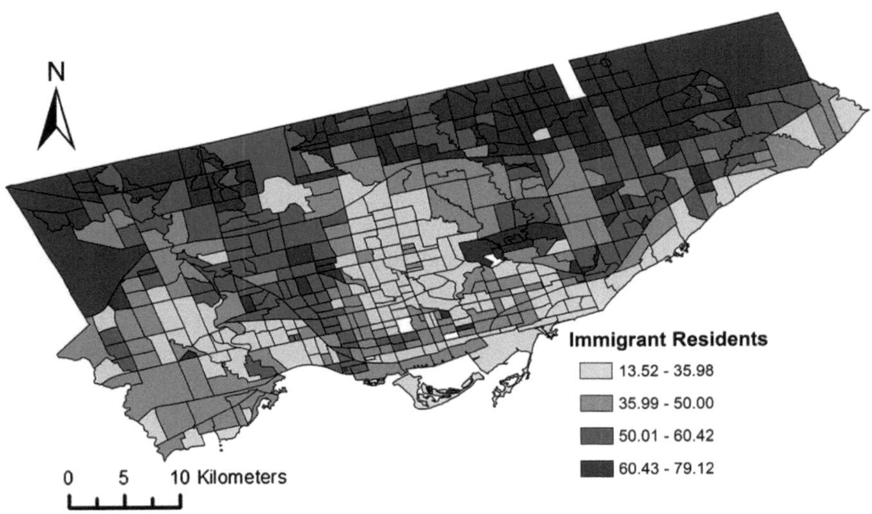

Fig. 14.3 Immigrant resident concentration at the census tract scale in Toronto

groups of census tracts where local Moran's I is positive and significant, or exhibit violent crime rates significantly greater than the average violent crime rate (Anselin 1995; Quick and Law 2013). Local Moran's I has been applied to explore spatial data and identify hotspots in past studies of crime (Baller et al. 2001; Murray et al. 2001).

The second stage of our analysis was regression modeling to identify significant associations between social disorganization characteristics and violent crime. We first used non-spatial ordinary least squares regressions to observe the relationship between each social disorganization covariate and violent crime rate (per 1000 people). Six covariates exhibited a significant relationship ($p < 0.05$): unemployment rate, percent of low-income families, one-year movers, five-year movers, and both concentration of immigrant residents and index of ethnic heterogeneity.

These covariates were included in one non-spatial multiple linear regression model to determine the simultaneous effect of these variables on violent crime. Insignificant covariates were removed until the final model contained only significantly associated covariates ($p < 0.05$). To ensure that this model did not exhibit multicollinearity, which could contribute to misleading parameter estimates (Farrar and Glauber 1967), we examined variance inflation factors and bivariate correlations for social disorganization characteristics (Warner and Rountree 1997). The only problematic variance inflation factor was between one-year movers and five-year movers. This was not unexpected because neighbourhoods with high short-term residential mobility likely have characteristics that lead to high long-term residential mobility (e.g., high density of rental housing). We kept one-year residential mobility because it exhibited more explanatory power in univariate regression models.

We tested residuals from the final multiple linear regression model for spatial autocorrelation using Moran's I. We anticipated spatial autocorrelation because violent crime rate exhibited significant clustering in exploratory analysis. Also, census tract boundaries are arbitrary and do not necessarily represent the spatial distribution of neighbourhood characteristics, so nearby neighbourhoods likely have similar social and demographic profiles (Baller et al. 2001).

Two simultaneous autoregressive models that account for spatial data were tested. These models, the spatial lag and spatial error models, are popular in social science research when analyzing continuous data for spatial areas (Tita and Radil 2010b). Briefly, the spatial lag model adds a covariate to the ordinary least squares regression model that represents the average of adjacent areas' violent crime rates. Superior fit of the spatial lag model can be used to interpret spatial effects in the dependent variable and suggests that small-area violent crime rates are influenced by violent crime rates in adjacent areas. The spatial error model modifies the ordinary least squares regression by incorporating an error term that averages the errors of adjacent areas. Superior fit of the spatial error model can be used to infer the presence of important and unmeasured covariates or unobservable social processes that do not match census tract boundaries (Tita and Radil 2010a, b; Anselin and Rey 1991).

A first-order row-standardized queen contiguity spatial weight matrix was used to specify spatial influence for exploratory spatial data analysis and spatial regression models (Anselin and Rey 1991; Murray et al. 2001). This spatial weight matrix specification assumes spatial interaction between a census tract and all adjacent census tracts that share any point including boundaries or vertices. We chose this specification because it follows past research observing that crime in one small-area is affected by crime in adjacent small-areas (Peterson and Krivo 2009) and there was no theoretical reasons to choose more particular specifications (i.e., rook-based). Construction of the spatial weight matrix, exploratory spatial data analysis, and spatial regression modeling was completed in GeoDa (Anselin 2005).

For modelling counts or binary outcomes rather than rates of crime, alternative spatial model types should be considered including binomial, Poisson, negative binomial (Haining et al. 2009), and Bernoulli (Law and Haining 2004). Further, spatio-temporal approaches should be considered when data containing both spatial and temporal components are to be modelled (Law et al. 2014; Chun 2014).

5 Results

Exploratory spatial data analysis using local Moran's I identified high violent crime rate hotspots in the downtown, northwest, and east of Toronto (Fig. 4). Specifically, these hotspots consist of census tracts with high violent crime that are surrounded by other high violent crime areas.

Residuals of the final non-spatial multiple linear regression model exhibited significant spatial autocorrelation ($I = 0.17, p < 0.05$). This model had an R-squared value of 0.12, suggesting that social disorganization characteristics explain a small amount of the variation in violent crimes and there are likely other important factors unaccounted for. For reference, fit of the non-spatial multiple regression model was indicated by a log likelihood value of −2090.0 and an AIC of 4190.07.

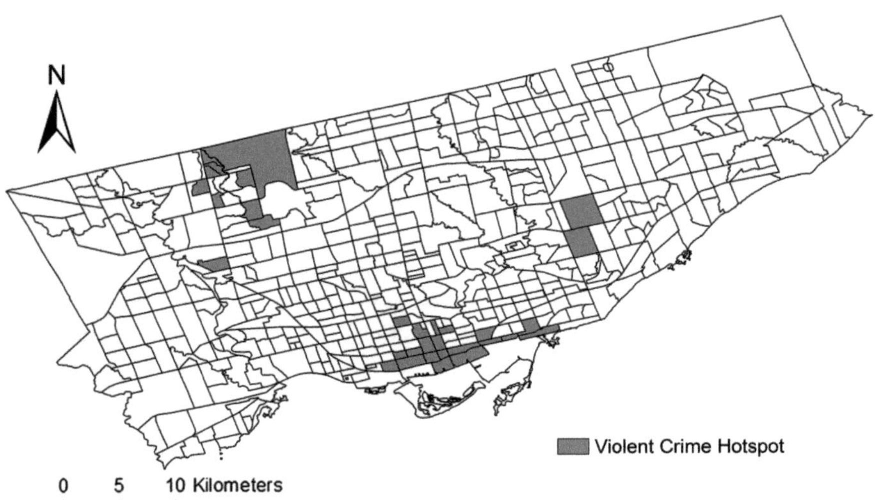

Fig. 14.4 Violent crime rate hotspots in Toronto based on local Moran's I

Table 2 Spatial lag regression results

Covariate	β	S.E.
One-year movers	0.51	0.09
Immigrant residents	−0.17	0.04
Index of ethnic heterogeneity	14.54	5.11
Spatial lag parameter	0.36	0.06

Evaluating model fit of spatial regression models, the spatial error model had a log likelihood of −2070.8 and an AIC of 4153.6 and the spatial lag model had a log likelihood of −2070.9 and an AIC of 4153.9. Models that have relatively smaller AIC values and relatively larger log likelihood values indicate better model fit, so both spatial regression models exhibited improved fit over the non-spatial multiple regression (Anselin 1998). Log likelihood and AIC values for spatial lag and spatial error models were not conclusive for either model type, so we evaluated alternatives through robust Lagrange Multiplier tests on the non-spatial linear regression model (Anselin 1998; Anselin 2005). To be clear, the robust Lagrange Multiplier for the spatial lag is robust to ignore spatial error dependence and the robust Lagrange Multiplier test for the spatial error is robust to ignore spatial lag dependence. The robust Lagrange Multiplier for the spatial lag model was highly significant and the robust Lagrange Multiplier for the spatial error was insignificant, suggesting that the spatial lag regression model was the appropriate model for this analysis (Anselin 2005; Anselin and Rey 1991). The results from the final spatial lag regression model can be seen in Table 2.

6 Discussion

According to the City of Toronto Neighbourhood Planning Areas (City of Toronto 2009), each of which consist of a number of census tracts, the violent crime hotspots shown in Fig. 4 roughly correspond to the neighbourhoods of York University Heights and Downsview-Roding-CFB in the northwest, the Bay Street Corridor and adjacent neighbourhoods in the downtown, and Wexford/Maryvale in east Toronto. Based on visual comparison of violent crime hotspots to maps of covariates (Hirschfield et al. 1995), in this case ethnic heterogeneity and immigrant residents (Figs. 2 and 3), it can be observed that hotspot location parallels the pattern of high ethnic heterogeneity neighbourhoods, but not high immigrant concentration neighbourhoods.

The results from the final spatial lag regression model found four covariates that were significantly associated with violent crime rate at the census tract scale. Three variables were positively related to violent crime; one-year movers, index of ethnic heterogeneity, and the spatial lag parameter. The large coefficient of the ethnic heterogeneity term is because it was measured on a scale of zero to one rather than as a percent. Immigrant resident concentration exhibited a negative relationship with violent crime rate. So, census tracts with high ethnic heterogeneity and a high percentage of one-year movers are likely to exhibit high violent crime, whereas census tracts with high immigrant concentrations are likely to exhibit low violent crime rates. The spatial lag parameter suggests that violent crime rate in a given census tract can, to a degree, be explained by violent crime rates in adjacent census tracts.

The negative association between immigrant resident concentration and violent crime supports recent research finding that immigrant populations are negatively associated with violent crime, perhaps due to decreased social disorganization (Lee and Martinez 2009). The immigrant revitalization thesis is one explanation, hypothesizing that immigrant residents contribute to economic revitalization and development and as a result, strengthen inform social control (Martinez et al. 2010; Reid et al. 2005; Velez 2009). In addition to economic forces, immigrant communities may have strong familial bonds and formal and informal neighbourhood institutions (Kubrin 2013). These characteristics, although likely only shared among immigrant residents, may contribute to entire neighbourhoods having distinct shared identities, strong social networks, and high collective efficacy, which is a combination of social cohesion among residents and the willingness of residents to intervene on behalf of the common good (Graif and Sampson 2009; Sampson et al. 1997).

Ethnic heterogeneity, on the other hand, was found to be positively associated with violent crime rate. This finding reinforces conventional interpretations of social disorganization that highlight the limited social interactions among ethnic groups and the role of heterogeneity in undermining social ties and weakening informal social control (Kubrin 2009). By finding a positive association between heterogeneity and crime and a negative association between immigrant concentration and crime, this research demonstrates the importance of distinguishing between these concepts in future spatial studies of crime. The results of this research suggest that recent spatial demographic changes in Toronto, and likely in other large, diverse, and multi-ethnic cities, are shaping the location and environments where crime occurs.

The superior fit of the spatial lag regression model, which has been applied in past crime research (Tita and Radil 2010b), and the positive association between the spatial lag parameter and violent crime rate also provides important insights into the dynamics of violent crime. Neighbourhoods adjacent to high-crime neighbourhoods are expected to have higher violent crime, even after accounting for social disorganization effects. One explanation of this is that violent crime may diffuse from high crime areas to adjacent areas through movement of criminals and offender knowledge of attractive targets in their activity and awareness spaces (Brantingham and Brantingham 2008). Also, the spatial distribution of violent crimes may follow social networks, which likely span a number of nearby census tracts (Peterson and Krivo 2005).

Two other social disorganization observations can be made from this research. First is that residential mobility, as operationalized through the percent of one-year movers in a census tract, was found to be positively associated with violent crime rate. High residential mobility, or when there is constant turnover of residents, is believed to make it difficult to establish social networks, disrupt existing community social networks, and reduce the capabilities of residents to identify and trust one another (Bursik 1988; Kubrin 2009). Second is that economic deprivation variables were found to be insignificantly associated with violent crime. This is in line with past research that has questioned the theoretical relevance of economic deprivation as it is not necessarily a cause of the social processes leading to social disorganization, but rather just a characteristic of neighbourhoods with high mobility and heterogeneity (Warner and Rountree 1997). It is possible, however, that the insignificance of economic deprivation is unique in this case study of Toronto, as poverty and ethnic concentration are not clustered to the same degree as in other cities or countries, including the United States (Walks and Bourne 2006).

In addition to the theoretical contributions of this study, this research provides evidence to inform practical operations including municipal public policy. Municipal land use planners have influence over residential settlement patterns (Siemiatycki and Isin 1997) and should carefully consider the location of land uses that inherently have high residential mobility (e.g., rental apartments) and their co-location with other neighbourhood characteristics including ethnic heterogeneity and criminogenic land uses such as bars (Ratcliffe 2012; Peterson and Krivo 2005; Bursik 1988). Further, in neighbourhoods that exhibit characteristics indicative of high social disorganization, policy makers and municipal agencies may attempt to strengthen informal social control through establishing neighbourhood groups or initiating public outreach and crime prevention efforts.

There are four limitations to this research. First, it is important for these results to be interpreted at the neighbourhood or census-tract scale and not at more precise (i.e., individual) or general (i.e., municipal or national) spatial scales (Openshaw 1984; Ouimet 2000; Ratcliffe and McCullagh 1999) because the neighbourhood-scale social processes hypothesized by social disorganization and inferred through structural characteristics are not interpretable at other spatial scales (Kubrin 2008; Mears 2001). For example, it is fallacious to assume that because high residential mobility is associated with high violent crime at the census tract scale, that individuals who move frequently commit violent crimes, or that cities with high mobility rates have high violent crime rates. It should be noted that census tracts are imperfect operationalizations of neighbourhoods, however it is the only spatial scale where Census and crime data are similarly aggregated (Sampson et al. 2002).

Second, official crime data may be under reported, particularly for immigrant-to-immigrant crime (Kubrin 2013) and for disadvantaged communities where residents may have negative perceptions of police and be less inclined to report crimes to law enforcement (Warner and Rountree 1997). Further, specific ethnic populations, and consequently specific neighbourhoods, may be overrepresented in crime as a result of discrimination of law enforcement and the criminal justice system (Wortley 2009).

Third, the mediating factors proposed by the social disorganization theory, for instance informal social control, are not directly measured through neighbourhood-scale structural characteristics (Sampson et al. 2002). For example, Warner and

Rountree (1997) and Sampson and Groves (1989) analyze structural characteristics including poverty, ethnic heterogeneity, and residential mobility, as well as mediating concepts such as community social ties, local friendship networks, and organizational participation in social disorganization models. While this was beyond the scope of this research, it would be beneficial for future studies to directly focus on the role of mediating concepts in this research context. It is important to remember that the social disorganization variables measured in this research are social constructs that must be considered in context of environmental, economic, and historical factors (Kaplan 2004).

Fourth, these results may only be representative of Toronto, Ontario and not generalizable to other cities. However, because our results support recent research that examines the link between immigrant residents and crime, we believe that this research is indicative of the influence of large-scale spatial demographic changes on violent crime and social disorganization.

7 Conclusion

This research investigates the influence of between ethnic heterogeneity and immigrant resident concentration on violent crime rate in Toronto, Ontario. Conventional interpretations of social disorganization theory propose that both immigrant concentration and ethnic heterogeneity contribute to increased social disorganization, weakened informal social control and increased crime. Recent spatial demographic changes over the last thirty years, including increasing immigration rates, increasing ethnic diversity of immigrants, and the movement of high-immigrant communities from the city centre to the inner suburbs, has motivated reconsiderations of the social disorganization theory and its relevance for explaining crime in contemporary cities.

Using exploratory spatial data analysis, we found that violent crime hotspots were located in downtown, northwest, and east areas of Toronto and closely parallel the location of neighbourhoods with high ethnic heterogeneity. Results from the best fitting spatial lag regression model indicated that the concentration of immigrant residents in a census tract was negatively associated with violent crime while ethnic heterogeneity was positively associated. Residential mobility was also positively associated with violent crime. The spatial lag parameter suggests that neighbourhood proximity to high crime areas influences violent crime rate after accounting for variations in socio-economic characteristics.

This research provides novel insights into the social disorganization theory, related social processes, and the spatial dynamics of crime that reflect spatial demographic changes in Toronto, Ontario at the neighbourhood scale. Future research should focus on the spatio-temporal effects of changes in immigration and ethnic composition on crime and investigate the influence of these demographic characteristics on crime at more precise spatial scales (i.e., census dissemination areas) to provide further insight into the social and spatial processes associated with violent crime. Additional research should investigate the role of crime in shaping neighbourhood characteristics (Morenoff and Sampson 1997) including ethnic composition and immigrant concentration. Methodologically, multi-level models that analyze city-scale and neighbourhood-scale characteristics would improve understanding of the spatial determinants of crime.

References

Andresen M (2006a) A spatial analysis of crime in Vancouver, British Columbia: a synthesis of social disorganization and routine activity theory. Can Geogr 50(4):487–502

Andresen M (2006b) Crime measures and the spatial analysis of criminal activity. Brit J Criminol 46:258–285

Anselin L (1995) Local indicators of spatial association—LISA. Geogr Anal 27(2):93–115

Anselin L (1998) Lagrange multiplier test diagnostics for spatial dependence and spatial heterogeneity. Geogr Anal 20(1):1–17

Anselin L (2005) Exploring spatial data with GeoDa: A Workbook. Available via Spatial Analysis Laboratory, University of Illinois and Centre for Spatially Integrated Social Science. https://www.csiss.org/clearinghouse/GeoDa/geodaworkbook.pdf. Accessed 14 Nov 2014

Anselin L, Rey S (1991) Properties of tests for spatial dependence in linear regression models. Geogr Anal 23(2):112–131

Baller RD, Anselin L, Messner SF et al (2001) Structural covariates of U.S. county homicide rates: incorporating spatial effects. Criminology 39(3):561–590

Bourne LS, Rose D (2001) The changing face of Canada: the uneven geographies of population and social change. Can Geogr 45(1):105–119

Brantingham PJ, Brantingham PL (1981) Environmental criminology. Sage Publications, Beverly Hills

Brantingham PJ, Brantingham PL (2008) Crime pattern analysis. In: Wortley R, Maerolle L (eds) Environmental criminology and crime analysis. Willan Publishing, Portland

Bursik RJ Jr (1988) Social disorganization and theories of crime and delinquency: problems and prospects. Criminology 26(4):519–551

Chun Y (2014) Analyzing space-time crime incidents using eigenvector spatial filtering: an application to vehicle burglary. Geogr Anal 46:165–184

City of Toronto (2009) Neighborhood planning areas. Available via city of Toronto Social Development, Finance, and Administration. http://www1.toronto.ca/wps/portal/contentonly?vgnextoid=1e68f40f9aae0410VgnVCM10000071d60f89RCRD. Accessed 14 Nov 2014

City of Toronto (2013) Toronto's racial diversity. http://www.toronto.ca/toronto_facts/diversity.htm. Accessed 14 Nov 2014

Farrar DE, Glauber RR (1967) Multicollinearity in regression analysis: the problem revisited. Rev Econ Stat 49(1):92–107

Graif C, Sampson RJ (2009) Spatial heterogeneity in the effects of immigration and diversity on neighbourhood homicide rates. Homicide Stud 13:242–260

Haining R, Law J, Griffith DA (2009) Modelling small area counts in the presence of overdispersion and spatial autocorrelation. Comput Stat Data Anal 53:2923–2937

Hirschfield A, Brown P, Todd P (1995) GIS and the analysis of spatially-referenced crime data: experiences in Merseyside, U.K. Int J Geogr Inf Syst 9(2):191–210

Hulchanski D (2010) Three cities within Toronto: income polarization among Toronto's neighbourhoods, 1970–2005. Available via University of Toronto Cities Centre. http://www.urbancentre.utoronto.ca/pdfs/curp/tnrn/Three-Cities-Within-Toronto-2010-Final.pdf. Accessed 14 Nov 2014

Kaplan GA (2004) What's wrong with social epidemiology, and how can we make it better? Empidemiol Rev 26(1):124–135
Kinney JB, Brantingham PL, Wuschke K et al (2008) Crime attractors, generators and detractors: land use and urban crime opportunities. Built Environ 34(1):62–74
Kubrin CE (2008) Racial heterogeneity and crime: measuring static and dynamic effects. Res Commun Sociol 10:189–218
Kubrin CE (2009) Social disorganization theory: then, now, and in the future. In: Krohn MD, Lizotte AJ, Hall GP (eds) Handbook on crime and deviance. Springer, New York
Kubrin CE (2013) Immigration and crime. In: Cullen FT, Wilcox P (eds) The Oxford handbook of criminological theory. Oxford University Press, Oxford
Kubrin CE, Weitzer R (2003) New directions in social disorganization theory. J Res Crime Delinq 40(4):374–402
Law J, Haining R (2004) A Bayesian approach to modeling binary data: the case of high-intensity crime areas. Geogr Anal 36(3):197–216
Law J, Quick M (2013) Exploring links between juvenile offenders and social disorganization at a large map scale: a Bayesian spatial modeling approach. J Geogr Syst 15:89–113
Law J, Quick M, Chan PW (2014) Bayesian spatio-temporal modeling for analyzing local patterns of crime over time at the small-area level. J Quant Criminol 30:57–78
Lee MT, Martinez R Jr (2002) Social disorganization revisited: mapping the recent immigration and black homicide relationship in Northern Miami. Sociol Focus 35(4):363–380
Lee MT, Martinez R Jr (2009) Immigration reduces crime: an emerging scholarly consensus. Sociol Crime Law Deviance 13:3–16
Martinez R Jr, Stowell JL, Lee MT (2010) Immigration and crime in an era of transformation: a longitudinal analysis of homicides in San Diego neighbourhoods, 1980–2000. Criminology 48 (3):797–829
Mears DP (2001) The immigration—crime nexus: toward an analytical framework for assessing and guiding theory, research, and policy. Sociol Perspect 44(1):1–19
Morenoff JD, Sampson RJ (1997) Violent crime and the spatial dynamics of neighborhood transition: Chicago, 1970–1990. Soc Forces 76(1):31–64
Morenoff JD, Sampson RJ, Raudenbush SW (2001) Neighbourhood inequality, collective efficacy, and the spatial dynamics of urban violence. Criminology 39(3):517–560
Murray AT, McGuffog I, Western JS, Mullins P (2001) Exploratory spatial data analysis techniques for examining urban crime. Brit J Criminol 41:309–329
Openshaw S (1984) Ecological fallacies and the analysis of areal census data. Environ Plann A 16 (1):17–31
Ouimet M (2000) Aggregation bias in ecological research: how social disorganization and criminal opportunities shape the spatial distribution of juvenile delinquency in Montreal. Can J Criminol 42:135–156
Peterson RD, Krivo LJ (2005) Macrostructural analyses of race, ethnicity, and violent crime: recent lessons and new directions for research. Annu Rev Sociol 31:331–356
Peterson RD, Krivo LJ (2009) Segregated spatial locations, race-ethnic composition, and neighborhood crime. Ann Am Acad Polit SS 623:93–107
Quick M, Law J (2013) Exploring hotspots of drug offences in Toronto: a comparison of four local spatial cluster detection methods. Can J Crim 55(2):215–238
Martinez Jr. R, Lee MT (2000) On immigration and crime. In: The nature of crime: continuity and change. Criminal Justice 2000. Available via National Criminal Justice Reference Service. https://www.ncjrs.gov/criminal_justice2000/vol_1/02j.pdf. Accessed 14 Nov 2014
Ousey GC, Kubrin CE (2009) Exploring the connection between immigration and violent crime rates in U.S. cities, 1980–2000. Soc Probl 56(3):447–473
Ratcliffe JH (2012) The spatial extent of criminogenic places: a changepoint regression of violence around bars. Geogr Anal 44(4):302–320
Ratcliffe JH, McCullagh MJ (1999) Hotbeds of crime and the search for spatial accuracy. J Geogr Syst 1:385–398

Reid LW, Weiss HE, Adelman RM et al (2005) The immigration-crime relationship: evidence across US metropolitan areas. Soc Sci Res 34:757–780
Sampson RJ, Groves WB (1989) Community structure and crime: testing social disorganization theory. Am J Sociol 94(4):774–802
Sampson RJ, Raudenbush SW (1999) Systematic social observation of public spaces: a new look at disorder in urban neighbourhoods. Am J Sociol 105(3):603–651
Sampson RJ, Raudenbush SW, Earls F (1997) Neighbourhoods and violent crime: a multilevel study of collective efficacy. Science 15(277):918–924
Sampson RJ, Morenoff JD, Gannon-Rowley T (2002) Assessing "neighbourhood effects": social processes and new directions in research. Annu Rev Sociol 28:443–478
Shaw C, McKay H (1942) Juvenile delinquency and urban areas: a study of rate of delinquents in relation to differential characteristics of local communities in American cities. University of Chicago Press, Chicago
Siemiatycki M, Isin E (1997) Immigration, diversity and urban citizenship in Toronto. Can J Reg Sci 20(1):73–102
Statistics Canada (2006) Suppression criteria, community profiles. http://www12.statcan.ca/census-recensement/2006/dp-pd/prof/92-591/help-aide/metadata/Suppression.cfm?Lang=E. Accessed 14 Nov 2014
Statistics Canada (2011) Data elements and violation coding structure for the Uniform Crime Reporting Survey. http://www23.statcan.gc.ca/imdb-bmdi/document/3302_D15_T9_V1-eng.htm. Accessed 12 Nov 2014
Statistics Canada (2012) Census tract. https://www12.statcan.gc.ca/census-recensement/2011/ref/dict/geo013-eng.cfm. Accessed 12 Nov 2014
Steenbeck W, Volker B, Flap H et al (2012) Local businesses as attractors or preventers of neighbourhood disorder. J Res Crime Delinq 49:213–248
Stowell JI, Messner SF, Mcgeever KF et al (2009) Immigration and the recent violent crime drop in the United States: a pooled, cross-sectional time-series analysis of metropolitan areas. Criminology 47(3):889–928
Tita GE, Radil SM (2010a) Making space for theory: the challenge of theorizing space and place for spatial analysis in criminology. J Quant Criminol 26:467–479
Tita GE, Radil SM (2010b) Spatial regression models in criminology: modeling social processes in the spatial weights matrix. In: Piquero AR, Weisburd D (eds) Handbook of quantitative criminology. Springer, New York
Townsley M (2009) Spatial autocorrelation and impacts on criminology. Geogr Anal 41:452–461
Velez MB (2009) Contextualizing the immigration and crime effect: an analysis of homicide in Chicago neighbourhoods. Homicide Stud 13(3):325–335
Veysey BM, Messner SF (1999) Further testing of social disorganization theory: an elaboration of Sampson and Groves's 'community structure and crime'. J Res Crime Delinq 36:156–174
Walks RA, Bourne LS (2006) Ghettos in Canada's cities? Racial segregation, ethnic enclaves and poverty concentration in Canadian urban areas. Can Geogr 50(3):273–297
Warner BD, Rountree PW (1997) Local social ties in a community and crime model: questioning the systemic nature of informal social control. Soc Probl 44(4):520–536
Wikstrom P-O, Dolmen L (1990) Crime and crime trends in different urban environments. J Quant Criminol 6(1):7–30
Wortley S (2009) The immigration-crime connection: competing theoretical perspectives. Int Migr 10:349–358

An Informed Virtual Geographic Environment Enhanced with Qualitative and Quantitative Information for the Geosimulations of Zoonosis Propagation

Mondher Bouden and Bernard Moulin

Abstract Public health decision makers need to better understand the propagation of zoonoses. The currently available zoonosis simulations are based on compartment models which do not integrate the influence of geographic features on the species' biological processes. In this context, we propose an approach that can generate an informed virtual geographic environment (IVGE) composed of a set of cells in which the evolution and interaction of the involved populations can be simulated plausibly. Since the number of these cells is huge, we propose a threshold-based merging algorithm that creates spatial subdivisions with the maximal size and suitability for a given biological phenomenon. Our approach also enhances each cell with qualitative and quantitative information such as the relative geographic orientation of the neighbors and the information about the distribution of individuals through trajectories. We used our IVGE to develop decision support tools that can simulate the spread of the West Nile Virus and Lyme disease.

Keywords Geosimulation · Zoonosis propagation · Merging process · Spatial distribution · Virtual environment

1 Introduction

Public health decision makers need to better understand the biological and geographic interactions of the different species involved in the propagation of zoonoses (infectious diseases transmitted from insects to animal and to humans) such as the

M. Bouden (✉) · B. Moulin
Department of Computer Sciences and Software Engineering,
Laval University, Quebec, Canada
e-mail: Mondher.Bouden@ift.ulaval.ca

B. Moulin
e-mail: Bernard.Moulin@ift.ulaval.ca

F. Harvey and Y. Leung (eds.), *Advances in Spatial Data Handling and Analysis*, Advances in Geographic Information Science,
DOI 10.1007/978-3-319-19950-4_15

West Nile Virus (WNV) and Lyme disease. They developed monitoring systems (Gosselin et al. 2005) which cannot be used to forecast the probable propagation of a zoonosis over a territory. Such a forecast, if it proved to be reliable, would allow public health authorities to initiate preventive actions at appropriate times and places. Besides, several approaches have been proposed to model and simulate the spread of zoonoses. However, these approaches such as mathematical modeling, cellular automata and traditional multi-agent systems have some weaknesses when trying to model and simulate the influence of geographic and climatic features on the disease spread and the spatial-temporal interactions of various kinds of actors (e.g. mosquitoes, ticks, birds, mammals, etc.). Indeed, the simulations based on mathematical models that generally use compartment models and differential equations (Wonham et al. 2004) do not take into account the characteristics of the geographical space in which populations operate, except in limited cases such as patchy models (Liu et al. 2006) which use an abstracted representation of space that is not based on data provided by Geographic Information Systems (GIS). In spite of the fact that a simulation based on cellular automata models the evolution of the spatial characteristics of a geographic area involved in the disease, it does not represent individuals and their mobility (White et al. 2009). On the other hand, agent-based simulations of epidemics represent the disease vectors (e.g. animals) as agents, but usually do not take advantage of data provided by GIS in order to properly locate agents in the geographic space (Emrich et al. 2007). Moreover, given the limitations of computational resources of computers and the lack of data, we cannot plausibly represent each individual by an agent, especially if we have to simulate a population composed of millions or even billions of individuals involved in zoonoses such as for example mosquito populations and tick populations transmitting the WNV and Lyme disease respectively.

In this context, we propose to use a multi-level population-based geosimulation approach (Bouden and Moulin 2012a) to remedy the shortcomings of current methods. We acquired some experience with the development of WNV-MAGS System (Bouden et al. 2008), a tool allowing public health decision makers to assess several intervention scenarios in order to understand and estimate the magnitude of the evolution of the WNV in a large territory. Furthermore, our team developed a generic solution (Zoonosis-MAGS) to be applied to a variety of zoonoses such as Lyme disease, with the aim to create realistic simulations at different levels of granularity. To develop such population-based geosimulations we introduced a new theoretical model (called MASTIM: Multi-Actor Spatio-Temporal Interaction Model) which can be used to simulate the interactions of various types of actors, including those representing populations containing a huge number of individuals (Bouden and Moulin 2012b).

In this paper we examine the critical issue of how to accurately represent and generate the virtual geographic environment (VGE) in which these geosimulations can be carried out, considering that the studied territories are huge and that we need to take into account the geographic characteristics that influence the biological cycles of the involved species and their behaviors. A good idea is to divide the VGE into basic cells that will contain sub-sets of individuals and in which the different stages of the involved sub-populations can be simulated. Such a space subdivision can be generated using either raster or vector data structures. In a raster-based system, the environment is divided into cells of the same size and shape. Such raster-based VGE are used by cellular automata (Emrich et al. 2007). However, such grids of cells are usually artificial and not related to the spatial characteristics of the studied phenomena. In a vector-based system, the environment is represented using geometrical primitives such as polygons (regions of space which are well defined using GIS data). Several kinds of polygons can be used to make a spatial subdivision of the VGE. The choice of the polygon types depends on the spatial characteristics that are important for the observation and analysis of the zoonosis propagation. In the case of WNV in which mosquitoes are spread over the territory, we used polygons representing either municipalities or census tracts, depending on the area of interest (municipalities are used to cover large areas such as the southern part of the province of Quebec whereas census tracks are used to characterize smaller areas such as the Ottawa metropolitan area). This administrative division fits the surveillance data which were available for the simulation. In the case of the simulation of the spread of Lyme disease, it does not make sense to use such an administrative subdivision because the cells representing municipalities or census tracks are too large and have no biological significance considering that tick populations can only survive in grass areas at the edge of (or in) forests. We therefore thought of using a spatial subdivision based on land cover characteristics. Land-cover data may be provided in vector-format in which polygonal cells cover large territories. However, another problem arises when it comes to simulate the propagation of zoonoses. Indeed, we need to compute for each cell the interactions of the involved species using some kind of compartment model. Given the huge number of land-cover cells it is impossible to carry out such computation for each time step in each individual cell of the land-cover subdivision. Consequently, we got the idea to merge land-cover cells having similar characteristics with respect to the phenomenon to be simulated. Hence, we could get the largest polygons possible in order to get plausible simulations while minimizing the needed computations. The merging method that we propose is based on criteria selected by the user who models the phenomenon. A very common criterion is related to the suitability of the habitat which may affect the survival, feeding, and/or breeding behaviors of species. In order to generate an efficient VGE, the cells resulting from such a merging process need to have the largest possible sizes with respect to the selected criteria. In this paper we propose an approach to create an informed VGE which is enhanced with qualitative and quantitative Information.

In Sect. 2 we present our new approach including the technique used to merge cells and to create the IVGE. In Sect. 3, we present an application of the IVGE which demonstrates the usefulness of the resulting IVGE for the geosimulation of the zoonosis propagation. Finally, we conclude the paper and evoke some future works.

2 Generation of the IVGE

We begin this section by presenting an overview of our new approach. Then, we present the criterion used to merge cells. After that, we discuss the different steps of our approach including the preprocessing of GIS data and deleting holes, the progressive merging of cells and the enrichment of the IVGE by adding data in cells.

2.1 *Overview of Our New Approach*

Figure 1 presents an overview of our approach that automatically generates the IVGE. The left part of this figure (i.e. transparent gray rectangles) shows how our system can generate the IVGE using an administrative subdivision. As previously mentioned this subdivision uses cells representing municipalities or census tracts. The user of our system can choose between the different types of cells. This choice is based on the selected spatial scale. The cells are then enriched with population data and/or climatic data. In this example, the IVGE obtained is used to geosimulate the propagation of the WNV.

The right side of Fig. 1 (blue rounded rectangles containing either numbers or letters) shows how our system can generate the IVGE using a subdivision based on

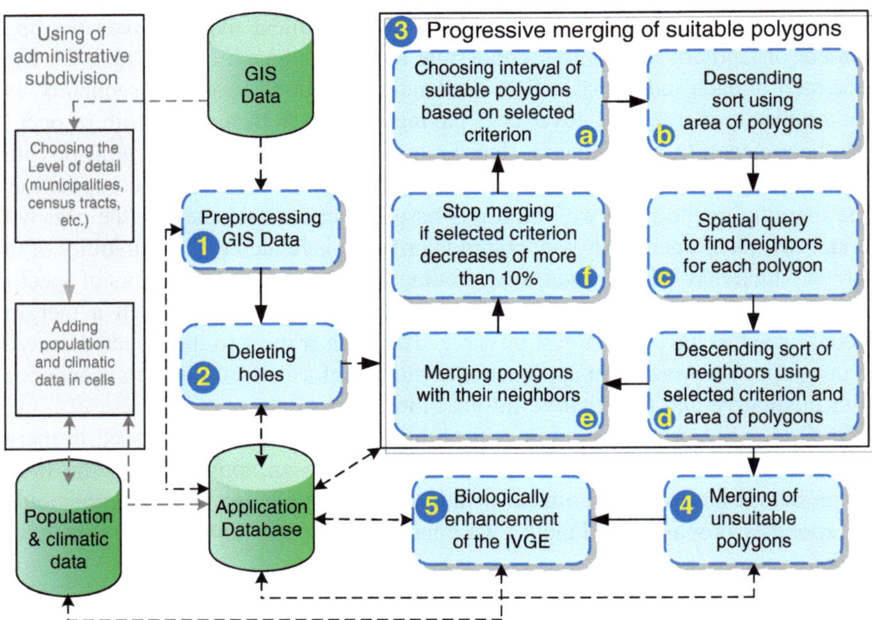

Fig. 1 Overview of our new approach

a criterion of favorability according to the land cover. This subdivision is more refined after merging of cells representing polygons provided by GIS data. Due to the large number of cells to be processed and because it is difficult to anticipate when to stop the merging process to get the largest polygons which satisfy the selected criterion, we propose an approach that is carried out in several steps. Indeed, in a first step (1 in Fig. 1) we preprocess the GIS data by selecting the region of interest and computing the suitability of all polygons with respect to the selected criterion. Then, we create plain polygons by removing their holes in order to reduce the number of basic cells (2). The result is stored in a new database which is updated as we go through the different processes. Besides, the most important step of our approach is the progressive merging of suitable polygons. We begin this process by selecting polygons having the best suitability (100 %) according to the selected criterion (3a). Then, we sort (3b) these polygons (the biggest is first processed) and get their neighbors (3c). These neighbors (3d) are also sorted (the biggest and the best one according to the selected criterion is first processed) in order to try to merge them to the original polygons (3e). Note that we choose different thresholds to stop the aggregation process if merging an additional neighbor decreases the suitability of the original polygon of more than 10 % (3f). In this way we preserve the suitability of the resulting polygons and thus we allow for a progressive merging process. We then iterate the merging process on a new set of polygons with a suitability value in the interval [90, 100 %] that we call the 'absorption threshold interval' (ATI). We therefore continue to apply this process several times by progressively increasing the ATI for the processed polygons and by simultaneously reducing the threshold used to stop merging. Finally, we do the opposite by selecting polygons having the worst suitability (0 %) according to the selected criterion and trying to merge them with other unsuitable polygons (4). As it is the case for suitable polygons, this process allows for generating homogeneous unsuitable areas (aggregation of unsuitable polygons) with maximal size. In addition, it further reduces the number of polygons used to generate the VGE. All the processes mentioned above are carried out until reaching a satisfactory result according to the user's appreciation. The final step (5) of our approach is the enrichment of the IVGE by adding data in cells (see Sect. 2.5).

2.2 Using Suitable Habitats as Merging Criterion

The biological phenomena that we study are closely tied to the characteristics of the landscape that is simulated by the VGE. Indeed, a zoonosis propagation greatly depends on the survival of the populations involved in the transmission of pathogens (i.e. virus, bacteria). For example, the capacity of tick populations to survive in suitable habitats (as for example 'sparse deciduous forests' according to Geobase terminology: www.geobase.ca) is an important factor that influences the spread of Lyme disease. In fact, a species can settle in an area only if it is suitable to its survival. Therefore, for each polygon of the GIS data presented in Sect. 2.3 we

Table 1 Suitability degrees (SD) of habitats for ticks in relation to land cover

Code	Type	SD	Code	Type	SD
0	No data	0	100	Herb	0.2
10	Unclassified	0	101	Tussock graminoid tundra	0.2
11	Cloud	0	102	Wet sedge	0.2
12	Shadow	0	103	Moist to dry non-tussock graminoid/dwarf shrub tundra	0.2
20	Water	0	104	Dry graminoid prostrate dwarf shrub tundra	0.2
30	Barren/non-vegetated	0	110	Grassland	0.2
31	Snow/ice	0	120	Cultivated agricultural land	0.2
32	Rock/rubble	0	121	Annual cropland	0.2
33	Exposed land	0	122	Perennial cropland and pasture	0.2
34	Developed	0	200	Forest/tree classes	0.8
35	Sparsely vegetated bedrock	0	210	Coniferous forest	0.8
36	Sparsely vegetated till-colluvium	0	211	Coniferous dense	0.6
37	Bare soil with cryptogam crust—frost boils	0	212	Coniferous open	0.9
40	Bryoids	0	213	Coniferous sparse	0.9
50	Shrubland	0.8	220	Deciduous forest	0.85
51	Shrub tall	0.75	221	Broadleaf dense	0.7
52	Shrub low	0.75	222	Broadleaf open	1
53	Prostrate dwarf shrub	0.2	223	Broadleaf sparse	1
80	Wetland	0.2	230	Mixed forest	0.8
81	Wetland—treed	0.2	231	Mixedwood dense	0.65
82	Wetland—shrub	0.2	232	Mixedwood open	0.95
83	Wetland—herb	0.2	233	Mixedwood sparse	0.95

propose to determine what we call a 'suitability degree' (SD) in order to estimate the quality of suitable habitats for a given species. To this end, we add a new attribute to the database table associated with each polygon. This attribute which represents the suitability degree of the polygon for a given species takes its values between 0 and 1 (see Table 1). The value 0 represents a place that is unsuitable for the species' individuals and the value 1 represents a very suitable place (100 % of suitability).

2.3 *Presentation and Preprocessing of the GIS Data*

We use land-cover shape files provided by the Geobase database. The land-cover information is the result of vectorization of raster thematic data originating from

Fig. 2 The six NTS regions of interest

classified Landsat 5 and Landsat 7 ortho-images, for agricultural and forest areas of Canada and for the Northern Territories. The land-cover data covers the totality of the Canadian territories and is divided in different regions using the index maps of the National Topographic System of Canada (NTS). Each region is identified by a unique number (e.g. 22 M, 30 K, etc.). Since we are interested in the southern part of Quebec and more specifically in the regions of *Montérégie* and *Estrie*, we consider six regions (i.e. 21E, 21L, 31G, 31H, 31I and 31 J) represented in Fig. 2.

The vector data are distributed as surface features (polygons) that have descriptive attributes such as the land-cover class which are based on the EOSD (Earth Observation for Sustainable Development) Land-Cover Classification. Indeed, the Coverage type attribute takes its value between 0 and 233 and represents different categories of land-covers (e.g. 20 for water, 34 for urban area, 50 for shrub land, 220 for deciduous forest, etc.). It is also worth mentioning that some polygons may have complex shapes as shown in Fig. 3.

We tested our approach using the shape file of Region 31H (its dimension is nearly 156 × 110 km: see Fig. 4). This file contains 133,780 polygons representing different kinds of land covers. Using Geomedia (Lisichenko 2008) we converted this file to an Access spatial database (read/write) in order to have the ability to modify it. For this conversion, we used the reference coordinate system of Canada

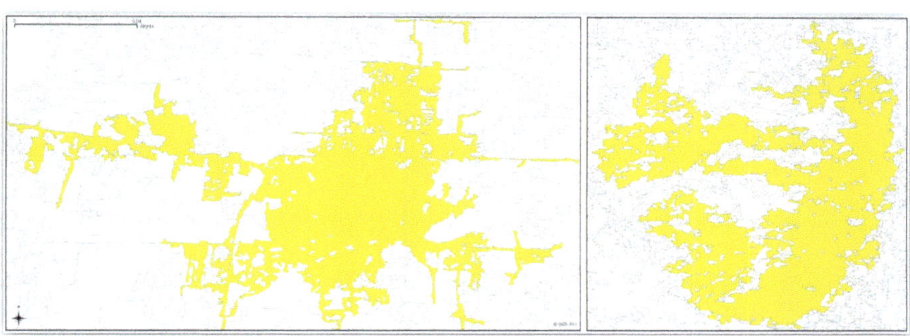

Fig. 3 Two complex polygons representing an urban area (*left*) and a forest (*right*)

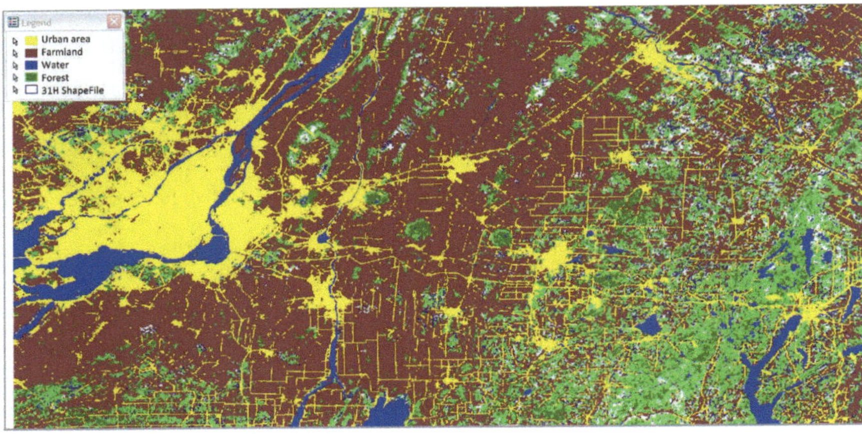

Fig. 4 Région 31H with different types of land cover

(NAD83-CSRS). Then, we added and computed the suitability attribute for each polygon as mentioned in the previous section. We also developed a new application which is able to query the Access database and modify its attributes.

We noticed that a large number of polygons have some holes. These holes are considered as polygons by Geobase with their own land cover attribute. To reduce the number of polygons, we decided to remove these holes and to create plain polygons. We assumed that the filling of holes will not greatly reduce the suitability degree of the polygon, since the size of most holes is a small portion of the polygon size. To explain our approach, let us take the example of the polygon shown in Fig. 5. Indeed, this polygon represents dense deciduous forest (SD = 0.7) and contains 20 holes. Some holes contain themselves other holes which brings the total number of holes to 35.

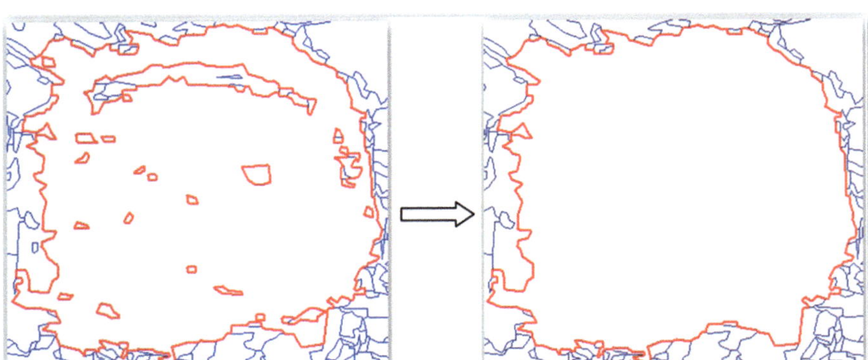

Fig. 5 Creating a solid polygon from a polygon that contains 20 holes and 15 sub-holes

Table 2 Computing the new suitability after deleting holes

Land cover	Nb	Code	Area (m^2)	SD	AP	SDP
Water	1	20	96,894.1	0	0.004534483	0
Shrub tall	3	51	35,420.8	0.75	0.001657635	0.001243226
Herb	1	100	11,284	0.2	0.000528072	0.000105614
Annual cropland	1	121	60,536.5	0.2	0.002833008	0.000566602
Perennial cropland and pasture	1	122	21,789.3	0.2	0.001019703	0.000203941
Coniferous dense	1	211	73,049.7	0.6	0.003418605	0.002051163
Broadleaf dense	2	221	19,914,816.3	0.7	0.931980327	0.652386229
Broadleaf open	1	222	13,574.2	1	0.00063525	0.00063525
Mixedwood dense	16	231	890,025.2	0.65	0.041651701	0.027073606
Mixedwood open	9	232	250,889.6	0.95	0.011741217	0.011154156
	36		21,368,279.7		1	0.695419786

Table 2 shows how we compute the new SD after deleting these holes. For example, there are 16 holes representing a dense mixed wood forest (SD = 0.65). The sum of areas of these holes is equal to 890,025.2 m^2. We compute an area proportion (AP) which is equal to the ratio between the sum of areas of the holes and the total area of the polygon without holes (890,025.2/21,368,279.7 = 0.041651701). Then, we compute the suitability degree proportion (SDP) which is equal to the multiplication between AP and SD (0.041651701 × 0.65 = 0.027073606). The sum of all SDP represents the new suitability of the solid polygon. We should mention that the total number of polygons listed in Table 2 is 36, including the polygon containing holes. Moreover, to delete holes, the system uses a spatial query which finds all the polygons located in the interior of the polygon containing holes. Then, we merge them to obtain a filled polygon which will have the new calculated SD (0.69542 instead of 0.7).

Moreover, we added a constraint to our application in order to block the removal of holes that exceed a threshold area and a threshold SD. This allows for a parameterization of our approach that can be customized to specific needs. After deleting the holes of polygons belonging to Region 31H (Fig. 6), we succeeded in removing 34,108 holes (25.5 % of the initial number of polygons). The remaining number of polygons is therefore 99,672 (74.5 %).

2.4 *Progressive Merging of Cells*

After deleting holes, we store the result in a new Access spatial database. Then, we start the cells merging process (by updating this database) in different steps using different intervals of suitability. In a first step, we query the database to get polygons having SD = 1 (100 % of suitability). Then, we apply to the 6261 obtained

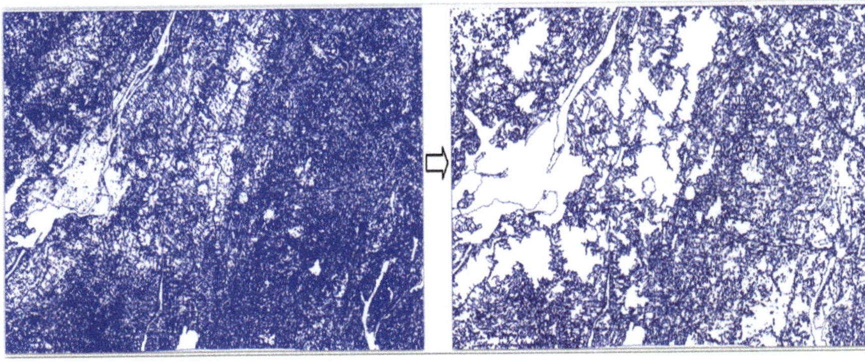

Fig. 6 The result of deleting holes of polygons of region 31H

polygons a descending sort using their area as an order criterion. Thus, the first processed polygon will be the one having the best SD and the biggest area. For each of these polygons, we apply a spatial query in order to find their immediate neighbors. We sort these neighbors in descending order according to their SD and area. Then, we try to merge to the selected polygon its neighbors one after the other, until we obtain a merged polygon that has a SD that does not drop below a chosen stopping-threshold of 0.9. After this first merging pass, 7137 polygons have been deleted from the database. Thus, until now we succeeded in removing 41,245 (30.8 %) polygons and the remaining number of polygons is 92,535 (69.2 % of the initial Geobase polygons). Figure 7 shows (on the left) a polygon (in blue) with three neighbors (red). This polygon is merged to one of its neighbors (on the right) and the resulting polygon keeps a SD higher or equal to the chosen threshold.

Then, we apply the merging process in successive steps by increasing the ATI (used to select the initial processed polygons) and decreasing the threshold of SD used to stop merging. In fact, we used in the next process the polygons that have a SD belonging to the interval [1, 0.9] and a stopping-threshold of 0.8. For the next

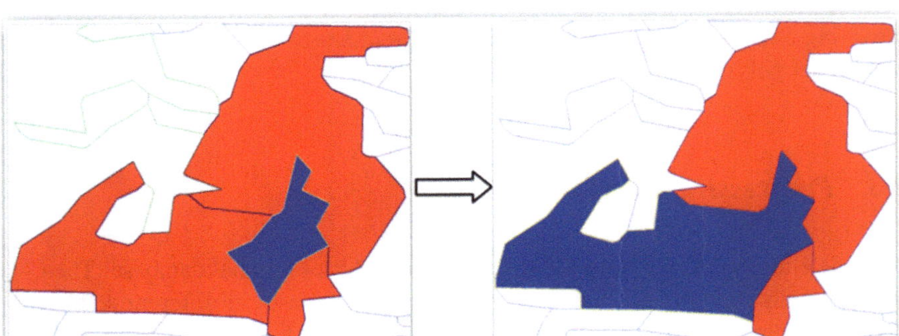

Fig. 7 Merging of cells using threshold of suitability degree

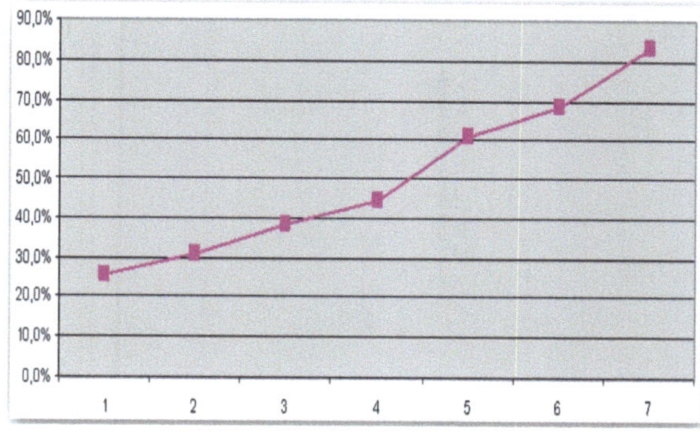

Fig. 8 Proportions of cells removed after the different merging iterations

merging iterations, we have used respectively the following intervals and thresholds: ([1, 0.8], 0.7), ([1, 0.7], 0.6), ([1, 0.6], 0.5).

After completion of the merging of suitable polygons, we apply another process in order to merge the unsuitable polygons as mentioned in Sect. 3. We select and sort by decreasing sizes polygons with a SD = 0 and we try to merge them with their neighbors which have a SD less than 0.5. We use this value since it represents the last stopping-threshold which is used by the merging process of suitable polygons. The proportions of cells removed after each of the applied merging processes are shown in Fig. 8 (the first process is the deletion of holes and the last process is the merging of unsuitable polygons).

Finally and after merging the GeoBase cells using the different processes mentioned above, it remains only 17 % (22,698) of the number of initial polygons belonging to the 31 H area which seems to be a good result.

2.5 Enrichment of the IVGE by Adding Data in Cells

For the purposes of the geosimulation of zoonosis propagation, we propose to add additional data to each polygon such as the identifiers of its neighbors, which is very useful for a variety of spatial functions. To this end, we add to each polygon a new attribute that contains the list of the IDs of its neighbors. Then, we apply a spatial query to the Access database in order to get the immediate neighbors of each polygon. We also add to each polygon its neighbors' orientation that is very useful for some processes that we want to simulate such as bird migrations in which the direction of displacement is very important. Thus, we associate two kinds of orientation data with each neighbor ID of a given polygon. The first one is quantitative information representing an angle between a virtual line (see Fig. 9a) and the North axis. These lines are drawn using the centroid of a given polygon and the centroid of each of its neighbors. To this end, we use a Geomedia function that returns the forward azimuth (FA) of a line segment. The returned value is normalized to fall

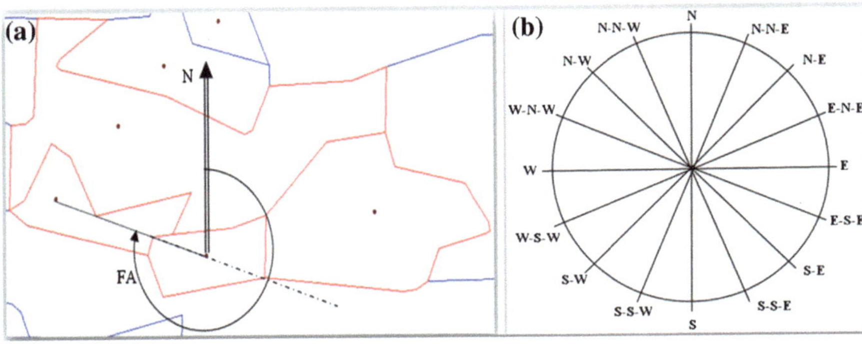

Fig. 9 Quantitative (**a**) and qualitative (**b**) information used for the neighbor's orientation

between zero and 2π and is measured clockwise from North. We use this quantitative information to also store qualitative information which represents either one of the four cardinal directions (i.e. North, East, South, West) or one of the four ordinal directions (i.e. North-East, South-East, South-West, North-West) or one of the eight further divisions represented in Fig. 9b. Besides, other types of data are added in cells and some of them will be presented in Sect. 3.

3 Application of the IVGE: Spatial Distribution of Flows of Individuals

We have to model the individuals' mobility in order to simulate the disease spread. Our approach facilitates this task because we do not need to use agent-based approaches which try to explicitly take into account the trajectories of each individual (i.e. agent) or group of individuals located in the VGE (Emrich et al. 2007) In the case of zoonoses, it is not feasible to use such approaches since we have to handle huge numbers of individuals. Besides, our IVGE allows for modeling a variety of processes influencing the geosimulation of the zoonosis propagation. As an example, we modeled the migration of birds that import juvenile ticks (some of them being infected) in Quebec from the US. We thought of modeling the Spring migrations as waves which distribute across cells birds of various species carrying ticks. Using a geo-referenced database of birds crossing the border on migration corridors at different periods of Spring, our simulation is initialized by associating the incoming birds to selected cells located at the border of Quebec and US (Francis and Hussell 1998). Then, bird groups spread to neighboring cells with respect to their attractiveness. This cell attractiveness for birds is computed using a qualitative attribute and associated with polygons (cells) in the same way as the suitability attribute. The process runs until all individuals are distributed. Indeed, such a process is facilitated by the identification of each neighbor of each cell which is available in our IVGE. Moreover, the distribution of individuals over cells should agree with the location of migration corridors. This is why it is important to know the orientation of neighboring cells with respect to a given cell in order to properly distribute individuals. This is possible thanks to our IVGE which stores in each cell the quantitative and qualitative orientation of its neighbors.

Figure 10 presents a screenshot of the main window of an application that we developed in order to test the distribution of the flows of individuals through some cells belonging to our IVGE. At the bottom left, this figure shows the number of individuals (set to 1500) representing the group of birds crossing the Canada/USA border. This figure shows also a trajectory having a direction to the north (individuals will try to follow the same direction when they travel). Furthermore, we introduce the concept of extension of the trajectory which represents a buffer area of the axis of this trajectory. In fact, we propose to extend the axis of the trajectory (i.e. line representing the direction of the studied species corridor) using a radius extension defined by the system user. Since this extension is spread over the cells as a shadow (see the grey area surrounding the trajectory curve in Fig. 10), it is possible to compute its proportion of overlapping with each of the cells. These proportions will be useful to determine the influence of the trajectory on the distribution.

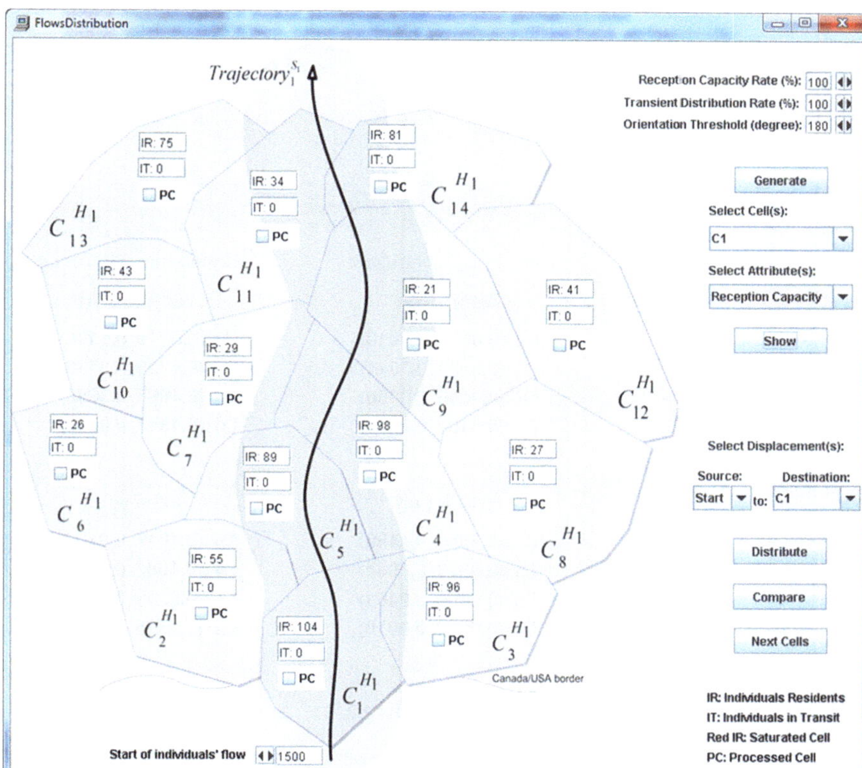

Fig. 10 Distribution of individuals through IVGE cells

In each cell, we store the number of resident individuals (i.e. IR) and the number of individuals in transit (i.e. IT). This information represents attributes attached to the cells. However, there are other cell's attributes used by our application. Indeed, Fig. 11 shows the reception capacity of cell C_1 which represents the maximum number of individuals that this cell can host. This attribute is computed according to the suitable habitat of the species and the land cover of the cell. Besides, the coefficient of attraction represents the percentage of individuals of a species that can be drawn by a cell. A cell containing suitable conditions for the survival and reproduction of a species may attract more individuals of this species then another cell containing unsuitable conditions.

Figure 11 also shows that C_1 has 4 neighbors: C_2, C_3, C_4 and C_5 and for example the cell C_2 has an orientation of West North-West (i.e. 5.0471 radians) from cell C_1. Moreover, the trajectory attribute is an important information that our system stores

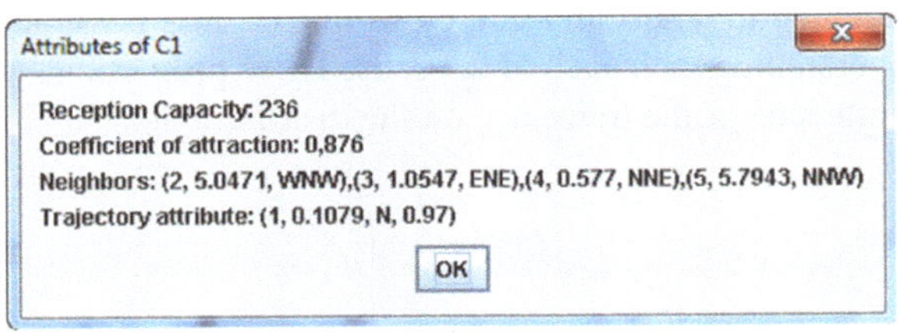

Fig. 11 Some attributes of the cell C_1

Fig. 12 Sorting the neighbors of the cell C_1

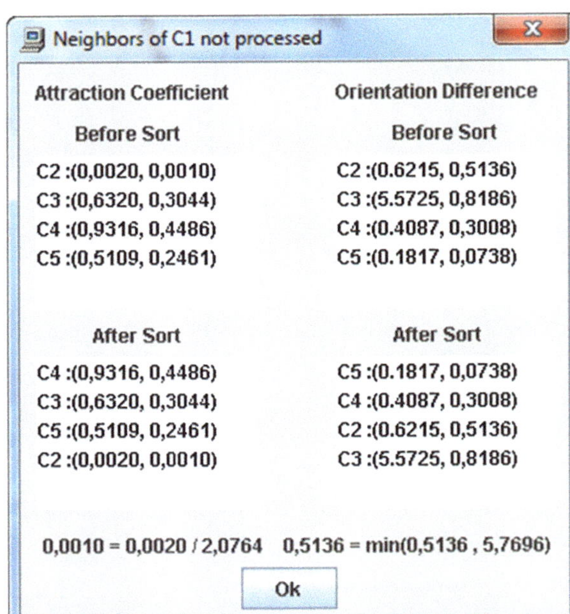

in each cell. This attribute contains mainly the orientation of the trajectory (i.e. North; 0.1079 rad) which is used to compute the direction to be taken by individuals in transit to follow this trajectory. It also contains the proportion of the extension of the trajectory (i.e. 0.97). This proportion is close to 100 % since the extension of the trajectory encompass almost the entire cell (see Fig. 10).

When the user clicks on the "Distribute" button, all the individuals are transferred to the cell C_1. The number of residents individuals in cell C_1 increases but it cannot exceed the reception capacity. Hence, this reception capacity is very useful to limit the number of resident individuals in order to not deplete the available resources in that cell. If the cell is saturated (it cannot accommodate more residents individuals), the system changes the color of the number of resident individuals (see Fig. 13).

Then, cell C_1 will act as a generator cell. Thus, non-resident individuals located in cell C_1 will be distributed to the neighboring cells (i.e. C_2, C_3, C_4 and C_5). Our system uses an algorithm that sorts the neighboring cells according to their attraction coefficients. For example, Fig. 12 shows on the left the list of neighboring cells before and after being sorted in descending order. The computed rates (for example 0.4486 for C4) represent the ratio between the attraction coefficient of each cell and the sum of

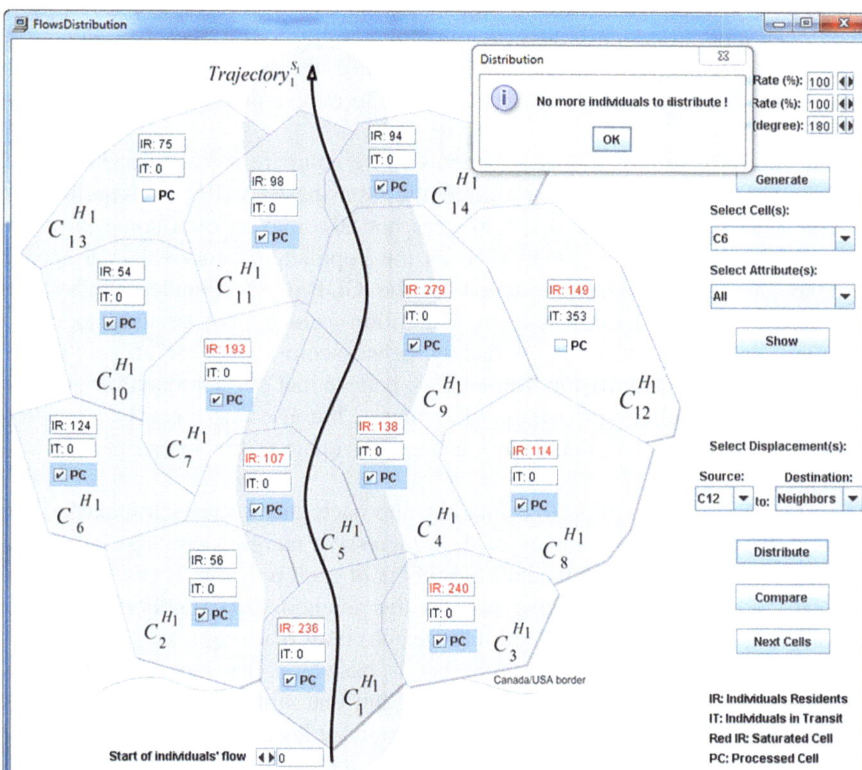

Fig. 13 End of the distribution process using the IVGE

all the attraction coefficients of all neighboring cells. These rates are used to compute the number of individuals in transit that have to be distributed to neighboring cells. However, the system has to determine which neighboring cell will be used to start the distribution. We propose to use the second sort presented in the right part of Fig. 12. Indeed, the neighboring cells are sorted this time in ascending order and using the absolute value of the difference between the orientation of a cell and the orientation of the trajectory of the generated cell. Thus, the first neighbor cell used by our system is the one having the orientation closest to the trajectory orientation of the generated cell. The distribution process continues until there are no more individuals to distribute or until all the cells become saturated. Figure 13 shows that there are no more individuals to distribute after the saturation of 8 cells.

4 Conclusion and Future Works

In this paper, we presented a new approach to generate an IVGE used to geosimulate the propagation of an infectious disease, taking into account the spatial-temporal characteristics of this phenomenon. We exploited vector-based land-cover data to progressively merge polygons according to their degree of suitability for selected biological processes. Hence, the reduction of the number of cells in which the populations of interest have to evolve and interact will improve the effectiveness of the simulation engine.

Our approach provides several benefits if we compare it to approaches which have been used up to now. For example, the raster-based VGE which are used by cellular automata (White et al. 2009) does not take into account some important factors related to disease spread such as the population's survival in a specific geographical space. Another example is the GDBSCAN (Sander et al. 1998) (generalizing density-based clustering algorithm) which may be used to cluster polygons, but it is based on extracting density-connected sets of neighboring objects within a circular region. Besides, we believe that our approach can not only be used to simulate the propagation of zoonoses, but also that it can be adapted to various other phenomena that do not necessarily relate to the spread of infectious diseases.

As future works, we plan to refine our approach and we are particularly interested in improving the efficiency of the progressive merging process. Indeed, we currently handle only the immediate neighbors of each polygon. It can be worth to take into account in the merging process the neighbors of neighbors and so on (considering a breadth-first or depth-first search) until reaching a satisfactory result according to the user's appreciation that is based on the selected criteria and thresholds. Besides, we investigate an algorithm that will be able to get the intersections between polygons of different IVGEs. Indeed, we need to create a VGE for each species involved in the phenomenon and then to compute the intersections between the corresponding polygons in order to obtain a VGE composed of cells which are qualified by the suitability parameters of all the species of interest.

Acknowledgments Many thanks to GEOIDE, the Canadian network of centers of excellence in geomatics (CODIGEOSIM Project), INSPQ (*Institut national de santé publique du Québec*) and the Saint-Hyacinthe Division of the Public Health Agency of Canada (PHAC) for their support (finance, expertise and data).

References

Bouden M, Moulin B (2012a) A interaction model used to geosimulate the zoonosis propagation. In: Symposium on theory of modeling and simulation (TMS'12), spring simulation multi-conference, the society for modeling and simulation international, 26–29 March, Orlando, USA

Bouden M, Moulin B (2012b) Zoonosis-MAGS: a generic multi-level geosimulation tool for zoonosis propagation. In: Global geospatial conference, spatially enabling government, industry and citizens. Quebec City, Canada, 14–17 May 2012

Bouden M, Moulin B, Gosselin P (2008) The geosimulation of West Nile virus propagation: a tool for risk management in public health. Int J Health Geogr 7(35):1–19

Emrich S, Suslov S, Judex F (2007) Fully agent based modelling of epidemic spread using anylogic. In: Proceedings of EUROSIM, September 2007, Ljubljana, Slovenia

Francis CM, Hussell DJT (1998) Changes in numbers of land birds counted in migration at long point bird observatory, 1961–1997. Bird Popul 5(6):37–66

Gosselin P, Lebel G, Rivest S, Douville-Fradet M (2005) The integrated system for public health monitoring of West Nile virus (ISPHM-WNV): a real-time GIS for surveillance and decision-making. Int J Health Geogr 4(21):1–12

Lisichenko R (2008) GIS using geomedia professional V6, 1st edn., Word Press

Liu R, Shuai J, Zhu H, Wu J (2006) Modeling spatial spread of West Nile virus and impact of directional dispersal of birds. Math Biosci Eng 3:145–160

Sander J, Ester M, Kriegel HP, Xu X (1998) Density-based clustering in spatial databases: the algorithm GDBSCAN and its applications. Data Min Knowl Discov 2(2):169–194

White SH, Martin del Rey A, Rodriguez Sanchez G (2009) Using cellular automata to simulate epidemic diseases. Appl Math Sci 3(20):959–968

Wonham MJ, De-Camino-Beck T, Lewis MA (2004) An epidemiological model for West Nile virus: invasion analysis and control applications. R Soc Lon Ser B Biol Sci 271(1538):501–507

Part IV
Statistical and Spatial Analysis

A Stochastic Method for the Generation of Optimized Building Layouts Respecting Urban Regulations

Shuang He, Julien Perret, Mickaël Brasebin and Mathieu Brédif

Abstract The development in an urban area normally has to obey planning regulations. In France, such regulations are specified in local urban planning schemes (LUPS or PLU in French) defining the right to build at municipal or inter-municipal level. Many aspects are concerned in a PLU. We address to the spatial aspect defining the rules for building development. Since these rules are stated in technical documents, it's not easy to comprehend or to assess their impacts. Driven by such issues, we propose to generate 3D building layouts that comply with the rules and have optimized indicators (e.g. floor area ratio), which is optional but useful. A building layout is a configuration of a number of buildings with various shapes (simplified as 3D boxes in this work). Thus, it can be seen as a realization of a marked point process (MPP) of 3D boxes, whose probability distribution can be defined through Gibbs energy with regard to a reference process. Its energy component reflects the compliance with the PLU rules in our problem. By maximizing this probability the optimal building layout can be found. The optimization task is realized by trans-dimensional simulated annealing (TDSA) coupled with a Reversible Jump Markov Chain Monte Carlo (RJMCMC) sampler. Several common types of the French PLU rules are studied and modeled into energy terms, and a case study is conducted to validate our approach.

Keywords Urban planning · Urban modelling · Scale · Building layout · 3D

S. He (✉) · J. Perret · M. Brasebin
IGN, COGIT, Université Paris-Est, 73 Avenue de Paris, 94165 Saint Mandé, France
e-mail: shuang.he@ign.fr

J. Perret
e-mail: julien.perret@ign.fr

M. Brasebin
e-mail: mickael.brasebin@ign.fr

M. Brédif
IGN, MATIS, Université Paris-Est, 73 Avenue de Paris, 94165 Saint Mandé, France
e-mail: mathieu.bredif@ign.fr

F. Harvey and Y. Leung (eds.), *Advances in Spatial Data Handling and Analysis*, Advances in Geographic Information Science,
DOI 10.1007/978-3-319-19950-4_16

1 Introduction

As urbanization is increasing, environmental, social and economic issues that cities are facing are becoming more and more critical. Thus, in order to consider these issues, urban planners design plans with the aim to regulate city development. These plans are various and differ by their scale (scale of a city or of a building) or by their preoccupations (displacement, social dwellings, etc.). The conception of these documents is a difficult task for two main reasons:

1. Information contained in plans is expressed with free text and their influence on a territory is difficult to assess;
2. As different city issues have complex relationships, the application of a plan supposed to improve an aspect of the city can decrease the performances of other aspects.

To support the design of such documents, it is necessary to provide tools that enable to check if the application of a given document has no side-effect according to the designers' wishes but also regarding other aspects of the city. Among different existing documents, this work is focused on constructability regulation. This kind of regulation is widely defined worldwide and describes constraints that new buildings have to respect to obtain a construction permit. The aim of this article is to propose a robust simulation through the generation of building layouts that respect constraints expressed by these regulations. In France, building regulation is specified in local urban planning schemes (LUPS or PLU in French). On the one hand, such schemes define the right to build at the scale of a parcel through 2D and 3D morphological rules (maximal height or floor area, for example) that, in particular, future buildings must respect. On the other hand, PLU documents require public survey to be effective. This is another issue because non-experts with such documents may have difficulties to understand its technical vocabulary and to interpret their effect on their district.

In Sect. 2, a state of the art describes several works that simulate similar constructability restrictions. Then, our proposed simulator that is based on the "trans-dimensional simulated annealing" optimization method is presented in Sect. 3. In order to produce desirable building layouts, this simulator integrates constraints from the regulation in its optimization function as described in Sect. 4. Some experiments are carried out with real world data using French urban regulations (Sect. 5).

2 State of the Art

Different works try to ease the comprehension of urban regulation through several types of approaches by: linking regulation to related geographic features in a 3D viewer (Métral et al. 2009); producing buildable hulls from geometric constraints

(El Makchouni 1987; Murata 2004; Brasebin et al. 2011); offering the possibility to explore a predefined set of parametric buildings respecting rules (Coors et al. 2009); generating buildings (Turkienicz et al. 2008; Parish and Müller 2001; Brasebin 2014) or proposing extensions to existing buildings (Laurini and Vico 1999). Among these works, building generation offers, in our opinion, the most promising results as it directly provides objects that can be built. Nevertheless, mentioned approaches use heuristics or procedural modeling that do not always fit with every terrain configuration and cannot integrate preferences of different builder-agents. Thus, we found necessary to adapt a building generation method to the specificity of urban regulations.

Building generation methods are explored through various fields such as architecture (Frazer 1995), urban planning (Rittel and Webber 1973), geosimulation (Ruas et al. 2011) or computer graphics (Wonka et al. 2003). The goal of the generation differs according to the domain. Vanegas (2013) distinguished geometric and behavior based approaches that are not always discordant. Behavior based approaches aim to produce buildings by integrating human processes and decisions whereas geometric approaches are designed to create fast and visually believable objects. As the objective of this paper is to simulate urban regulations, it is necessary to integrate the preferences of different agents that can construct buildings in order to assess the influence of the rules on different actors (for example, households or building promoters). Generally, these preferences are translated into utility functions that agents try to maximize.

To maximize such functions, optimization methods are used such as Multi-Agent Systems (Ruas et al. 2011) or meta-heuristics like evolutionary algorithms (Frazer 1995) or simulated annealing (Bao et al. 2013) combined to geometric generative methods like primitive instancing (Perret et al. 2010; Kämpf et al. 2010) or shape grammars (Talton et al. 2011). In order to integrate constraints, a large set of methods and their comparison are described in (Coello Coello 2010) including rejection, penalization of the optimization function or fixing automatically solutions that do not respect constraints.

3 Proposed Approach

Brasebin (2014) proposed a generic approach to generating buildings according to a model of French regulations, based on the approach presented in (Tournaire et al. 2010) for extracting building footprints from digital elevation models which relies on marked point processes, a class of random processes which realizations represent a set of objects in a certain parameter space. In (Tournaire et al. 2010), a configuration represents a collection of rectangle footprints in image space, whereas in (Brasebin 2014), a configuration consists of rectangular parallelepipeds or cuboids (we will from now on refer to as boxes), each representing a building placed inside a land parcel. The authors propose to optimize a given criteria (such as building volume) and a constraint management by rejection during the optimization process.

Using rejection allows for generality in this approach, but might deteriorate generated optimized configurations: the creation of a non-convex configuration search space and "barrier effects" due to strict constraints (such as "a building must be aligned with the roads"). Our objective is to propose a more robust simulator, based on the principles in (Brasebin 2014), that overcomes these issues by using an alternative constraint handling approach: (a) geometric constraints on individual buildings are configured as search space parameters; (b) constraints on spatial relations (e.g. distance to parcel borders, to other buildings) are expressed by an energy function as objective function to be minimized through stochastic optimization.

3.1 Marked Point Process

A marked point process (van Lieshout 2000) is a stochastic model defined by a probabilized space (Ω, π) with $\Omega = \bigcup_{n=0}^{\infty} K^n$, where K denotes the set of possible values of a single object and n is the number of objects in a configuration. A simple probabilization of Ω may be given by the probabilization of K and a discrete probability over $\mathbb{N}$ that samples the number n of objects. Assuming that the probabilization of K may be sampled directly, this yields a direct sampling of Ω according to this reference process. What is needed is a sampler of Ω according to a target density π that encodes the target objective function to be maximized. We will present in the next sections how this can be achieved by using Reversible Jump Markov Chain Monte Carlo (RJMCMC) and simulated annealing.

In our problem, each realization of the marked point process is a building layout (a configuration of n buildings). Extending the parameterization of 2D rectangles introduced in (Tournaire et al. 2010), each building is described by its 2D center $c_i = (x_i, y_i)$, its 2D semi-major axis vector $\overrightarrow{v_i} = (\rho_i, \theta_i)$, its aspect ratio $r_i \leq 1$ and its height h_i, using $r_i = \frac{l_i}{L_i}$ where $L_i = 2 \parallel \overrightarrow{v_i} \parallel$ and l_i are respectively the horizontal box dimensions along and across $\overrightarrow{v_i}$.

For efficiency, we prevent the sampling of buildings which centers c_i lie outside the considered land parcel. This generation of centers c_i within the land parcel only is performed using a triangulation of its defining polygon and a parameterization $c'_i = (x'_i, y'_i) \in [0, 1]^2$ that enables the uniform sampling of points inside this set of triangles. K is therefore defined as Eq. (1). Furthermore, r_i is limited to $[r_{\min}, 1]$ in order to better control building shapes, h_i is sampled in $[h_{\min}, h_{\max}]$ with $h_{\min}$ the minimum height of a building and $h_{\max}$ the maximum height depending on the local urban planning rules. Figure 1 illustrates the search space of the location and two realizations of this stochastic process.

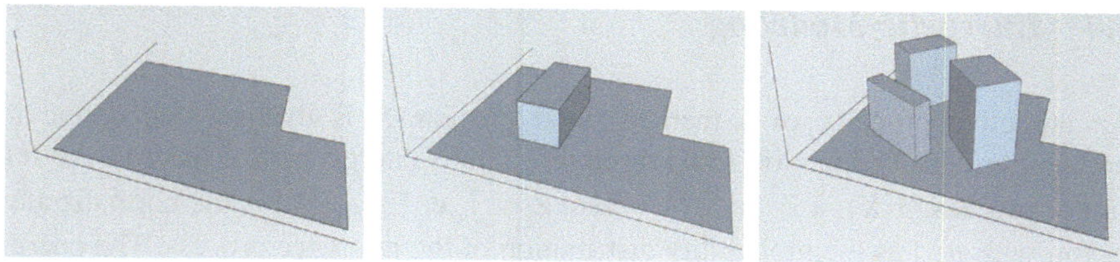

Fig. 1 A target parcel and two realizations of the marked point process of 3D box on the given parcel ($n = 1$ and $n = 3$)

$$K = \overbrace{\underbrace{[0,1]\times[0,1]}_{c_i'}}^{points} \times \overbrace{\underbrace{[\rho_{\min},\rho_{\max}]\times[0,\pi]}_{\overrightarrow{v_i}} \times \underbrace{[r_{\min},1]}_{r_i} \times \underbrace{[h_{\min},h_{\max}]}_{h_i}}^{mark} \tag{1}$$

3.2 *Reversible Jump Markov Chain Monte Carlo*

RJMCMC is an extension of MCMC that allows sampling from a configuration space Ω of varying dimension that only requires the unnormalized evaluation of its target distribution. This approach consists in using reversible kernels Q_i associated with probabilities $q(i|X_t)$ modeling a random modification of the current configuration $\mathrm{X_t}$ to successively propose a new configuration and evaluate its acceptance probability. The algorithm operated by repeating such steps in order to build a series of configurations which stationary distribution is the desired target distribution π. A generic implementation of this algorithm is available and detailed in (Brédif and Tournaire 2012). In the proposed approach, the reversible kernels used are: birth and death kernel, edge translation kernel, and height scaling kernel.

3.3 *Trans-Dimensional Simulated Annealing*

In order to drive the RJMCMC sampler to the target probability distribution, we use simulated annealing (SA), a widely used local metaheuristic (Salamon et al. 2002). This approach of coupling a RJMCMC sampler within a simulated annealing may be referred to as Trans-Dimensional Simulated Annealing (TDSA) (Singh et al. 2008). The idea of the approach is to sample increasingly more selectively from the RJMCMC sampler until a convergence criteria is reached (usually a given number of iterations or a maximum variation of energy during a certain number of iterations).

3.4 Energetic Modeling

An attractive property of a marked point process X is that we can define its probability density function (pdf) through a Gibbs energy with respect to the reference process: $\pi(X) = Z^{-1}e^{-E(X)}$, where $Z = \int_X e^{-E(X)}d\mu(X)$ is the normalization factor, with $\mu(.)$ as the probability distribution of the reference process. The energy function $E(X)$ can express the quality of a configuration of marked points. In our case, it indicates the conformity with urban planning rules as well as desired criteria to optimize (the total building volume for instance). We introduce a global energy composed as the sum of finite weighted energy terms, each of which is formed according to a specific urban planning rule. Therefore, the problem of finding the optimal building layout that complies with urban planning rules can be translated into finding the configuration of building boxes $\tilde{X}$ by maximizing the probability $\pi(.)$ using a RJMCMC sampler under a TDSA framework: $\tilde{X} = \arg\max \pi(.)$. Urban planning rules vary by countries and cities. The French urban planning scheme is studied in this paper. Common rules are extracted and modeled by energy terms. The detail is presented in the following section.

4 Urban Rules to Energy Terms

A PLU (French urban planning scheme) document provides a zoning plan dividing an urban territory into several zones, and specifies applicable regulations for each type of zone. The regulations may include all or some of the 16 articles provided by the urban planning code. Each of these articles describes a fixed theme. For instance, article 1 describes prohibited land use and article 16 describes electronic communication infrastructures and network. This work only focuses on the articles that regulate the spatial aspects of buildings. The studied articles are:

- Article 6: building position in relation to public roads,
- Article 7: building position in relation to separative limits,
- Article 8: building position in relation to other buildings,
- Article 9: building footprint,
- Article 10: building height,
- Article 14: floor area ratio.

Articles 6 and 7 can be considered as one theme: building position in relation to parcel borders (front, side, and back borders). Front borders are the borders adjacent to public roads (or at least public space). Side and back borders are separative limits adjacent to private roads or other parcels. Regarding the buildings position, the most common rules are for orientation and location. Therefore, we take into account the following types of rules when generating optimal building layouts:

- Rule A1: distance to parcel borders,
- Rule A2: distance between buildings,
- Rule A3: parcel coverage ratio,
- Rule A4: floor area ratio,
- Rule B1: angle to parcel borders,
- Rule B2: building height.

Rules B1 and B2 can be satisfied by directly setting constraints on geometry parameters ρ and h before MPP sampling if the rules do not depend on dynamic building attributes (e.g. location and/or height of itself and/or of its adjacent buildings). For example, if the angle rule is that all buildings should be parallel to a fixed border, which means it does not depend on dynamic building attributes, then the constraint on ρ can be known before sampling. If the angle rule requires a building to be parallel to its nearest border, which means it depends on its own location, then the rule cannot be configured before sampling. Similar for rule B2, the actual rule may have constraints on height differences between adjacent buildings. In such cases, rules B1 and B2 should be modeled by energy functions, along with rules A1 to A4. We discuss the energetic modeling of these rules by categorizing them into three types: unary, binary, and global energy.

4.1 Unary Energy

If the calculation of a type of energy for one building does not depend on the attributes of other buildings, we refer to such kind of energy as *unary energy*. Among all types of rules that we studied in this work, rule A1 can be modeled by *unary distance energy* E^{u_d}, and rule B1 can be modeled by *unary angle energy* E^{u_θ}, when it needs energetic modeling. Thus the total *unary energy* can be defined as their weighted sum:

$$E^u = w^{u_d} E^{u_d} + w^{u_\theta} E^{u_\theta} \tag{2}$$

4.1.1 Unary Distance Energy

There are generally three types of borders: front, side, and back borders. One parcel can consist of zero or more than one borders of each type. Regardless of the border type, we refer to the distance from a building to a border as a *unary distance*. A PLU rule for unary distance could be a simple constraint involving only one given constant, for instance, $d < 2m$. It could also depend on external and/or internal parameters. For instance, $d > \max(c/2, h)$, where c is the largest width of the road adjacent to the border considered as an external parameter, and h is the building height considered as an internal parameter. The valuation of external parameters can be done before launching the optimization process, whereas the valuation of internal parameters has to be performed during the optimization process. Regardless of the timing of valuation, the left value of the constraint has to be calculated before energetic modeling.

Therefore, all PLU rules for unary distance can eventually be described by the union of finite disjoint real intervals: $d^u_{ij} \in \bigcup_{k=1}^{n} \{I_k(a_k, b_k) | I_k \cap \forall I_{p \neq k} = \emptyset\}$, where d^u_{ij} is the distance from the ith building to the jth border, and I_k is an real interval with endpoints a_k and b_k. The corresponding energy can be defined as:

$$E^{u_d}_{ij} = \begin{cases} 0 & d^u_{ij} \in \bigcup I_k \\ g\left(d^u_{ij}\right) & d^u_{ij} \notin \bigcup I_k \end{cases} \tag{3}$$

where $g(.)$ is a penalty function that penalizes non-acceptable distance values.

If the rules for all borders are of logical conjunction: $d^u_{i1} \in A_1$ $\& \, d^u_{i2} \in A_2 \ldots \& \, d^u_{im} \in A_m$, then the unary distance energy for ith building is $E^{u_d}_i = \sum_{j=1}^{m} E^{u_d}_{ij}$. If there exist rules of logical disjunction like $d^u_{ij} \in A_j || d^u_{ik} \in A_k$, then $E^{u_d}_{i_{jk}} = \min(E^{u_d}_{ij}, E^{u_d}_{ik})$. There could also be of logical conjunction in each operand of logical disjunction, so the final $E^{u_d}_i$ can hardly be described by a generic formula. Example of such complex rules can be found in Sect. 5. Eventually, the overall unary distance energy for a parcel with n buildings is:

$$E^{u_d} = \sum_{i=1}^{n} E^{u_d}_i \tag{4}$$

4.1.2 Unary Angle Energy

We refer to the angle between a building and a parcel border as *unary angle*. All PLU rules for unary angle can be eventually described by the union of finite disjoint real intervals: $\theta^u_{ij} \in \bigcup_{k=1}^{n} \{I_k(a_k, b_k) | I_k \cap \forall I_{p \neq k} = \emptyset\}$, where θ^u_{ij} is the angle from the ith building to the jth border, and I_k is an real interval with endpoints a_k and b_k. The corresponding energy can be defined as:

$$E^{u_\theta}_{ij} = \begin{cases} 0 & \theta^u_{ij} \in \bigcup I_k \\ g(\theta^u_{ij}) & \theta^u_{ij} \notin \bigcup I_k \end{cases} \tag{5}$$

where $g(.)$ is a penalty function that penalizes non-acceptable angle values.

In practice, a PLU rule for unary angle normally only involves one border, for example, each building should be parallel to a fixed border or to its nearest border.

The involved Jth border can be determined before energetic modeling. Therefore, the unary angle energy for the ith building is:

$$E_i^{u_\theta} = E_{iJ(i)}^{u_\theta} \tag{6}$$

The overall unary angle energy for a parcel with n buildings is:

$$E^{u_\theta} = \sum_{i=1}^{n} E_i^{u_\theta} \tag{7}$$

4.2 Binary Energy

If the calculation of a type of energy involves two buildings in the same parcel, we refer to such kind of energy as *binary energy*. Among all types of rules that we studied in this work, rule A2 can be modeled by *binary distance energy* E^{b_d}, and rule B2 can be modeled by *binary height energy* E^{b_h}, when it concerns the height difference between two adjacent buildings. Thus the total *binary energy* can be defined as their weighted sum:

$$E^b = w^{b_d} E^{b_d} + w^{b_h} E^{b_h} \tag{8}$$

4.2.1 Binary Distance Energy

We refer to the distance between two buildings on the same parcel as *binary distance*. All PLU rules for binary distance can be eventually described by the union of finite disjoint real intervals: $d_{ij}^b \in \bigcup_{k=1}^n \{I_k(a_k, b_k) | I_k \cap \forall I_{p \neq k} = \emptyset\}$, where d_{ij}^b is the distance between the ith and the jth building, and I_k is an real interval with endpoints a_k and b_k. The corresponding energy can be defined as:

$$E_{ij}^{b_d} = \begin{cases} 0 & d_{ij}^b \in \bigcup I_k \\ g(d_{ij}^b) & d_{ij}^b \notin \bigcup I_k \end{cases} \tag{9}$$

where $g(.)$ is a penalty function that penalizes non-acceptable values.

The overall binary distance energy for a parcel with n buildings is therefore:

$$E^{b_d} = \sum_{i=1, j > i}^{n-1, n} E_{ij}^{b_d} \tag{10}$$

4.2.2 Binary Height Energy

We refer to the height difference between two buildings on the same parcel as *binary height*. All PLU rules for binary height can be described by the union of finite disjoint real intervals: $h_{ij}^b \in \bigcup_{k=1}^n \{I_k(a_k, b_k) | I_k \cap \forall I_{p \neq k} = \emptyset\}$, where h_{ij}^b is the height difference between the ith and the jth buildings, and I_k is an real interval with endpoints a_k and b_k. The corresponding energy can be defined as:

$$E_{ij}^{b_h} = \begin{cases} 0 & h_{ij}^b \in \bigcup I_k \\ g(h_{ij}^b) & h_{ij}^b \notin \bigcup I_k \end{cases} \tag{11}$$

where $g(.)$ is a penalty function that penalizes non-acceptable values.

The overall binary height energy for a parcel with n buildings is therefore:

$$E^{b_h} = \sum_{i=1, j>i}^{n-1, n} E_{ij}^{b_h} \tag{12}$$

4.3 Global Energy

If the calculation of a type of energy involves all buildings in the same parcel, we refer to such kind of energy as *global energy*. Among all types of rules that we studied in this work, rule A3 can be modeled by *global coverage energy* E^{g_c}, and rule A4 can be modeled by *global builtup energy* E^{g_f}. Thus the total *global energy* can be defined as their weighted sum:

$$E^g = w^{g_c} E^{g_c} + w^{g_f} E^{g_f} \tag{13}$$

4.3.1 Global Coverage Energy

Global coverage energy is used to model the PLU rule for lot coverage ratio (LCR). Normally, a constant value maxlcr is given as the upper limit of LCR, and the implicit lower limit of PCR is zero (no building on the parcel). Therefore, the global coverage energy can be defined as:

$$E^{g_c} = \begin{cases} 0 & x_c \in [0, \max lcr] \\ g(x_c) & x_c \in (\max lcr, +\infty) \end{cases} \tag{14}$$

where $g(.)$ is a penalty function that penalizes non-acceptable values.

In order to help urban planners to access the effect of the PLU rule for LCR, an optimization task for LCR is often demanded, which is to generate building layouts with LCR values approaching to maxpcr. Therefore, an alternative global coverage energy function can be defined as:

$$E^{g_c} = \begin{cases} g_1(x_c) & x_c \in [0, \max lcr] \\ g_2(x_c) & x_c \in (\max lcr, +\infty) \end{cases} \tag{15}$$

where $g_1(.)$ and $g_2(.)$ can be different penalty functions that penalizes less favorable values and non-acceptable values respectfully.

4.3.2 Global Builtup Energy

Global builtup energy is used to model the PLU rule for floor area ratio (FAR). Normally, a constant value max*far* is given as the upper limit of FAR, and the implicit lower limit of FAR is zero (no buildings on the parcel). Therefore, the global builtup energy can be defined as:

$$E^{g_f} = \begin{cases} 0 & x_f \in [0, \max far] \\ g(x_f) & x_f \in (\max far, +\infty) \end{cases} \tag{16}$$

where $g(.)$ is a penalty function that penalizes non-acceptable values.

Similarly, optimization for FAR is also frequently demanded, and an alternative global builtup energy function can be defined as:

$$E^{g_f} = \begin{cases} g_1(x_f) & x_f \in [0, \max far] \\ g_2(x_f) & x_f \in (\max far, +\infty) \end{cases} \tag{17}$$

where $g_1(.)$ and $g_2(.)$ can be different penalty functions that penalizes less favorable values and non-acceptable values respectfully.

5 Case Study

This section presents a case study realized on an actual parcel from the French city of La Courneuve illustrated in Fig. 2. The objective is to generate optimized building layouts on the given parcel of land with maximal floor area ratio, meanwhile satisfying all the PLU rules concerning this parcel. The PLU rules associated to this parcel are:

- Article 6: (1) Distance from each building to the front border should be greater than 3 meters; (2) All buildings should be parallel to the front border,
- Article 7: Distances from each building (with height h) to the two side borders d_1, d_2 should satisfy: $(d_1 = 0, d_2 = 0) || (d_1 = 0, d_2 \geq \max(6, h)) || (d_1 \geq \max(6, h), d_2 = 0)$. Distance from each building to the back border should be greater than 4 meters,

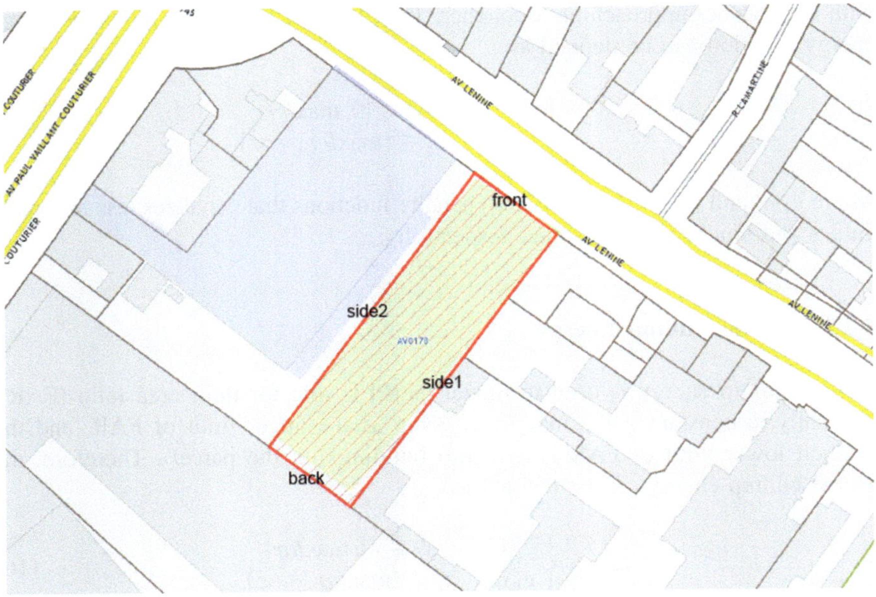

Fig. 2 The target land parcel for the case study. See Fig. 3 for **a** Building layout, **b** Footprints, **c** Energy plot

- Article 8: Distance between each two buildings should be greater than 4 meters.
- Article 9: Maximal lot coverage ratio is 0.6.
- Article 10: Minimal and maximal building heights are 6 and 18 m.
- Article 14: Maximal floor area ratio is 3.

The strategy for handling these rules is:

- Rule A1: Article 6(1) and Article 7 → unary distance energy,
- Rule A2: Article 8 → binary distance energy,
- Rule A3: Article 9 → global coverage energy,
- Rule A4: Article 14 → global builtup energy,
- Rule B1: Article 6(2) → constraint on parameter ρ,
- Rule B2: Article 10 → constraint on parameter h.

5.1 Energy Terms

The total energy for each building layout is defined as the weighted sum of unary distance energy E^{u_d}, binary distance energy E^{b_d}, global coverage energy E^{g_c}, and global builtup energy E^{g_f}.

$$E = w^{u_d} E^{u_d} + w^{b_d} E^{b_d} + w^{g_c} E^{g_c} + w^{g_f} E^{g_f} \tag{18}$$

The Gaussian error function $erf(.)$ is used as penalty function to form all the energy terms, with parameters $a, b (a > 0, b > 0)$ for tuning the magnitude and gradient. The square function is used for reaching optimum floor area ratio, with parameters $a', b' (a' > 0, b' > 0)$ for tuning the magnitude and gradient.

5.1.1 Unary Distance Energy

1. Distance from ith building to 1st border (front border) should satisfy $d_{i1}^u \geq 3$. The energy is defined as:

$$E_{i1}^{u_d} = \begin{cases} -a \cdot erf(b \cdot (d_{i1}^u - 3)) & d_{i1}^u \in (-\infty, 3) \\ 0 & d_{i1}^u \in [3, +\infty) \end{cases} \tag{19}$$

2. Distance from ith building to 2nd and 3rd borders (side borders) should satisfy: $(d_{i2}^u = 0 \,\&\&\, d_{i3}^u = 0) || (d_{i2}^u = 0 \,\&\&\, d_{i3}^u \geq \max(6, h_i)) || (d_{i2}^u \geq \max(6, h_i) \,\&\&\, d_{i3}^u = 0)$. We first define four atomic energy terms:

   - For rule $d_{i2}^u = 0$:

$$E_{i2_a}^{u_d} = \begin{cases} -a \cdot erf(b \cdot d_{i2}^u) & d_{i2}^u \in (-\infty, 0) \\ a \cdot erf(b \cdot d_{i2}^u) & d_{i2}^u \in [0, +\infty) \end{cases} \tag{20}$$

   - For rule $d_{i3}^u = 0$:

$$E_{i3_a}^{u_d} = \begin{cases} -a \cdot erf(b \cdot d_{i3}^u) & d_{i3}^u \in (-\infty, 0) \\ a \cdot erf(b \cdot d_{i3}^u) & d_{i3}^u \in [0, +\infty) \end{cases} \tag{21}$$

   - For rule $d_{i2}^u \geq \max(6, h_i)$:

$$E_{i2_b}^{u_d} = \begin{cases} -a \cdot erf(b \cdot (d_{i2}^u - R_i)) & d_{i2}^u \in (-\infty, \max(6, h_i)) \\ 0 & d_{i2}^u \in [\max(6, h_i), +\infty) \end{cases} \tag{22}$$

   - For rule $d_{i3}^u \geq \max(6, h_i)$:

$$E_{i3_b}^{u_d} = \begin{cases} -a \cdot erf(b \cdot (d_{i3} - R_i)) & d_{i3} \in (-\infty, \max(6, h_i)) \\ 0 & d_{i3} \in [\max(6, h_i), +\infty) \end{cases} \tag{23}$$

Then the energy for side border rule is defined as:

$$E^{u_d}_{i_{(2,3)}} = \min(\max(E^{u_d}_{i2_a}, E^{u_d}_{i3_a}), \max(E^{u_d}_{i2_a}, E^{u_d}_{i3_b}), \max(E^{u_d}_{i2_b}, E^{u_d}_{i3_a})) \tag{24}$$

3. Distance from ith building to 4th border (front border) should satisfy $d^u_{i4} \geq 4$. The energy is defined as:

$$E^{u_d}_{i4} = \begin{cases} -a \cdot erf(b \cdot (d^u_{i4} - 4)) & d^u_{i4} \in (-\infty, 4) \\ 0 & d^u_{i4} \in [4, +\infty) \end{cases} \tag{25}$$

Then, the unary distance energy for the ith building is:

$$E^{u_d}_i = E^{u_d}_{i1} + E^{u_d}_{i_{(2,3)}} + E^{u_d}_{i4} \tag{26}$$

And the total unary distance energy for a configuration with n buildings is:

$$E^{u_d} = \sum_{i=1}^{n} E^{u_d}_i \tag{27}$$

5.1.2 Binary Distance Energy

Distance from ith and jth building should satisfy $d^b_{ij} \geq 4$. The energy is:

$$E^{b_d}_{ij} = \begin{cases} -a \cdot erf(b \cdot (d^b_{ij} - 4)) & d^b_{ij} \in (-\infty, 4) \\ 0 & d^b_{ij} \in [4, +\infty) \end{cases} \tag{28}$$

Then, the total binary distance energy for a configuration with n buildings is:

$$E^{b_d} = \sum_{i=1, j>i}^{n-1, n} E^{b_d}_{ij} \tag{29}$$

5.1.3 Global Coverage Energy

The lot coverage ratio should satisfy $x_c \leq 0.6$, so the global coverage energy is defined as:

$$E^{g_c} = \begin{cases} 0 & x_c \in [0, 0.6] \\ a \cdot erf(b \cdot (x_c - 0.6)) & x_c \in (0.6, +\infty) \end{cases} \tag{30}$$

5.1.4 Global Builtup Energy

The floor area ratio should satisfy $x_f \leq 3$, and there is an optimization task for the final configuration approaching to the upper limit $x_f = 3$. Instead of assigning the same energy (e.g. zero) for all the values that satisfy the rule, we can favorite the values closer to the optimum value. Thus, we define the energy function as:

$$E^{g_f} = \begin{cases} a' \cdot (b' \cdot (x_f - 3))^2 & x_f \in [0, 3] \\ a \cdot erf(b \cdot (x_f - 3)) & x_f \in (3, +\infty) \end{cases} \tag{31}$$

5.2 *Implementation and Results*

The proposed approach is implemented using the open source C++ library librjmcmc (Brédif and Tournaire 2012), which provides a framework for stochastic optimization using RJMCMC sampler and simulated annealing. 50 experiments are conducted on a machine (HP workstation Z210) with 3.3 GHz dual core CPU (Intel Core i5-2500) and 8 GB memory under 32 bit Linux environment (Ubuntu 13.04). The average CPU computation time is 56,413.4 ms (about 1 min) per experiment with 1.5 million iterations for simulated annealing. The weights of the energy terms used in these experiments are as in Eq. (32).

$$E = 30E^{u_d} + 20E^{b_d} + 20E^{g_c} + 50E^{g_f} \tag{32}$$

The parameters used from Eqs. (19) to (29) are: $a = 100$, $b = 0.2$; The parameters used in Eqs. (30) and (31) are: $a = 100$, $b = 1$, $a' = 10$, $b' = 1$. The statistics of the parcel coverage ratio and floor area ratio are given in Table 1. The results can be evaluated from two aspects: (1) the optimization of the floor area ratio and (2) the conformity with all the PLU rules reflected by energy values. Therefore, we verify the two results with highest and lowest floor area ratio and the two results with highest and lowest energy.

Table 1 Statistics of 50 experiments

Floor area ratio (FAR)			Lot coverage ratio (LCR)		
Min	Max	Average	Min	Max	Average
1.60993	2.99672	2.21107	0.310011	0.61312	0.469303

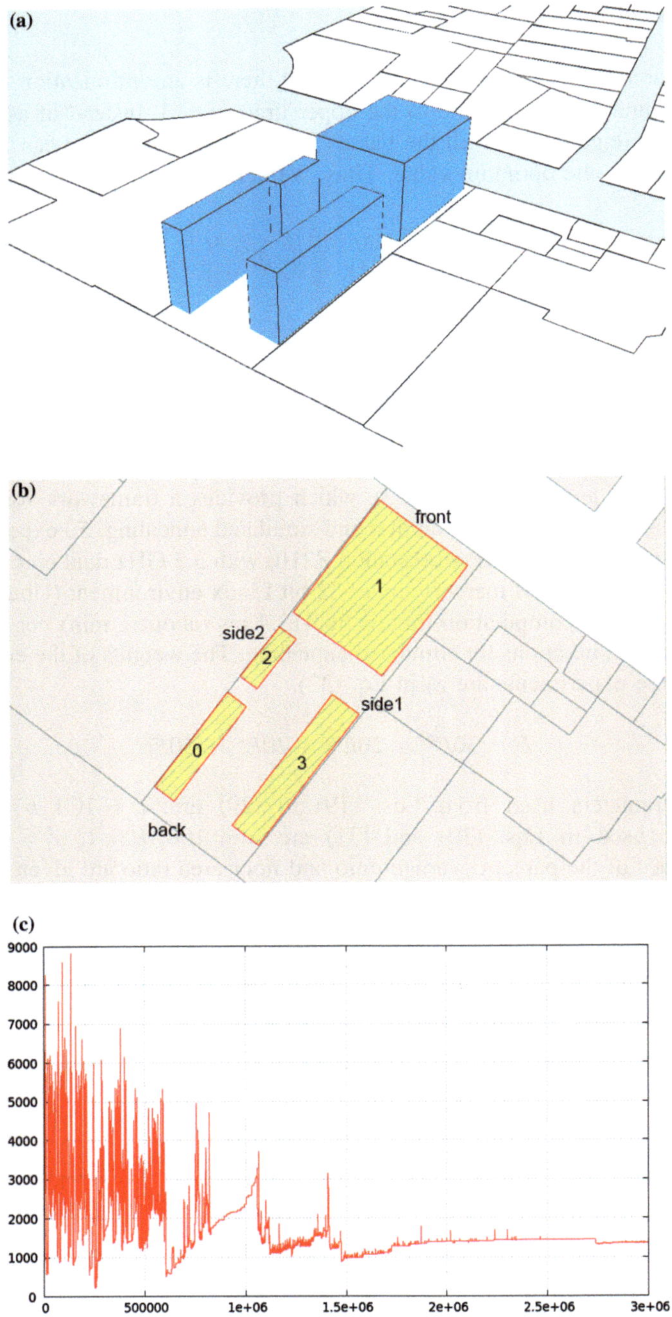

Fig. 3 Result with the highest floor area ratio. **a** Building layout, **b** Footprints, **c** Energy plot

Table 2 Properties of the result with the highest floor area ratio

	dFront	dSide1	dSide2	dBack	Height
$bldg_0$	50.3493	18.2548	0.0004	4.1528	18
$bldg_1$	2.8052	1.5012	0.2108	47.5014	18
$bldg_2$	33.9305	18.8940	0.0129	31.1415	15
$bldg_3$	38.6350	0.2542	16.1756	7.2704	15
Binary distance					
d_{01}	d_{02}	d_{03}	d_{12}	d_{13}	d_{23}
20.3253	3.9695	11.6544	3.9066	8.6111	12.1260
Lot coverage ratio = 0.5230					
Floor area ratio = 2.9967					
Energy					
$30E^{u_d}$	$20E^{b_d}$	$20E^{g_c}$	$50E^{g_f}$	Total energy	
1299.39	55.9319	0	0.00538	1355.32	

5.2.1 Results with the Highest and Lowest Floor Area Ratio

Since the optimization task of this case study is to maximize the floor area ratio within an upper limit of 3 m, we verify these two extreme outcomes. The result with the highest floor area ratio is shown in Fig. 3, and its properties are given in Table 2. Some rules can be satisfied within acceptable tolerance of error (in blue). There is one violation of the rule (in red) may not be accepted in reality, but it can be easily fixed by postprocessing transformation. The result with the lowest floor area ratio is shown in Fig. 4, and its properties are given in Table 3. All the PLU rules are satisfied within 0.2 m tolerance of error, and the floor area ratio reached to 53.66 % of its maximum. This result validates our approach by showing that even the worst possible result can still be satisfying.

5.2.2 Results with the Highest and Lowest Energy

Since the compliance of the PLU rules and the optimization of floor area ratio are reflected by energy, we verify these two extreme outcomes. The result with the highest energy value is shown in Fig. 5, and its properties are given in Table 4. There is one violation of the rule (in red) may not be acceptable in practice, but can be easily rectified by transformation in postprocessing. The result with the lowest

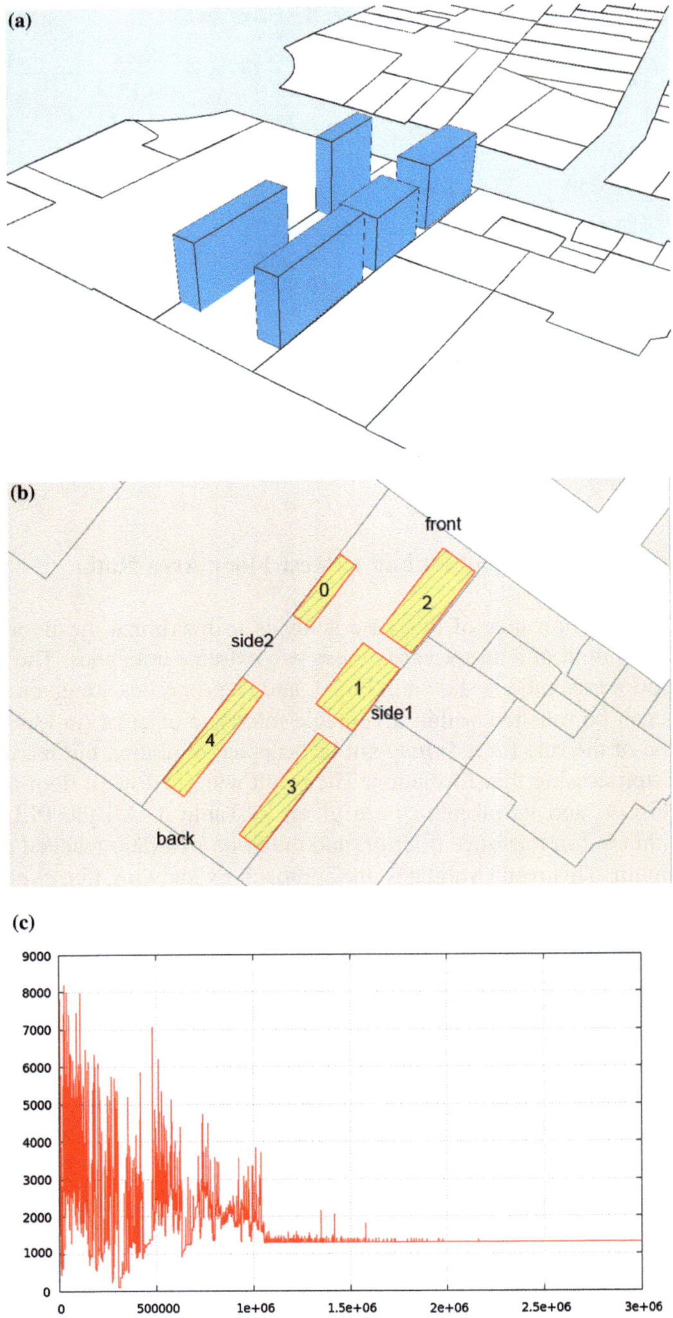

Fig. 4 Result with the lowest floor area ratio. **a** Building layout, **b** Footprints, **c** Energy plot

Table 3 Properties of the result with the lowest floor area ratio

	dFront	dSide1	dSide2	dBack	Height
$bldg_0$	15.5205	18.8965	6.529e-07	47.1977	18
$bldg_1$	26.7261	0.0796	14.062	36.0455	12
$bldg_2$	2.9182	0.1945	14.9991	54.9106	15
$bldg_3$	45.9468	0.0518	17.8391	7.0341	15
$bldg_4$	45.9698	17.808	0.0003	6.38574	15
Binary distance					
d_{01}	d_{02}	d_{03}	d_{04}	d_{12}	d_{13}
9.98308	11.0886	20.7292	15.6475	3.92361	4.29164
d_{14}	d_{23}	d_{24}	d_{34}		
10.1291	23.1443	25.3415	12.8055		
Lot coverage ratio = 0.329935					
Floor area ratio = 1.60993					
Energy					
$30E^{u_d}$	$20E^{b_d}$	$20E^{g_c}$	$50E^{g_f}$	Total energy	
276.123	34.4739	0	966.147	1306.74	

energy is regarded as the best solution. It is shown in Fig. 6, and its properties are in Table 5. All the PLU rules are satisfied within 0.2 m's tolerance of error, and the floor area ratio reached to 81.18 % of its maximum.

6 Conclusion and Perspectives

This paper proposed an original approach in order to provide better understanding of urban planning rules, especially the rules that regulate spatial aspects in building development (e.g. individual geometry, spatial relationship with other buildings and parcel borders). For this purpose, we propose to simulate legal scenarios by generating 3D building layouts in conformity with these rules using a stochastic optimization approach. Moreover, our approach also allows to generate acceptable building layouts with optimized urban indicators (e.g. floor area ratio), so as to help urban planners in the assessment of the impacts of the rules on constructability. As a result, the generated building layouts can also be used for preliminary urban design and for presenting new construction projects.

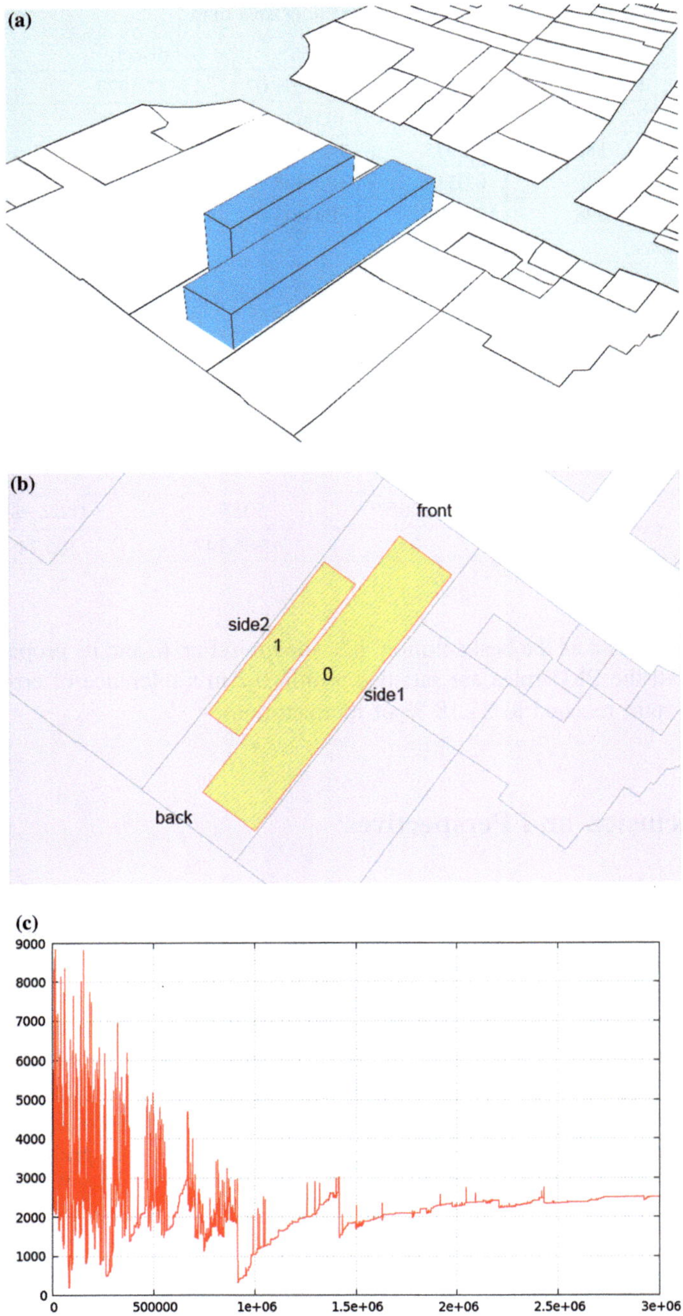

Fig. 5 Result with the highest energy. **a** Building layout, **b** Footprints, **c** Energy plot

Table 4 Properties of the result with the highest energy

	dFront	dSide1	dSide2	dBack	Height
$bldg_0$	7.32	0.558052	9.31182	6.81932	9
$bldg_1$	20.1219	14.4473	0.793807	19.9008	15
Binary distance d_{01} = 1.1065					
Lot coverage ratio = 0.60879					
Floor area ratio = 2.1413					
Energy					

$30E^{u_d}$	$20E^{b_d}$	$20E^{g_c}$	$50E^{g_f}$	Total energy
909.201	1173.75	39.6696	368.681	2491.3

Taking French PLU rules as example, several common rules are studied and modeled into energy terms which eventually form an overall energy function for optimization. Lower value of an energy term indicates less violation of a rule or closer to the optimum value of a target urban indicator. The minimization of the overall energy function is realized by stochastic optimization using RJMCMC sampler in a simulated annealing framework. We developed a simulation tool named BuildUP (Building generator for Urban Planning), and the prototype is available as open source (https://github.com/IGNF/BuildUP).

With this simulator, we carried out a case study using real data and rules. The results proved that our method is capable of accomplishing the required task. Due to the diversity of the rules, the robustness and the generality of our method are still to be assessed. More complex rules and use cases will be studied in the near future. Since the total energy function is the sum of weighted energy terms, we also plan to work on determining the weight coefficients.

Some extensions of the simulator will be realized such as the possibility of managing several types of objects associated to future buildings. This includes objects such as parking spaces, whose dimensions are linked, in the regulation, to the parcel floor area, or building architectural elements, that can be generated with procedural grammars [as it is realized in Talton et al. (2011)]. A last perspective is to link this tool with other simulators that consider other phenomena such as solar radiation or house prices in order to assess the influence of the regulation on these phenomena.

This work is part of the FEDER e-PLU project (http://www.e-plu.fr/) whose aim is to propose a web platform dedicated to territorial engineering. This platform will provide services such as 3D city navigation, right to build consultation and co-design of urban regulations and will be tested on Plaine-Commune inter-communality by the end of 2015.

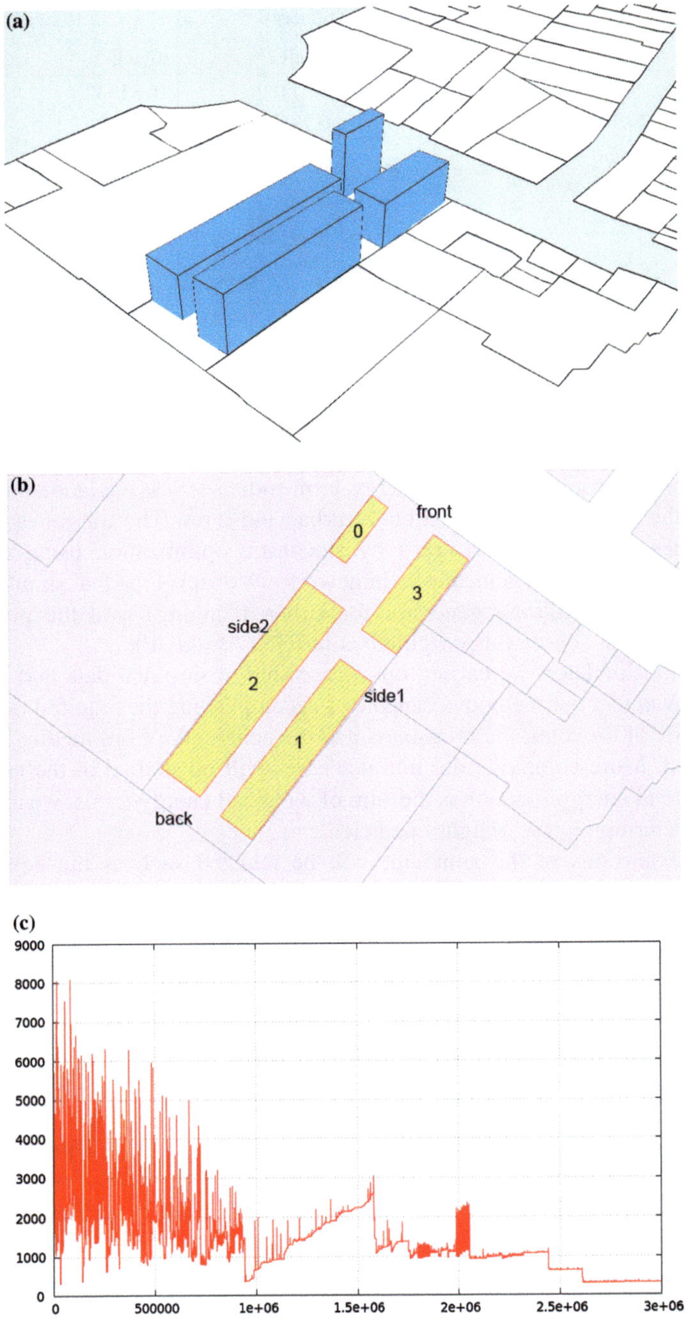

Fig. 6 Result with the lowest energy

Table 5 Properties of the result with the lowest energy

	dFront	dSide1	dSide2	dBack	Height
$bldg_0$	3.98306	18.9307	5.364e−06	59.772	18
$bldg_1$	33.7205	0.20625	14.8007	5.28895	15
$bldg_2$	25.1778	13.3745	1.194e−05	5.76251	12
$bldg_3$	2.99916	0.02221	14.2492	51.3647	12

Binary distance

d_{01}	d_{02}	d_{03}	d_{12}	d_{13}	d_{23}
19.163	7.42981	10.303	5.42371	7.38893	5.08948

Lot coverage ratio = 0.55157

Floor area ratio = 2.43554

Energy

$30E^{u_d}$	$20E^{b_d}$	$20E^{g_c}$	$50E^{g_f}$	Total energy
155.179	0	0	159.305	314.484

References

Bao F, Yan D-M, Mitra NJ, Wonka P (2013) Generating and exploring good building layouts. ACM Trans Graph 32

Brasebin M (2014). Les données géographiques 3D pour simuler l'impact de la règlementation urbaine sur la morphologie du bâti., Université Paris-Est

Brasebin M, Perret J, Haëck C (2011) Towards a 3D geographic information system for the exploration of urban rules: application to the French local urban planning schemes. In: 28th urban data management symposium (UDMS 2011), Delft, Netherlands

Brédif M, Tournaire O (2012) Librjmcmc: an open-source generic C++ library for stochastic optimization. The XXII congress of the international society of photogrammetry and remote sensing, Melbourne, Australia

Coello Coello CA (2010) Constraint-handling techniques used with evolutionary algorithms. In: Proceedings of the 12th annual conference companion on genetic and evolutionary computation. ACM, Portland, Oregon, USA

Coors V, Hünlich K, On G (2009) Constraint-based generation and visualization of 3D city models. In: 3rd International Workshop on 3D Geo-Information. Seoul, Korea, pp 365–378

El Makchouni M (1987) Un système graphique intelligent d'aide à la conception des plans d'occupation des sols: SYGRIPOS. In: 12th urban data management symposium, Blois, France

Frazer J (1995) An evolutionary architecture. Architectural Association

Kämpf JH, Montavon M, Bunyesc J, Bolliger R, Robinson D (2010) Optimisation of buildings' solar irradiation availability. Sol Energy 84:596–603

Laurini R, Vico F (1999) 3D symbolic visual simulation of building rule effects in urban master plans. In: The second international workshop on urban 3d/multi-media mapping (UM3'99), Japan

Métral C, Falquet G, Cutting-Decelle A (2009) Towards semantically enriched 3D city models: an ontology-based approach. GeoWeb, Vancouver

Murata M (2004) 3D-GIS application for urban planning based on 3D city model. In: 24th annual ESRI international user conference

Parish YIH, Müller P (2001) Procedural modeling of cities. In: Proceedings of the 28th annual conference on computer graphics and interactive techniques. ACM, New York, NY, USA

Perret J, Curie F, Gaffuri J, Ruas A (2010) A multi-agent system for the simulation of urban dynamics. In: 10th European conference on complex systems (ECCS 2010), Lisbon, Portugal

Rittel HWJ, Webber MM (1973) Dilemmas in a general theory of planning. Policy Sci 4:155–169

Ruas A, Perret J, Curie F, Mas A, Puissant A, Skupinski G, Badariotti D, Weber C, Gancarski P, Lachiche N, Lesbegueries J, Braud A (2011) Conception of a GIS-platform to simulate urban densification based on the analysis of topographic data. Lect Notes Geogr Cartography. Springer, Heidelberg 413–430

Salamon P, Sibani P, Frost R (2002) Facts, conjectures, and improvements for simulated annealing. Soc Ind Appl Math

Singh HK, Isaacs A, Ray T, Smith W (2008) A simulated annealing algorithm for single objective trans-dimensional optimization problems. In: Eighth international conference on hybrid intelligent systems, 2008. HIS'08. School of Aerospace, University of New South Wales, Canberra, ACT

Talton JO, Lou Y, Lesser S, Duke J, Měch R, Koltun V (2011) Metropolis procedural modeling. ACM Trans Graph 30

Tournaire O, Brédif M, Boldo D, Durupt M (2010) An efficient stochastic approach for building footprint extraction from digital elevation models. ISPRS J Photogram Remote Sens 65:317–327

Turkienicz B, Gonçalves BB, Grazziotin P (2008) CityZoom: a visualization tool for the assessment of planning regulations. Int J Architectural Comput 6:79–95

van Lieshout MNM (2000) Markov point processes and their applications. Imperial College Press, London

Vanegas CA (2013) Modeling the appearance and behavior of urban spaces. Purdue University

Wonka P, Wimmer M, Sillion F, Ribarsky W (2003) Instant architecture. ACM Trans Graph 22:669–677

Spatial Variation of Privacy Measured Through Individual Uniqueness Based on Simple US Demographics Data

Allen Lin and Francis Harvey

Abstract Previous studies reveal that, using solely U.S. census data, over 60 % population could be uniquely identified with (gender, zip code, date of birth) in 1990 and 2000. This paper extends these studies to examine spatial variation of uniqueness in 2010. We provide (1) the comparison of national level uniqueness between 2000 and 2010, and (2) the investigation of spatial variation of uniqueness in different regions and at different scales. The comparison between 2000 and 2010 reveals that, although overall uniqueness changes little, the uniqueness of middle-age group has significantly decreased. Spatial variation studies show that similar characteristics in age-group uniqueness exist in different regions. Finally, the analysis at different scales discloses that overall uniqueness decreases, and the difference between age-group uniqueness reduce, when geographical scales focus on the cores of urban area. This study offers contributions to geographic information privacy, particularly relevant to reverse geocoding and related spatial aggregation techniques.

Keywords Spatial statistics · Census · Privacy · Spatial analysis · Accuracy

1 Introduction

Privacy has been one of the top priorities in the release of the U.S. census data (US Census Bureau 2013). In general, the anonymity of individual in the aggregated census data is measured by k-anonymity (Sweeney 2002), which, in the context of census data, ensures that in a data release at least k people share the same demo-

A. Lin (✉)
Department of Geography, Environment and Society, University of Minnesota, Twin Cities, 414 Social Sciences 267 19th Ave S, Minneapolis, MN 55455, USA
e-mail: linx0344@umn.edu

F. Harvey
Leibniz Institute for Regional Geography and Leipzig University, Schongauerstr. 9, 04238 Leipzig, Germany
e-mail: f_harvey@ifl-leipzig.de

F. Harvey and Y. Leung (eds.), *Advances in Spatial Data Handling and Analysis*, Advances in Geographic Information Science,
DOI 10.1007/978-3-319-19950-4_17

graphic characteristics. A k value of 5 means that a group of 5 people with the same demographic characteristics is identifiable from the population. Using this measurement, research based on the past two census (i.e. 1990 and 2000) reveals that, in the United States, 87 % of the population in 1990, and 63 % of the population in 2000 can be uniquely identified (i.e. 1-anonymity of k-anonymity), given the simple demographics (gender, zip code, and date of birth) (Golle 2006; Sweeney 2000). These studies have influenced privacy protection guidelines for numerous public releases of microdata[1] (El Emam et al. 2008; US Census Bureau 2013), and inspired a series of algorithms to optimize the process of k-anonymization (Bayardo et al. 2005; Winkler 2002).

A related research area is geographic information privacy, which studies the identification and the anonymization of individuals' location information. The techniques to protect geographic confidentiality are referred to as geographic masking (Allshouse et al. 2010; Armstrong et al. 1999; El Emam et al. 2005; Zimmerman et al. 2008). With regards to census data, the specific protecting technique being used is called aggregation (Armstrong et al. 1999), which aggregates people found in a region and uses a regionalized attribute to represent the attributes of individuals who reside in it. A typical example is aggregating individuals' addresses to zip code areas, which is then used as the geographic attributes to represent all people in this zip code zone. This method prevents the linkability (Pfitzmann et al. 2010) between the resulting dataset and the original dataset. In this paper, we consider the spatial variation of k-anonymity in relation to privacy concerns.

Compared with research in geo-privacy, previous studies of uniqueness in census data fail to take into account the data's spatial nature. In census data, demographic information are aggregated according to various levels of geographic enumeration areas, including zip code, county, city, metropolitan statistical areas (MSA) and states. As a result, the spatially heterogeneous population distributions of different geographic enumeration areas would impact the variation of uniqueness. For instance, as shown in the 2010 population distribution map (US Census Bureau 2010), the population density in the eastern part of U.S. is prominently higher than that in the western part of U.S. Likewise the state population pyramids also show disparities between states. As a result, we are interested to examine how privacy in the census data, measured through individual uniqueness, varies spatially. This paper presents the first published analysis of spatial variation of k-anonymity privacy in U.S. census data according to regional differences and scale differences. Specifically, the paper makes the following contributions:

- **Provides the national level individual uniqueness using the simple demographics in the latest 2010 census data**. We compare the 2010 results with the results from 2000 to confirm that, on the national level, the individual uniqueness remains the same (62 % in 2010 census and 63 % in 2000 census). Nevertheless, in a finer grained analysis of age groups uniqueness, we found

[1]Microdata are data on the characteristics of units of a population, such as individuals, households, or establishments, collected by a census, survey, or experiment.

significantly more (about 10 % more) people of middle-age group (45–70) are uniquely identifiable in 2010 census data.

- **Explores the impact of regional differences**. Using the 2010 census data, we choose four exemplary states: Florida, Connecticut, Iowa and Minnesota and calculate both their state-wise overall uniqueness and age group uniqueness. These four states represent important aspects of demographic diversity in the U.S. The results indicate some remarkable differences among the four states. More importantly, besides the differences, the results also suggest some regionally invariant characteristics of uniqueness that deserve further investigation.
- **Investigates the impact of spatial scales**. For these four states, we choose three spatial scales: state, metropolitan statistical area (MSA) and county. We interpret scale in this paper from a human geographical perspective that is the size of the study area (Sheppard et al. 2008). We calculate both the overall and the age group uniqueness. The results suggest that although general trends persist at different scales, inter-age-group variation subsides and the overall uniqueness decreases as the scales become smaller and more focused on the core of the region.

The rest of the paper is organized as follows: Sect. 2 describes the data collection and analysis techniques used in this paper. Section 3 presents the result and discusses its implication. Section 4 offers the summary, conclusion and future direction for this research.

2 Data and Methods

2.1 Data

In order to be comparable with previous studies of the 1990 and the 2000 census data, the same data sources are used for this study. Specifically, we use table PCT 12 (Sex by Age) from 2010 census data release, which documents the respective number of males and females, of the specific age (1 year interval between 0 and 99 years old, 5 year interval between 100 and 110 years old and over 110 years old) in specific ZIP Code Tabulation Areas (ZCTAs). According to U.S. Census Bureau, ZCTAs are the generalized areal representations of USPS Zip Code service areas (US Census Bureau 2012). Their construction involves (1) assigning each census block a zip code that the majority of its containing addresses use, and (2) aggregating census block with the same majority zip code. ZCTA is the census geographic identity that most approximates zip code and is the same with ZIP code in most cases (US Census Bureau 2012). ZCTAs data is downloaded from National Historical Geographic Information System (NHGIS) project (Minnesota Population Center 2011). Other auxiliary data are retrieved to translate zip code areas to larger geopolitical boundaries, including city, metropolitan statistical area and state. Specifically, the Zip Code Lookup table is downloaded from U.S. Department of Labor (Department of Labor 2010).

2.2 Method

2.2.1 Examine National Level Individual Uniqueness

To examine the national level uniqueness, we calculate the uniquely identifiable population percentage based on two sets of demographic characteristics: (zip code, gender, date of birth) and (zip code, gender, age). These two sets of the attributes are chosen because they are commonly requested by most online registration process ranging from social media such as Facebook to the financial agency such as auto insurance company.

Since the table PCT12 (Sex by Age) is aggregated on age, it needs additional transformation to estimate the number of people born on a specific date. This estimation can be reduced into the question of estimating the expected number of days in a year on which k people are born in a given geographic region and calculated by:

$$E(X_n^k) = C_k^n(\frac{1}{365})^{k-1}(\frac{364}{365})^{n-k} \quad (1)$$

where X_n^k is the number of days in a specific year on which exact k individual are born, given n as the total number of people in the given gender, zip code and year. Golle offers a proof of the solution in his paper (Golle 2006).

As a result, the generic form of estimation of population under k-anonymity of i-anonymity, given (zip code, gender, date of birth) is as follows:

$$P_{i-anonymity} = \sum_{k=1}^{i}(k \cdot E(X_n^k)) \quad (2)$$

2.2.2 Exploring Spatial Variation of Uniqueness

We explore the spatial variation of uniqueness in two dimensions: regional differences and scale differences.

Regional Differences:
We choose these four states, Florida (FL), Iowa (IA), Minnesota (MN) and Connecticut (CT), due to their distinct age-gender population constitution, seen in the respective population pyramids (Fig. 1).

The population pyramids of the four states show the regional differences of the simple demographics used in this paper. For example, Florida stands out with a dominant percentage of the people above 60 years old while Minnesota exhibits a higher percentage of middle age (30–40 years old) people. In this paper, both the state-wise overall uniqueness and age group uniqueness of the four states are computed based on (date of birth, gender, zip code) using Eq. 2.

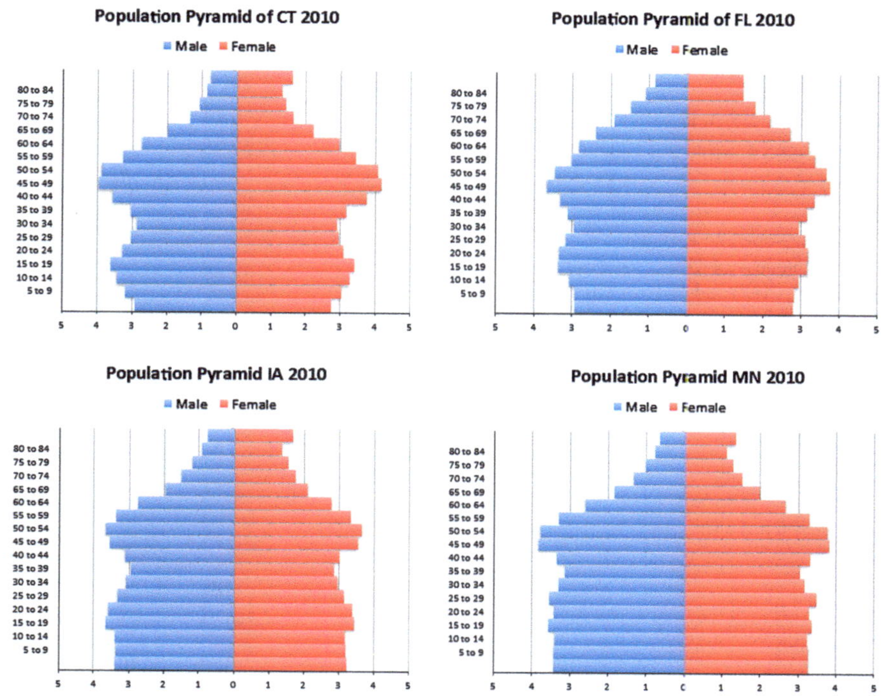

Fig. 1 Population pyramids of Florida, Minnesota, Iowa and Connecticut

Scale Differences:
Scale has been a central topic in almost every facet of geographical research and held diverse meanings for geographers working in different domains (Sheppard et al. 2008). As mentioned before, in this paper we adopt a human geographical concept of scale, meaning that it refers to the scope of the study area. (That is, a large scale refers to a large area.) For each of the four selected states, we select one metropolitan statistical area (MSA) within that state and one county within that MSA. Similarly, both the overall uniqueness and the age specific uniqueness are computed using Eq. 2.

3 Results and Analysis

3.1 2010 National Level Privacy

First, we compute the national level percentage of the uniquely identifiable population using (gender, date of birth, zip code) and (gender, age, zip code), based on 2010 census data. The results are compared with those of 2000 census data (See Table 1).

Table 1 National level uniquely identifiable population percentage, 2000 and 2010 (2000 result is excerpted from (Golle 2006))

Given attributes	2000 census data (%)	2010 census data (%)
Gender, age, zip code	0.20	0.14
Gender, date of birth, zip code	63.30	62.48

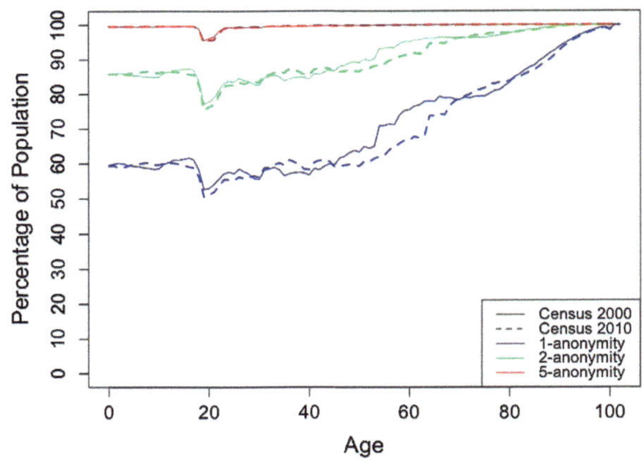

Fig. 2 Age-specific national level uniqueness of U.S population, 2000 and 2010

Second, following Golle' paper (Golle 2006), we provide a fine-grained characterization of the national level uniqueness of different age group population. We compute the population percentages that are (1) under 1-anonymity (uniqueness), (2) under 2-anonymity and (3) under 5-anonymity and compared with those of 2000 (See Fig. 2).

Table 1 and Fig. 2 shed light on the changes of national level uniqueness in the past ten years. In general, the results confirm the conclusions in Golle's paper that significant amount of people can be uniquely identified by the simple and frequently used demographic attributes. It additionally shows that the level of anonymity has not improved between 2000 and 2010. However, when breaking down to anonymity of different age-group population, there is remarkably fewer people in middle age (from age 45 to 70) to be unique identified from their peers, thus showing an improvement of individual anonymity of that particular group of people. This change is of particular interest since it cannot be completely explained by matching the uniqueness of this group to that of the corresponding group (age 35–60) in 2000 result. Therefore, it indicates more complex social-economic changes of this group of people during these ten years, which is beyond the scope of this paper, but deserves further investigation.

3.2 Spatial Variation of Uniqueness

The impacts of regional differences and scale differences to uniqueness are analyzed using the selected states and scales based on the explanation in Sect. 2.2. We compute only the population percentage of 1-anonymity (i.e. individual uniqueness) given gender, date of birth and zip code for the easiness of comparison.

Table 2 Regional differences of overall uniqueness, using MN, FL, IA, CT

	Minnesota	Florida	Iowa	Connecticut
Population	5,304,141	18,801,226	3,046,945	3,574,333
Pct. of uniquely identified people	70.18 %	60.99 %	77.43 %	66.76 %

Table 2 and Fig. 3 allow us to consider how regional differences impact individual uniqueness. Table 2 presents the total population and the overall uniquely identifiable population percentage of the four states. Figure 3 offers a more fine-grained comparison of the age-specific uniqueness in these states and the national average. Although the overall uniqueness differs markedly between states, some similarities show in the age-group uniqueness comparison. First, despite the different youth population percentages, the uniquely identifiable percentage of the college-age population in the four states drop to approximately the same level. Golle attributed this plunge to the concentration of campus student housing, which makes the people in these areas more homogeneous. However, the results of this paper further imply that this could be a spatially invariant feature in most part of U.S. Second, the same extent (about 10 %) increase from age 63 to 69 is observed from all four states. Unfortunately, we find no clear explanation to this trend. These two similarities are of special interests, considering the disparate state-wise demographics (see Fig. 1) and might imply some common spatially distinct characteristics to explore further.

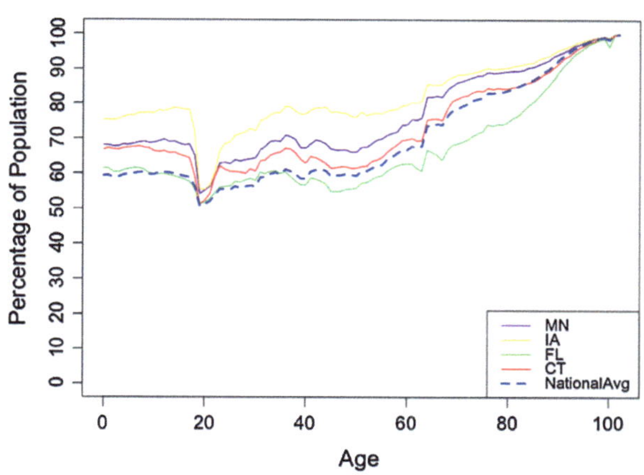

Fig. 3 Regional differences of age-group uniqueness, using MN, FL, IA, CT

Table 3 Scale differences of overall uniqueness, using state, MSA, county

	State level (%)	MSA level (%)	County level (%)
Minnesota	70.18	62.89	65.12
Iowa	77.43	71.5	67.68
Florida	60.99	53.49	51.90
Connecticut	66.76	67.32	65.15

Table 3 and the horizontal interpretations of Fig. 4 explain the scale differences of uniqueness. The legend of Fig. 4 indicates the specific MSA and counties we selected. It is worth reiterating that these MSAs and counties are chosen to ensure that the smaller scale regions reside in bigger scale regions. This nesting of scales provides a comparison between the core of the region and the region overall. Table 3 reveals the general trend in which the percentages of uniquely identifiable population decrease in the core area of smaller scales, meaning proportionally more people are anonymous. The horizontal reading of Fig. 4 verifies this trend and

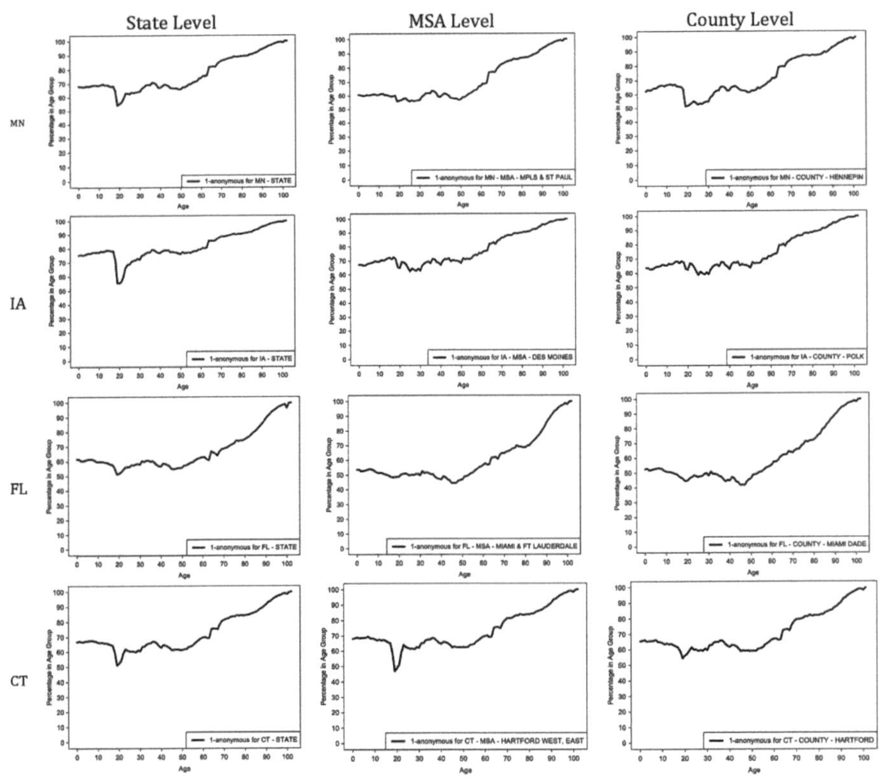

Fig. 4 Scale differences of age-group uniqueness, using MN, FL, IA, CT

further suggests that the age-group differences also tend to subside in core regions at smaller scales.

4 Conclusions and Future Work

In this paper, we provide the first empirical analysis of the spatial variation of uniqueness in the census data considering simple demographics. We first compare the national level uniqueness between the 2010 census and the 2000 census. The findings suggest that the overall uniqueness does not change noticeably, yet there is a significant decrease to the uniquely identifiable population of the middle-aged group. This decrease reveals that people from middle-age group (age 45–60) are more spatially aggregated in 2010 comparing to 2000. We then break down the data into regions of different demographic constitutions and into scales of different areas. The analysis of the regional differences indicates some spatial invariants of the uniqueness among the four states (i.e. FL, IA, MN and CT). Although both the overall uniqueness and the age-group uniqueness vary remarkably between states, the college-age group tends to reach the same level of uniqueness. Furthermore, in the group between the age of 63 and the age of 70, a similar increase (approximately 10 %) of uniquely identifiable population percentage exists, despite the different average uniqueness percentage. The study of the scale difference suggests that the uniquely identifiable population percentage decreases as the scales become smaller and more focus on the core of the region. In addition, the age-group variation also diminishes in smaller scale regions.

As noted before, in this study, we consider only four states and three scales to show the spatial variation of individual uniqueness. One direction for future work is to conduct a thorough comparison of uniqueness values in all 50 states, which should provide solid verification of some specific patterns of uniqueness such as the spatial invariants across regions and the decreasing variation across spatial scales. It will also support an exploration of spatial autocorrelation of uniqueness. Second, another direction of research is to investigate what social-demographic attributes explain the spatial distribution patterns of individual uniqueness found in this paper. Finding this correlation will help improvement of privacy protection strategy in census data by applying place-specific aggregation limits (El Emam et al. 2005). Third, the marked decrease of individual uniqueness on national level between 2010 and 2000 indicates a more spatially aggregated distribution of middle-aged group population. However, the reason of the increasing spatial aggregation remains uninvestigated and deserves exploration with additional evidence from population studies. Regarding scale differences, the multi-level relations between different scales need be investigated with a specific research designs. Finally, given the potential of the simple demographic to contravene common geostatistical measures for protecting individual privacy, coupling census data and other publicly available online social media data for geo-privacy combination attack is of considerable interest for future research.

Acknowledgments This research is partially supported by the Summer Research Scholarship of Department of Geography, Environment and Society, University of Minnesota, Twin Cities, provided through the Abler Foundation.

References

Allshouse WB, Fitch MK, Hampton KH, Gesink DC, Ia Doherty, Pa Leone, Miller WC (2010) Geomasking sensitive health data and privacy protection: an evaluation using an E911 database. Geocarto Int 25(6):443–452

Armstrong MP, Rushton G, Zimmerman DL (1999) Geographically masking health data to preserve confidentiality. Stat Med 18(5):497–525

Bayardo RJ, Agrawal R (2005) Data privacy through optimal K-anonymization. In: ICDE 2005. Proceedings. 21st international conference on data engineering, pp 217–228

Department of Labor (2010) Geographic practice cost index values by ZIP code US. Available at. http://www.dol.gov/owcp/regs/feeschedule/fee/fee10/fs10gpci Accessed 1 Feb 2014

El Emam K, Brown A, AbdelMalik P (2005) Evaluating predictors of geographic area population size cut-offs to manage re-identification risk. J Am Med Inform Assoc JAMIA 16(2):256–266

El Emam K, Dankar FK (2008) Protecting privacy using K-anonymity. J Am Med Inform Assoc 15(5):627–637

Golle P (2006) Revisiting the uniqueness of simple demographics in the US population. In: Proceedings of the 5th ACM workshop on privacy in electronic society ACM, New York, pp 77–80

Minnesota Population Center (2011) National historical geographic information system: version 2.0. https://www.nhgis.org/. Accessed 1 Feb 2014

Pfitzmann A, Hansen M (2010) A terminology for talking about privacy by data minimization: anonymity, unlinkability, undetectability, unobservability, pseudonymity, and identity management. Available via. http://dud.inf.tu-dresden.de/Anon_Terminology.shtml. Accessed 1 Nov 2014

Sweeney L (2000) Uniqueness of simple demographics in the US population. In: LIDAPWP4. Carnegie Mellon University, laboratory for international data privacy, Pittsburgh, PA

Sweeney L (2002) K-anonymity: a model for protecting privacy. Int J Uncertainty Fuzziness Knowl Based Syst 10(05):557–570

Sheppard E, McMaster RB (2008) Introduction: scale and geographic inquiry. In: Sheppard E, McMaster RB (ed) Scale and geographic inquiry: nature, society, and method. Wiley, New York

U.S. Census Bureau (2010) Population distribution in the United States and Puerto Rico [map]. 1:7,500,000. https://www.census.gov/geo/maps-data/maps/2010popdistribution.html. Accessed 14 Sep 2014

United States Census Bureau (2012) ZIP Code Tabulation Areas (ZCTAs). https://www.census.gov/geo/reference/zctas.html. Accessed 20 Feb 2014

United States Census Bureau (2013) About data protection and privacy. Available http://www.census.gov/privacy/. Accessed 1 Feb 2014

Winkler W (2002) Using simulated annealing for K-anonymity. Research Report 2002–07, US Census Bureau Statistical Research Division

Zimmerman DL, Pavlik C (2008) Quantifying the effects of mask metadata disclosure and multiple releases on the confidentiality of geographically masked health data. Geogr Anal 40(1):52–76

Monte Carlo-Based Analysis of the Effect of Positional and Thematic Uncertainties on Biodiversity Models

Patrick J. Kirby and Scott W. Mitchell

Abstract Monte Carlo methods are a common approach to quantifying uncertainty propagation. We used Monte Carlo simulation to quantify the effects of positional and thematic uncertainties in a set of landscape maps on model averaged regression coefficients that were based on metrics derived from these maps. Results indicate that the uncertainty estimates from model averaging outweigh the effects of positional and thematic uncertainties in the landscape maps. Shifts between reference and simulated coefficients indicate a need for further research into simulation approaches that account for spatial autocorrelation.

Keywords Monte Carlo · Positional uncertainty · Thematic uncertainty · Biodiversity

1 Introduction

With declining biodiversity levels and food security issues related to a growing global population, questions regarding how to maintain both biodiversity and agricultural yield are being raised (Phalan et al. 2011). To better understand how we can preserve biodiversity without reducing the amount of land in agricultural production, the relationship between agricultural landscape heterogeneity and biodiversity is being studied (Duro et al. 2014; Fahrig et al. 2015). The influence of landscape configuration and composition is a topic of particular interest (Fahrig et al. 2011).

Geographic Information Systems (GIS) and remote sensing products can be useful for studying the relationship between biodiversity and landscape

P.J. Kirby (✉) · S.W. Mitchell
Geomatics and Landscape Ecology Laboratory, Department of Geography and Environmental Studies, Carleton University, Ottawa, Canada
e-mail: Patrick.Kirby@carleton.ca

S.W. Mitchell
e-mail: Scott.Mitchell@carleton.ca

F. Harvey and Y. Leung (eds.), *Advances in Spatial Data Handling and Analysis*, Advances in Geographic Information Science,
DOI 10.1007/978-3-319-19950-4_18

characteristics, though an understanding of the positional and thematic uncertainties in these products is important (Shi et al. 1999). Positional uncertainty in vector datasets can arise from the error in the positioning of individual vertices and the interpretation of boundaries when digitizing (Edwards and Lowell 1996; Shi et al. 1999), while thematic uncertainty in maps derived through image classification can arise due to spectral overlap between the defined classes (Lillesand et al. 2004; Foody et al. 1992).

Uncertainty analysis (UA) is a tool through which the influence of these uncertainties can be better understood. It involves modelling uncertainty in the input data and quantifying its propagation through a model (Crosetto and Tarantola 2001; Crosetto et al. 2001). Monte Carlo (MC) simulation, a numerical, black box technique to quantifying uncertainty propagation, has become widely used for UA (Crosetto and Tarantola 2001). It involves generating multiple error-sensitized inputs, referred to as realizations, by repeatedly sampling probability density function (PDF) representations of model inputs (Crosetto and Tarantola 2001; Crosetto et al. 2001). The model is then re-evaluated with the perturbed inputs to obtain an output distribution that is representative of the propagated input uncertainty (Crosetto and Tarantola 2001; Crosetto et al. 2001).

Realizations can be sensitized based on multiple types of uncertainty. For example, Delmelle et al. (2014) sensitized disease outbreak reports with positional and temporal error to visualize and compare the relative stability of spatio-temporal outbreak clusters. The influence of individual uncertainty sources can be compared by analyzing the difference in model output variance when one source is isolated (Benke and Pelizaro 2010).

Finally, MC simulation is not limited to the analysis of model predictions. For example, Richardson and Hollinger (2005) used MC simulation to sensitize data with measurement error and repeatedly re-fit models to obtain a PDF for model parameters. These approaches can be useful in cases where much of the focus is placed on the relationships established by the model, rather than solely on model predictions.

1.1 Aim

We build on a study wherein several forms of biodiversity had been modelled on landscape metrics derived from agricultural landscape maps. Our aims were to: (1) use MC simulation to quantify the uncertainty in the regression coefficients resulting from positional and thematic uncertainties in the landscape maps (and corresponding uncertainty in landscape metrics derived from these maps); and (2) to compare the variance in regression coefficients associated with these uncertainties to the variance associated with the modelling approach, which had already been considered.

2 Methods

2.1 Background and Data

This section outlines the combined work of Pasher et al. (2013), Duro et al. (2014), and Fahrig et al. (2015), which our study builds on. Together, these studies involved the statistical modelling of alpha, beta, and gamma diversity, as well as abundance, for a series of species groups (including beetles, birds, butterflies, plants, spiders, and syrphids) in agricultural landscapes.

Biodiversity data were recorded for several 1 km^2 landscapes in eastern Ontario (46 in 2011 and 47 in 2012). Maps of agricultural fields in each landscape had been created by manually digitizing high resolution air photos. Digitizers had some previous digitizing experience, though not extensive. Crop type data for each landscape were recorded through ground referencing.

Landscape metrics were extracted from each landscape map, including the proportion of land in agricultural production (*P.AG*), the logged mean field size (*MFS*), and the Shannon Index of crop diversity (*SHDI*). These three sets of landscape metrics were predictor variables in the biodiversity models. An indicator variable representing the data collected in 2011 versus 2012 was also included.

Seven linear candidate models, each consisting different combinations of the three predictors (all three, all possible pairs, and all individually) and all containing the year indicator, were produced for each diversity type and species group. For each candidate model set, second-order Akaike's information criterion (AICc) model averaging was used to establish the general relationship between the three predictors and biodiversity.

The biodiversity models discussed in this paper refer to the results of this model averaging approach. Of these biodiversity models 23 were considered by this study. This study was performed while the Fahrig et al. (2015) models were being developed, so the models used here are earlier versions of those final, published models.

2.2 Imagery

A 2008, 50 cm resolution, colour orthophoto, covering much of eastern Ontario, was available (Ministry of Natural Resources 2009). 50 cm resolution, colour air photos from 2011 and 2012 were also available for each landscape (except one). Image registration error for the 2011 air photos was not available, but was for most of the 2012 photos. The 2012 photos, georectified to the 2008 orthophoto, had an average root mean squared error (RMSE) of 3.06 m.

Landsat-5 Thematic Mapper (TM) scenes from 3 June, 19 June, 5 July, and 9 October in 2011 were acquired (NASA Landsat Program 2011). A linear shift was applied to the TM imagery to better align it with the 2008 orthophoto, resulting in RMSEs of 8.19 m (East-West) and 8.56 m (North-South). Landsat 7 ETM+ scenes were not used for 2012 due to a combination of SLC-off gaps and cloud cover.

2.3 *Extended Work*

Our study builds on the same dataset used to generate the biodiversity models, with the exception of the crop types for the 2011 landscapes. In order to consider thematic uncertainty, we assumed a scenario where the 2011 crop types were recorded through image classification rather than ground referencing. Therefore, the models analyzed here are not identical to the biodiversity models described in Sect. 2.1.

Uncertainties were modelled for: (1) field boundary interpretation; (2) vertex position; and (3) thematic class assignment.

2.3.1 Uncertainty in Field Boundary Interpretation

Uncertainty associated with identifying, via photo interpretation, features that constitute the boundaries of a field was modelled at the field level. This uncertainty was assumed to be null in cases with reasonably well-defined boundaries. In 339 of the 1916 fields, however, there was considerable uncertainty in boundary interpretation (Fig. 1). For these cases, alternate field representations were produced manually. Reference field polygons were replicated and altered as necessary, usually involving splitting or merging fields along potential boundary features.

Alternate field representations (polygons) were assigned probability estimates of being the true representation, based on the visibility of the potential boundary and photo interpretation. For each potential merge or split, 25 % probability was assigned in favour of the reference boundary representation, with an additional 0–75 % assigned based on the results of three runs of a Canny edge detector at

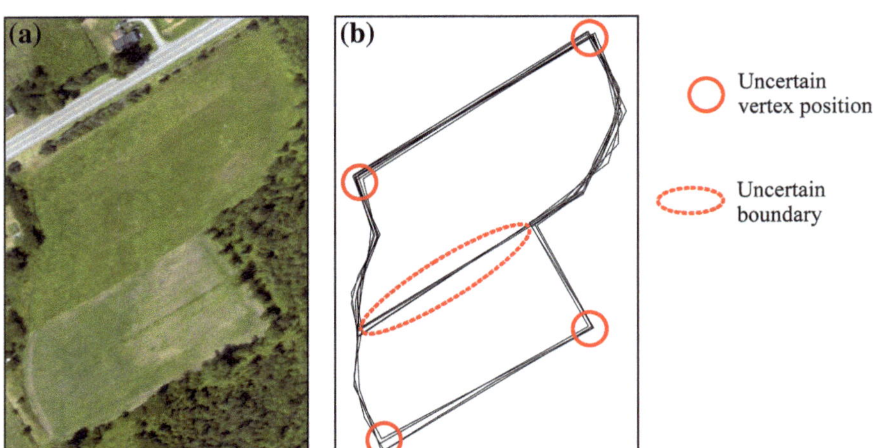

Fig. 1 **a** An air photo of a field with **b** re-digitized boundaries showing uncertainty in vertex position and boundary interpretation

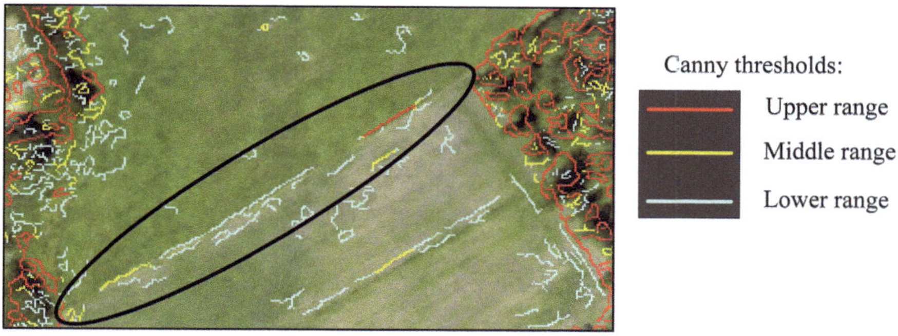

Fig. 2 An uncertain edge (*circled*) with edges detected by the Canny detector, at three threshold ranges

varying detection thresholds (up to 25 % for each, based on proportion of edge detected) (Fig. 2). Leeway of ±25 % was assigned based on visual interpretation.

For some alternatives, fields were either taken out of or entered into consideration. These were usually small features with uncertainty regarding whether they were agricultural fields, lawns, or other semi-natural areas. Probabilities for these cases were estimated by interpretation alone.

2.3.2 Vertex Position Uncertainty

The uncertainty associated with vertex positioning along field boundaries was assigned a global estimate in the East-West and North-South directions. A subset of 100 fields from the 2011 landscapes were manually re-digitized 24 times (half on the 2008 air photos, half on the 2011 air photos).

Two to four distinct corners were selected per subset field. The East-West and North-South distances between re-digitized vertices and the corresponding reference vertex were recorded for each selected corner. These distances were approximately normally distributed, with standard deviations of 3.89 m (East-West) and 4.39 m (North-South). Vertex position uncertainty in all vertices was modelled on these distributions.

2.3.3 Thematic Uncertainty

A supervised classification of the 2011 TM imagery was performed. Classes included corn, soybean, hay (>75 % grassy hay), legumes (>25 % legume hay), pasture, cereals, and fallow. These classes accounted for 99.2 % of the area of known crop types in the 2011 landscape centers.

A Random Forest (RF) classifier of 500 trees was trained on pixels outside of the landscape maps. All TM bands, except for the thermal infrared band, were included.

The normalized difference vegetation index (NDVI) from each date was also included. Some landscapes did not have coverage by the 5 July scene due to cloud cover. For these landscapes, a separate RF was trained without that scene.

Estimates of class membership probability were assigned to each pixel based on normalized class assignment votes across the 500 trees (Breiman et al. 2012). Pixel-specific class membership probabilities were aggregated to the field level so that individual fields were each represented by a single PDF of class membership.

2.3.4 Simulation

A total of 2000 realizations of alternate landscape configuration and composition were generated. For each realization, the vertex positions, including those in alternate boundary polygons, were simulated using the Brown and Heuvelink (2007) Data Uncertainty Engine, based on the vertex position uncertainty models (Sect. 2.3.2). This was followed by a recalculation of field area. All fields were then independently assigned a simulated class based on MC sampling of the class membership probabilities (Sect. 2.3.3). Finally, for fields that had alternate boundary representations, a single representation was selected based on MC sampling of their probability estimates (Sect. 2.3.1).

For each landscape realization, the three landscape metrics were recalculated. Conditional AICc model averaging, as described in Anderson (2008) and Burnham and Anderson (2002), was performed for each biodiversity model, with each recalculated (perturbed) set of landscape metrics. Regression coefficients and standard error estimates were averaged across all realizations. The variance in each of the regression coefficients, across all realizations, was representative of the uncertainty resulting from positional and thematic uncertainties in the landscape maps.

3 Results and Discussion

The overall accuracy of the image classification was reasonable, at 83.4 % (Table 1). Corn and soybean were accurately classified, though high confusion among the grass-like classes (hay, pasture, legumes and fallow) resulted in lowered the overall accuracy. Fallow was very poorly classified, though it was the rarest class.

Figure 3 shows model averaged regression coefficient estimates ($\widehat{\overline{\theta}}$) for the gamma diversity models, along with 95 % confidence intervals (CIs). The hard coefficients ($\widehat{\overline{\theta}}_{hard}$) are based on the reference landscape maps, but with the 2011 crop types from the hardened image classification (each field assigned its most-likely class). Coefficients based on the official landscape maps (crops ground-referenced in

Table 1 Classification accuracy

	Accuracy by class (%)						
	Corn	Soybean	Hay	Pasture	Legumes	Cereals	Fallow
User's	93.0	92.0	72.6	73.1	52.6	63.1	29.4
Producer's	92.3	94.8	66.7	75.9	45.1	93.9	30.3
Overall accuracy		83.4 %	Cohen's kappa coefficient			0.779	

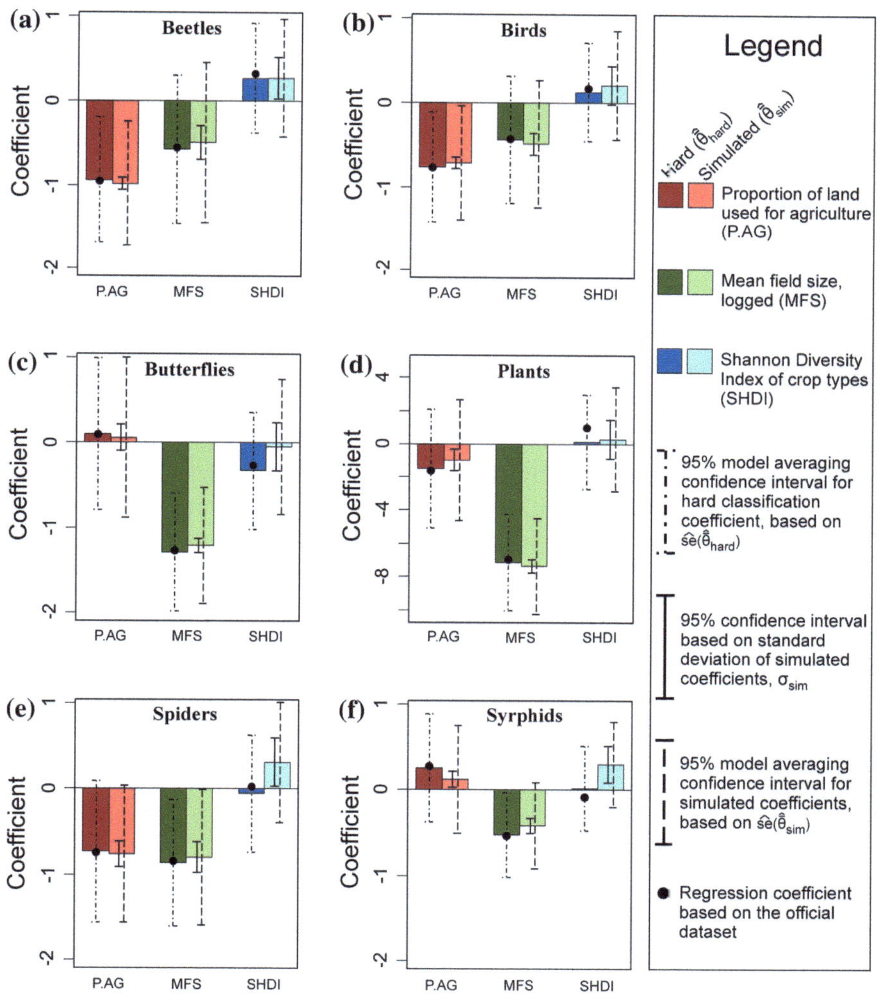

Fig. 3 Hard and simulated model averaged regression coefficients for **a** beetle, **b** bird, **c** butterfly, **d** plant, **e** spider, and **f** syrphid gamma diversity models

both years) are also shown. The simulated coefficients ($\widehat{\overline{\theta}}_{sim}$) are included, wherein field boundaries for both years and the 2011 crop types were simulated ($\widehat{\overline{\theta}}$ was averaged across all realizations).

The additional, smaller set of CIs encompassing $\widehat{\overline{\theta}}_{sim}$ are based on the standard deviation of $\widehat{\overline{\theta}}$, across all of the simulation realizations (σ_{sim}). Unlike $\widehat{se}(\widehat{\overline{\theta}}_{hard})$ and $\widehat{se}(\widehat{\overline{\theta}}_{sim})$, which consider sampling variance (given the candidate models) and variation arising from model selection uncertainty (Anderson 2008), σ_{sim} is an estimate of the variation in $\widehat{\overline{\theta}}_{sim}$ resulting from positional and thematic uncertainties alone.

For all 23 models, σ_{sim} was less than $\widehat{se}(\widehat{\overline{\theta}}_{sim})$. The average proportion (across all 23 models) of σ^2_{sim} to $\widehat{se}(\widehat{\overline{\theta}}_{sim})^2$ for the *P.AG*, *MFS*, and *SHDI* coefficients were 0.022, 0.033, and 0.148 respectively. This indicates that variance in the regression coefficients that can be attributed to positional and thematic uncertainties is quite low, relative to the uncertainty that can be attributed to the sample and model selection.

The hard and simulated model averaging standard error estimates were similar, with average proportions of $\widehat{se}(\widehat{\overline{\theta}}_{sim})^2$ to $\widehat{se}(\widehat{\overline{\theta}}_{hard})^2$ for the *P.AG*, *MFS*, and *SHDI* coefficients at 1.008, 1.067, and 1.148 respectively. However, there were cases of clear offset between $\widehat{\overline{\theta}}_{sim}$ and $\widehat{\overline{\theta}}_{hard}$, particularly for the *SHDI* coefficients (Fig. 3c, e, f).

One potential source of the *SHDI* coefficient offset is the fact that fields were assigned crop types independently, despite the presence of spatial autocorrelation in crop type between fields. Moran's I for the class membership probabilities ranged from 0.03 (cereals) to 0.22 (soybean), significantly greater than zero for all classes (at $\alpha = 0.05$), indicating some degree of positive spatial autocorrelation (O'Sullivan and Unwin 2010). While this spatial autocorrelation is not extremely high, failure to account for it in the simulation may have resulted in some degree of underprediction of the variance in the landscape metrics.

Further research into simulation approaches that consider spatial autocorrelation could be beneficial for simulations of landscape composition in similar landscapes. The approach demonstrated by Canters (1997) and Canters et al. (2002), where nearby features with similar characteristics are simulated with a common random number, may be useful in this regard.

Another potential source of offset may come from the fact that the accuracy of an output PDF obtained through MC simulation is dependent on the quality of the uncertainty models used to sensitize the inputs (Crosetto et al. 2001). For example, here we have used class membership probability estimates from the classifier, which itself can be biased due to the nature of the training data, to model the uncertainty in the *SHDI* metrics. The use of simulation approaches that condition probabilities on local reference data may have produced more accurate results (De Bruin 2000), but would require reference data distributed throughout each landscape.

4 Conclusions

MC methods were useful for quantifying the influence of positional and thematic uncertainties on the regression coefficients. The effect of positional and thematic uncertainties had little influence on these models, relative to the uncertainties associated with the sample and model selection. *SHDI* coefficients were the most influenced by the consideration of these uncertainties. Further research into simulation techniques that consider spatial autocorrelation, without reliance on ground reference data, would be beneficial for future MC analyses in similar landscapes.

Acknowledgments Assistance and guidance from Andrew Davidson, Dennis Duro, Lenore Fahrig, Jude Girard, Steve Javorek, Doug King, Kathryn Lindsay, Jon Pasher, Murray Richardson, Adam Smith, Lutz Tischendorf, Jessica van den Berg as well as others involved in the Farmland Biodiversity Project. Ground reference data from Agriculture Canada was used to supplement training samples. Funding was provided through Environment Canada. Additional thanks to Dan Patterson.

References

Anderson DR (2008) Model based inference in the life sciences: a primer on evidence. Springer, New York

Benke KK, Pelizaro C (2010) A spatial-statistical approach to the visualisation of uncertainty in land suitability analysis. J Spat Sci 55(2):257–272. doi:10.1080/14498596.2010.521975

Breiman L, Cutler A, Liaw A, Wiener M (2012) Breiman and Cutler's random forests for classification and regression. CRAN. http://cran.r-project.org/web/packages/randomForest/randomForest.pdf. Accessed 16 Jul 2013

Brown JD, Heuvelink GBM (2007) The data uncertainty engine (DUE): a software tool for assessing and simulating uncertain environmental variables. Comp Geosci 33(2):172–190. doi:10.1016/j.cageo.2006.06.015

Burnham KP, Anderson DR (2002) Model selection and multimodel inference: a practical information-theoretic approach, 2nd edn. Springer, New York

Canters F (1997) Evaluating the uncertainty of area estimates derived from fuzzy land-cover classification. Photogram Eng Remote Sens 63(4):403–414

Canters F, De Genst W, Dufourmont H (2002) Assessing effects of input uncertainty in structural landscape classification. Int J Geogr Inf Sci 16(2):129–149. doi:10.1080/13658810110099143

Crosetto M, Tarantola S (2001) Uncertainty and sensitivity analysis: tools for GIS-based model implementation. Int J Geogr Inf Sci 15(5):415–437. doi:10.1080/13658810110053125

Crosetto M, Ruiz JAM, Crippa B (2001) Uncertainty propagation in models driven by remotely sensed data. Remote Sens Environ 76(3):373–385. doi:10.1016/S0034-4257(01)00184-5

De Bruin S (2000) Predicting the areal extent of land-cover types using classified imagery and geostatistics. Remote Sens Environ 74(2):387–396. doi:10.1016/S0034-4257(00)00132-2

Delmelle E, Dony C, Casas I, Jia M, Tang W (2014) Visualizing the impact of space-time uncertainties on dengue fever patterns. Int J Geogr Inf Sci 28(5):1107–1127. doi:10.1080/13658816.2013.871285

Duro DC, Girard J, King DJ, Fahrig L, Mitchell S, Lindsay K, Tischendorf L (2014) Predicting species diversity in agricultural environments using landsat TM imagery. Remote Sens Environ 144:214–225. doi:10.1016/j.rse.2014.01.001

Edwards G, Lowell KE (1996) Modeling uncertainty in photointerpreted boundaries. Photogram Eng Remote Sens 62(4):337–391

Fahrig L, Baudry J, Brotons L, Burel FG, Crist TO, Fuller RJ, Sirami C, Siriwardena GM, Martin JL (2011) Functional landscape heterogeneity and animal biodiversity in agricultural landscapes. Ecol Lett 14(2):101–112. doi:10.1111/j.1461-0248.2010.01559.x

Fahrig L, Girard J, Duro D, Pasher J, Smith A, Javorek S, King D, Lindsay KF, Mitchell S, Tischendorf L (2015) Farmlands with smaller crop fields have higher biodiversity. Agric Ecosyst Environ 200:219–234. doi:10.1016/j.agee.2014.11.018

Foody GM, Campbell NA, Trodd NM, Wood TF (1992) Derivation and applications of probabilistic measures of class membership from the maximum-likelihood classification. Photogram Eng Remote Sens 58(9):1335–1341

Lillesand TM, Kiefer RW, Chipman JW (2004) Remote sensing and image interpretation, 5th edn. Wiley, Hoboken

Ministry of Natural Resources (2009) Digital raster acquisition project eastern ontario (DRAPE). Metadata available via https://www.javacoeapp.lrc.gov.on.ca/geonetwork/srv/en/main.home. Accessed 20 Aug 2014

NASA Landsat Program (2011) Landsat-5 TM scenes LT50150292011154EDC00, LT50150292011170EDC00, LT50150292011186EDC00, LT50150292011282EDC00. L1T. USGS, Sioux Falls. 3 June, 19 June, 5 July, 9 October, 2011

O'Sullivan D, Unwin D (2010) Geographic information analysis, 2nd edn. Wiley, Hoboken

Pasher J, Mitchell SW, King DJ, Fahrig L, Smith AC, Lindsay KE (2013) Optimizing landscape selection for estimating relative effects of landscape variables on ecological responses. Land Ecol 28(3):371–383. doi:10.1007/s10980-013-9852-6

Phalan B, Balmford A, Green RE, Scharlemann JPW (2011) Minimising the harm to biodiversity of producing more food globally. Food Policy 36(1):S62–S71. doi:10.1016/j.foodpol.2010.11.008

Richardson AD, Hollinger DY (2005) Statistical modeling of ecosystem respiration using eddy covariance data: Maximum likelihood parameter estimation, and Monte Carlo simulation of model and parameter uncertainty, applied to three simple models. Agric For Meteorol 131(3–4):191–208. doi:10.1016/j.agrformet.2005.05.008

Shi WZ, Ehlers M, Tempfli K (1999) Analytical modelling of positional and thematic uncertainties in the integration of remote sensing and geographical information systems. Trans GIS 3(2):119–136. doi:10.1111/1467-9671.00011

Examining Tail Distributions of Moran's *I* Statistic through Intensive Simulations

Ikuho Yamada and Atsuyuki Okabe

Abstract Moran's *I* statistic is arguably the most widely applied method for testing spatial autocorrelation in areal datasets. This study examines its probability distribution through intensive simulation experiments with a specific focus on its tails, which are the most important for statistical testing. While Moran's *I* statistic is known to be asymptotically normal when the number of zones in a study region is sufficiently large, results of this study indicate that the normality is hardly achievable even when the number of zones is increased up to 2,500. Our results also suggest that discrepancies from the normality are more serious for target variables with larger skewness and kurtosis. Simulation-based testing, which does not rely upon the assumed normality of *I*, is thus recommended. This study proposes to carry out multiple sets of simulations and use the mean of simulated percentage point values as critical values to reduce instability inherent in simulation-based approaches.

Keywords Moran's *I* statistic · Spatial autocorrelation · Statistical testing · Spatial statistics · Probability distributions

1 Introduction

Moran's *I* statistic (Moran 1948) is arguably the most widely applied method for testing spatial autocorrelation in areal datasets. When the number of zones, *N*, in a study region is sufficiently large and a variable, *X*, to be analyzed satisfies certain conditions, it is known that the probability distribution of Moran's *I* statistic can be assumed to follow the normal distribution asymptotically and statistical testing is generally carried out based upon that assumption. However, previous studies (Cliff

I. Yamada (✉)
Chuo University, 1-13-27 Kasuga, Bunkyo-ku, Tokyo 112-8551, Japan
e-mail: iyamada.87s@g.chuo-u.ac.jp

A. Okabe
Aoyama Gakuin University, 4-4-25 Shibuya, Shibuya-ku, Tokyo 150-8366, Japan

F. Harvey and Y. Leung (eds.), *Advances in Spatial Data Handling and Analysis*, Advances in Geographic Information Science,
DOI 10.1007/978-3-319-19950-4_19

and Ord 1971; Anselin 1995; Tiefelsdorf and Boots 1995; Tiefelsdorf 1998, 2002) pointed out that the probability distribution of Moran's I statistic was likely to deviate from the normal distribution especially around its tails, which are the most important for statistical testing. Furthermore, whether or not X satisfies the required conditions is barely scrutinized in application studies.

By extending Cliff and Ord's work (1971) that examined potential impacts of target variable X that did not satisfy the required conditions, this study closely investigates properties of the tail distributions of Moran's I statistic for varying distributions of X as well as varying sizes of a study region through intensive computational simulations.

2 Moran's *I* Statistic—Basics and Related Literature

Moran's I statistic (Moran 1948) is defined as:

$$I = \frac{N\sum_{i=1}^{N}\sum_{j=1}^{N} w_{ij}(X_i - \bar{X})(X_j - \bar{X})}{\left(\sum_{i=1}^{N}\sum_{j\neq i}^{N} w_{ij}\right)\sum_{i=1}^{N}(X_i - \bar{X})^2}, \tag{1}$$

where X_i is the observed value of the target variable X in zone i (i = 1, …, N), $\bar{X}$ is the mean of the observed X_i (i.e., $\bar{X} = \sum_{i=1}^{N} X_i/_N$), and w_{ij} is an i-j element of the spatial proximity matrix, **W**, that indicates proximity between zones i and j. Moran's I statistic ranges approximately $-1 \leq I \leq 1$ and can be interpreted by analogy with Pearson's correlation coefficient.

The sampling distribution of Moran's I statistic is known to approach a normal distribution asymptotically under the null hypothesis of no spatial autocorrelation when N is sufficiently large and either of the following two assumptions is justifiable.

1. Normal variate spatial randomness assumption: values of X observed in individual zones are generated from an identical and independent normal distribution.
2. Permutation spatial randomness assumption: all possible permutations of the observed X values among the N zones are equally likely.

When normality of the sampling distribution of Moran's I statistic can be assumed, one may carry out statistical testing using a normal distribution of which mean and variance can algebraically be determined. When the normality cannot be assumed, a Monte Carlo simulation that generates a large number of possible permutations of the observed X_i values (i = 1, …, N) is generally conducted to approximate the sampling distribution of I. While the latter approach implicitly presumes the second assumption above, it is often utilized in practical applications without explicit discussion about which assumption is appropriate for the variable X under study.

Cliff and Ord (1971) pointed out the possibility of serious inferential errors caused by wrongly assuming the normality for a non-normal sampling distribution. Based upon the results of their simulation experiments, they recommended statistical testing that utilized Monte Carlo simulations over testing that assumed the normality of Moran's I statistic and also proposed a numerical adjustment for the statistic to handle its discrepancy from the normality.

Anselin (1995) also illustrated non-normality of a local version of Moran's I statistic through a simulation study and recommended a Monte Carlo simulation approach to statistical testing. *GeoDa* (Anselin et al. 2005), which is a free software program for spatial data analysis developed by Anselin and his colleagues, implements an intensive simulation process to provide users with a quick tool to test spatial autocorrelation both globally and locally with Moran's I statistic.

The series of studies by Tiefelsdorf and his colleague (Tiefelsdorf and Boots 1995; Tiefelsdorf 1998, 2002) also dealt with the problem associated with the potential non-normality of Moran's I statistic. Instead of relying upon Monte Carlo simulation, their studies focused on numerical statistical solutions. Tiefelsdorf and Boots (1995) and Tiefelsdorf (1998) discussed derivation of the exact distribution of I and Tiefelsdorf (2002) proposed the saddlepoint approximation, but both were specifically designed for X that represented normally distributed regression residuals.

Whereas several approaches have already been proposed to handle the non-normality of the sampling distribution of Moran's I statistic as seen above, these approaches have limitations. The numerical approaches proposed by Tiefelsdorf and his colleague are not only mathematically complex but also rather restrictive for practical applications due to its selective handling of regression residuals. On the other hand, the Monte Carlo simulation approach is easily accessible with the help of software programs such as *GeoDa*, and places no restrictions on the distribution of the target variable X. However, the use of random simulations inevitably leads to some level of instability of testing results. An easy-to-understand approach that provides stable testing results is still needed.

The present study therefore closely investigates properties of the tail distributions of Moran's I statistic with the aim of developing an alternative method for statistical testing with I. We extend Cliff and Ord's work (1971) and conduct intensive simulation experiments to examine potential influences of N and the distribution of the target variable X on the tail distributions of Moran's I statistic.

3 Method

The simulation experiments conducted in this study are based upon the permutation spatial randomness assumption, the second assumption described in Sect. 2. Compared with the first assumption, which assumes that observed values of X come from a normal distribution, this second assumption is more flexible and thus widely applicable to practical situations. Cliff and Ord (1971) also based their experiments upon this assumption.

For a study region consisting of N zones, the simulation procedure is as follows:

1. Generate a set of N observed values of X, one for each of the N zones, by randomly deriving values from a given probability distribution, D;
2. Compute Moran's I statistic, I_{obs}, for the X values generated in Step (1);
3. Permute the observed X values randomly in the study region. In other words, assign the N observed values randomly to each of the N zones;
4. Compute Moran's I statistic, I_k, for the permuted X values;
5. Repeat Steps (3)–(4) K times and create a frequency distribution of Moran's I statistic based upon I_k ($k = 1, \ldots, K$);
6. Test the significance of I_{obs} in relation to the frequency distribution created in Step (5);
7. Repeat Steps (1)–(6) L times.

Steps (2)–(6) correspond to a common testing procedure with the permutation spatial randomness assumption. The procedure here repeats that procedure multiple times in order to capture instability inherent in simulation-based testing. In this study, we use $K = 100{,}000$ and $L = 1{,}000$. That is, we conduct 1,000 sets of hypothesis testing, each of which generates 100,000 of permutations to obtain a sampling distribution of Moran's I statistic. As the probability distribution D, we examine a normal distribution, $N(0, 1)$, and two log-normal distributions, $LN(0, 1)$ and $LN(0, 0.5)$. The two parameters of the log-normal distribution represent, respectively, the mean and standard deviation of its natural logarithm, which by definition is normally distributed. Theoretical moments of the three probability distributions are summarized in Table 1.

The study regions examined are $m \times m$ regular grid systems with $m = 5, 7, 10, 20, 30, 40$, and 50. This choice of the study regions may appear too restrictive, but it allows us to assess the potential influence of N in a controlled manner. Furthermore, the availability of remote sensing data has rapidly been expanding in terms of both volume and areas covered, so that regular grid data are likely to be the mainstream in spatial data analysis in the near future.

Row-standardized Queen's connectivity matrix is used as the spatial proximity matrix $\mathbf{W}$. In Queen's connectivity matrix, $w_{ij} = 1$ when zones i and j share a boundary or a corner, and $w_{ij} = 0$ otherwise. The matrix is row-standardized as:

Table 1 Theoretical moments of the probability distribution examined in this study

	Normal distribution $N(0, 1)$	Log-normal distribution $LN(0, 1)$	Log-normal distribution $LN(0, 0.5)$
Mean	0.00	1.65	1.13
Variance	1.00	4.67	0.36
Skewness	0.00	6.18	1.75
Kurtosis	3.00	113.94	8.90

$$w'_{ij} = {w_{ij}}\Big/{\sum_{j=1}^{N} w_{ij}} \quad (2)$$

so that every zone i has $\sum_{j=1}^{N} w'_{ij} = 1$ and thus the same magnitude of influence in Eq. 1 regardless of the number of zones to which it is connected.

4 Results

In our preliminary experiments with $K = L = 999$, we found that the sampling distribution of Moran's I statistic hardly followed the normal distribution even when N was increased up to 2,500 and the target variable X satisfied the permutation spatial randomness assumption. Figure 1 shows the percentage of times out of 999 where the normality of Moran's I statistic was rejected by Shapiro-Wilk test (Shapiro and Wilk 1965) at the 5 % significance level. This result clearly indicates unsuitability of statistical testing that relies upon the assumed normality of Moran's I statistic. Q-Q plots of the simulated sampling distributions (not presented here) suggested that this discrepancy from the normality was mostly attributable to the tails of the distributions.

Based on the finding above, the present study focuses on the tails of the sampling distributions of Moran's I statistic. More specifically, we examines 5 and 1 % points of the distributions, which are often used as critical values in hypothesis testing.

Figure 2 shows the observed variability in the upper 5 % point values of 1,000 simulated sampling distributions. Results for the lower 5 % point and the upper and lower 1 % points are qualitatively the same and not presented here. The variability in the 5 % point values is larger for the target variable X with higher skewness and

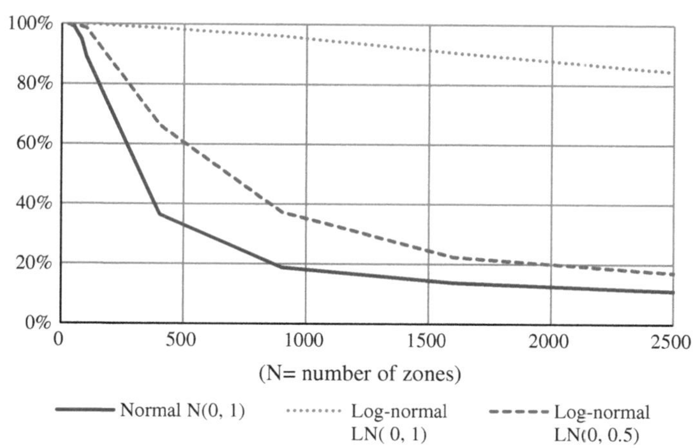

Fig. 1 Percentage of rejecting the normality of Moran's I statistic by Shapiro-Wilk test (number of simulations = 999; significance level = 5 %)

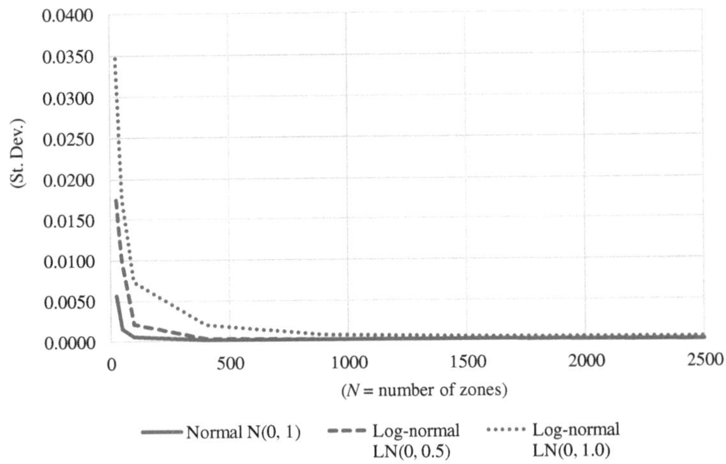

Fig. 2 Standard deviations of simulated upper 5 % point values (number of simulations = 1,000)

kurtosis as well as for the study region with smaller N. It appears that the variability becomes negligibly small when N is greater than about 30×30.

Statistical properties of the simulated 5 % point values are summarized in Table 2. "Max," "min," "mean," and "S.D." in the tables indicate the maximum, minimum, mean, and standard deviation, respectively, of the 1,000 simulated values of the 5 % point. "Theoretical" indicates a value of the 5 % point calculated algebraically assuming that the sampling distribution of Moran's I statistic follows a normal distribution. The tables suggest that observed discrepancies between the mean of the simulated 5 % point values and the theoretical 5 % point values become smaller and smaller as N increases. However, even for $N = 50 \times 50$, the discrepancies are more than triple of the standard deviations of the simulated 5 % point values in many cases.

The last three rows of each table show the number of Type I errors out of the 1,000 simulations when the mean of the simulated 5 % point values is used as the critical value. Note that usually a simulated 5 % point value is used for each simulation. The use of the mean proposed here would diminish the concern about instability inherent in simulation-based procedures, but it might potentially bias testing results. The results in Table 2 imply that, as far as Type I errors are concerned, such a bias seems to be minimal. Tables for the 1 % points are not presented here due to space limitation, but qualitatively the same discussion applies.

Table 2 Statistical properties of the simulated 5 % point values

		5 × 5	7 × 7	10 × 10	20 × 20	30 × 30	40 × 40	50 × 50
(a) Normal distribution $N(0, 1)$								
Upper 5 %	Max	0.2024	0.1483	0.1039	0.0516	0.0339	0.0253	0.0202
	Mean	0.1927	0.1439	0.1018	0.0504	0.0334	0.0249	0.0199
	S.D.	0.0057	0.0016	0.0006	0.0002	0.0002	0.0001	0.0001
	Theoretical	0.1668	0.1299	0.0947	0.0486	0.0326	0.0245	0.0196
Lower 5 %	Mean	−0.2227	−0.1570	−0.1076	−0.0518	−0.0340	−0.0253	−0.0201
	Min	−0.2271	−0.1595	−0.1092	−0.0525	−0.0343	−0.0256	−0.0204
	S.D.	0.0026	0.0009	0.0004	0.0002	0.0001	0.0001	0.0001
	Theoretical	−0.2501	−0.1716	−0.1149	−0.0537	−0.0348	−0.0257	−0.0204
Type I errors	Upper[a]	29	25	23	18	30	28	21
	Lower[b]	21	25	23	21	22	27	19
	Total	50	50	46	39	52	55	40
(b) Log-normal distribution $LN(0, 1)$								
Upper 5 %	Max	0.2028	0.1507	0.1193	0.0590	0.0385	0.0282	0.0221
	Mean	0.1675	0.1342	0.1001	0.0520	0.0349	0.0260	0.0207
	S.D.	0.0347	0.0177	0.0074	0.0020	0.0008	0.0005	0.0003
	Theoretical	0.1384	0.1135	0.0866	0.0465	0.0318	0.0240	0.0193
Lower 5 %	Mean	−0.1908	−0.1334	−0.0928	−0.0461	−0.0309	−0.0233	−0.0188
	Min	−0.2270	−0.1581	−0.1057	−0.0518	−0.0336	−0.0248	−0.0197
	S.D.	0.0278	0.0180	0.0100	0.0041	0.0018	0.0012	0.0008
	Theoretical	−0.2218	−0.1552	−0.1068	−0.0516	−0.0340	−0.0253	−0.0201
Type I errors	Upper[a]	33	21	28	30	26	16	25
	Lower[b]	32	23	26	25	29	26	22
	Total	65	44	54	55	55	42	47
(c) Log-normal distribution $LN(0, 0.5)$								
Upper 5 %	Max	0.2035	0.1518	0.1043	0.0518	0.0343	0.0255	0.0203
	Mean	0.1856	0.1405	0.1017	0.0509	0.0337	0.0251	0.0200
	S.D.	0.0174	0.0098	0.0021	0.0003	0.0002	0.0001	0.0001
	Theoretical	0.1584	0.1239	0.0927	0.0483	0.0325	0.0244	0.0196
Lower 5 %	Mean	−0.2134	−0.1485	−0.1039	−0.0507	−0.0335	−0.0250	−0.0199
	Min	−0.2271	−0.1580	−0.1082	−0.0519	−0.0341	−0.0255	−0.0202
	S.D.	0.0133	0.0109	0.0033	0.0007	0.0002	0.0001	0.0001
	Theoretical	−0.2417	−0.1656	−0.1129	−0.0533	−0.0347	−0.0257	−0.0204
Type I errors	Upper[a]	26	19	19	28	23	28	34
	Lower[b]	30	25	29	17	23	25	18
	Total	56	44	48	45	46	53	52

[a]Number of times that the observed value of Moran's I statistic I_{obs} is greater than the mean upper 5 % point out of 1,000 simulations
[b]Number of times that I_{obs} is less than the mean lower 5 % point out of 1,000 simulations

5 Summary

In this study, we examined the 5 and 1 % points of the sampling distributions of Moran's I statistic that were obtained through intensive Monte Carlo simulations. The results supported the prior studies by Cliff and Ord (1971) etc. in that the normality of Moran's I statistic could hardly been realized even when N was considerably large. Especially around the tails of the sampling distribution, discrepancies between the simulated percentage point values and the theoretically calculated values were not negligible. In addition, discrepancies tended to be large when the target variable X came from a probability distribution with large skewness and kurtosis. Cliff and Ord (1971) were concerned with $10 \leq N \leq 50$ considering such N as moderate size and most important in practice. These days we have datasets with much larger N, but our study indicates that it is still insufficient to assume the normality for Moran's I statistic. It is of great interest to see if and how extremely large values of N such as those found in remote sensing data would change the results.

We further looked into the possibility of using the mean of the simulated percentage point values as critical values in order to diminish the concern about unstable testing results that are inevitable in simulation-based testing. The proposed procedure performed well in terms of Type I errors. The result implies that it may be possible to construct standard tables of percentage point values for a given combination of N and X through large-scale simulations. Such tables would offer a great opportunity not only for quick and precise testing of new datasets but also for re-examination of past analyses.

So far, we have worked only with synthetic data to have better control over our experiments. We are currently preparing for a case study with remote sensing data to investigate the effectiveness of the proposed approach in the real-world setting. Remote sensing data is selected because it has the same regular grid system as used in the present synthetic experiments. Remote sensing data also offers the opportunity to deal with a very large N.

There are also a few other follow-up studies we plan to conduct. One is the power analysis of the proposed approach with varying levels of spatial autocorrelation. The other is to extend the proposed approach to local analysis. Especially when analyzing data with a large N, it is unlikely that the whole study region has a consistent spatial pattern and it is of greater use and interest to identify local hotspots of spatial autocorrelation. These follow-ups would considerably increase the practical applicability of the proposed approach.

Acknowledgements This work was supported by JSPS KAKENHI Grant Number 24650606. The authors greatly appreciate helpful comments and information from anonymous reviewers as well as attendees of the Joint International Conference on Geospatial Theory, Processing, Modelling and Applications in Toronto in 2014.

References

Anselin L (1995) Local indicators of spatial association—LISA. Geogr Anal 27(2):93–115

Anselin L, Syabri I, Kho Y (2005) *GeoDa*: an introduction to spatial data analysis. Geogr Anal 38 (1):5–22

Cliff AD, Ord K (1971) Evaluating the percentage points of a spatial autocorrelation coefficient. Geogr Anal 3(1):51–62

Moran PAP (1948) The interpretation of statistical maps. J Roy Stat Soc B Met 10(2):243–251

Shapiro SS, Wilk MB (1965) An analysis of variance test for normality (complete samples). Biometrika 52(3–4):591–611

Tiefelsdorf M (1998) Some practical applications of Moran's *I*'s exact conditional distribution. Pap Reg Sci 77(2):101–129

Tiefelsdorf M (2002) The saddlepoint approximation of Moran's *I*'s and local Moran's I_i's reference distributions and their numerical evaluation. Geogr Anal 34(3):187–206

Tiefelsdorf M, Boots B (1995) The exact distribution of Moran's *I*. Environ Plann A 27 (6):985–999

Index

F. Harvey and Y. Leung (eds.), *Advances in Spatial Data Handling and Analysis*, Advances in Geographic Information Science,
DOI 10.1007/978-3-319-19950-4

MIX
Papier aus verantwortungsvollen Quellen
Paper from responsible sources
FSC® C105338

If you have any concerns about our products,
you can contact us on
ProductSafety@springernature.com

In case Publisher is established outside the EU,
the EU authorized representative is:
Springer Nature Customer Service Center GmbH
Europaplatz 3, 69115 Heidelberg, Germany

Printed by Libri Plureos GmbH
in Hamburg, Germany